ADAPTIVE THERMAL COMFORT: FOUNDATIONS AND ANALYSIS

There has been widespread dissatisfaction with accepted models for predicting the conditions that people will find thermally comfortable in buildings. These models require knowledge about clothing and activity, but can give little guidance on how to quantify them in any future situation. This has forced designers to make assumptions about people's future behaviour based on very little information and, as a result, encouraged static design indoor temperatures.

This book is the second in a three-volume set covering all aspects of adaptive thermal comfort. The first part narrates the development of the adaptive approach to thermal comfort from its early beginnings in the 1960s. It discusses recent work in the field and suggests ways in which it can be developed and modelled. Such models can be used to set dynamic, interactive standards for thermal comfort which will help overcome the problems inherited from the past. The second part of the volume engages with the practical and theoretical problems encountered in field studies and in their statistical analysis, providing guidance towards their resolution so that valid conclusions may be drawn from such studies.

Michael Humphreys is known for his pioneering work on the adaptive approach to comfort. He was Head of Human Factors at the Building Research Establishment, and has been a Research Professor at Oxford Brookes University. His scientific interests are the methodology of field studies of environmental comfort, the structure and statistical modelling of human adaptive behaviour, and the interactions between the several aspects of the indoor environment.

Fergus Nicol has led a number of important research projects on comfort, which have influenced thinking internationally. He has authored numerous journal articles and other publications including guidance on comfort and overheating. Fergus convenes the Network for Comfort and Energy Use in Buildings and organises their regular international Windsor Conferences.

Susan Roaf did her PhD on comfort and the windcatchers of Yazd, and, after a decade working with Nicol and Humphreys at the Oxford Thermal Comfort Unit, she is now Professor of Architectural Engineering at Heriot Watt University in Edinburgh. She is a teacher, researcher, designer and author or editor of 21 books, including *Ecohouse: A Design Guide* and *Adapting Buildings and Cities for Climate Change*.

ADAPTIVE THERMAL COMFORT: FOUNDATIONS AND ANALYSIS

Michael Humphreys, Fergus Nicol and Susan Roaf

Routledge
Taylor & Francis Group

LONDON AND NEW YORK

First published 2016
by Routledge

2 Park Square, Milton Park, Abingdon, Oxfordshire OX14 4RN
52 Vanderbilt Avenue, New York, NY 10017

Routledge is an imprint of the Taylor & Francis Group, an informa business

First issued in paperback 2020

British Library Cataloguing-in-Publication Data
A catalogue record for this book is available from the British Library

Library of Congress Cataloging-in-Publication Data
Humphreys, Michael A. (Michael Alexander), 1936-
Adaptive thermal comfort : foundations and analysis / Michael
Humphreys, Fergus Nicol and Susan Roaf.
pages cm
Includes bibliographical references and index.
1. Buildings—Environmental engineering. 2. Buildings—Thermal
properties. 3. Architecture—Human factors. 4. Architecture and
climate. I. Nicol, Fergus, 1940- II. Roaf, Susan. III. Title.
TH6025.H86 2016
697—dc23
2015007799

ISBN: 978-0-415-69161-1 (hbk)
ISBN: 978-0-367-59824-2 (pbk)

Typeset in Bembo Std
by Swales & Willis Ltd, Exeter, Devon, UK

For Mary

CONTENTS

ILLUSTRATIONS

Tables

PREFACE

The preface to Volume 1 of our trilogy applies also to this volume, so rather little further is needed by way of preface.

Volume 1 was in two parts. This second volume is divided into two corresponding parts:-

In Volume 1, Part I, we presented the principles of the adaptive approach in a logical order, giving reasons for the importance of the adaptive approach and explaining the underlying behavioural processes. It explained how to conduct a field survey to gather quantitative information, and how to relate its findings to the adaptive processes. It explained the thermal physiological processes by which comfort is achieved, and also noted how the adaptive approach is expressed in various national and international standards and guidelines. Part I of Volume 2 considers this same material, but from a different point of view. We have adopted a narrative style as a powerful method for explaining the growth of the concepts behind adaptive thermal comfort by setting them in their historical context. The impersonal style in which much of science is reported does rather obscure the fact that science is a very human activity. This is especially true of thermal comfort field work, where the researcher and their respondents ('subjects' in the jargon of experimental psychology) are intelligent, creative and interactive human beings. The field study method, essential to the development and quantification of adaptive concepts, deals with real people in real buildings and responding to real situations. The narrative method makes this abundantly clear and in Volume 2, Part I the story, as it is developed, relies to a large part on the recollections of Michael Humphreys, based on his unpublished memoirs. We have used a wide range of original sources to build what we hope is a clear picture of the development of the adaptive approach to thermal comfort at the Building Research Station between the years 1965–1978, and the subsequent work we undertook at Oxford Brookes University to further develop the approach between 1993 and 2006.

Volume 1, Part II covered the practical matters entailed in the conduct of a thermal comfort survey and the statistical analysis of the data acquired. Part II of this volume covers this same area, but treating selected topics in greater depth and detail. We show how the properties of subjective scales of warmth and comfort can be examined, and discuss the considerations necessary for the construction of new scales. In Volume 1 we took a particular dataset and showed how it might best be analysed. In this volume we build a statistical 'model dataset' whose properties are known, and then consider the effect upon the statistics derived from it, should it be subjected to different approximations and analytical procedures. This leads us to consider how important statistics may best be estimated from the available international databases of thermal comfort surveys, and to show the patterns that emerge from them. All this brings together the work of many years of research effort by numerous teams of researchers across the world. The pace of gathering field-data is accelerating, and the patterns and outcomes will need periodic updating to retain their relevance. We regret that the pace of acquisition field data has outstripped our ability to collect it, and that consequently rather little of the more recent survey data from the Far East and from South America has contributed to our summaries.

We conclude the volume by pointing out that the adaptive approach applies not only to the thermal environment but also to other aspects such as lighting, acoustics and air quality. The numerous and varied interactions among these aspects can influence the acceptability of the indoor climate. Here is a whole new field for collaborative adaptive research.

We hope that this volume will be of interest to all who are concerned with adaptive thermal comfort, whether researchers, building engineers or architects. Part II will be of particular interest to those whose task is to collect and analyse thermal comfort survey data, to those planning the acquisition of such data, and to those using the information to form standards and guidelines for the thermal environment in buildings.

Michael Humphreys
Fergus Nicol
Susan Roaf
September 2014

ACKNOWLEDGEMENTS

This is the second book in our trilogy on adaptive thermal comfort and the lead author of Volume 2 is Michael Humphreys. The book takes the themes of Volume 1 and both treats them from a different perspective and takes them to a more advanced level.

In the acknowledgements in Volume 1, we noted the help of our many colleagues and friends across the world, and need not repeat their names here. There are however some extra names we should add for this volume.

During the writing of Part I, which recounts the history of the adaptive approach, we were reminded of colleagues who worked at the UK Building Research Station (BRS) in the 1960s and 1970s. Without them, the early research work would perhaps not have been started. They include Teddy Danter, Alec Loudon, Neil Milbank, John Harrington-Lyn and Peter Petherbridge whose work on summer overheating of buildings showed the need for a better understanding of thermal comfort. Alan Wise helped Michael Humphreys shape his first published work, and Stan Leach saw the potential of a behavioural approach in comfort research. David Fisk quickly grasped the adaptive approach and used it in his study of the control of buildings. Particular appreciation goes to those who at one time or another worked alongside Michael Humphreys and Fergus Nicol in the BRS Thermal Comfort Section headed by Charles Webb, including Dorothy Hume, Beryl England and Jackie English. Later, the Human Factors Section was headed by John Langdon with Margaret Gidman, Barbara Buller, Mike Crook, Pat Denison, John Hart, Liz Mansbridge, Ed Bywater, Frank Ball, Mark Porcheret and David Hunt in his team. Our interaction with Don McIntyre and Ian Griffiths was always stimulating. It was Professor R. K. MacPherson and Dr Andris Auliciems who independently suggested that Michael Humphreys explore the effect of the outdoor climate on temperatures for comfort indoors, leading to Humphreys developing the understanding of the relationship between the two, as described in Chapter 9. We have drawn freely on BRS (later BRE) publications from the 1960s and 1970s, and are grateful to IHS BRE Press, the current copyright holders, for permission to reproduce extracts and photographs from these publications and other research records.

For their comments on the content and text of various chapters, our thanks go to Mary Hancock, Alicia Montarzino, Iftikhar Raja and Hom Rijal. Jane Galbraith generously spent time with Michael Humphreys discussing and commenting on the statistical chapters and their context, and the chapters have benefitted from her critique. Our special thanks go to Mary Humphreys and Hilary de Visme for their reading of the final manuscript. Kevin Bowe and Abdulrahman Alshaikh helped us prepare the volume for the publisher, formatting text and extracting and tabulating the figures and plates. The responsibility for any remaining errors lies of course with the authors.

We thank the publishers for their support during the writing and completion of this second volume in our thermal comfort trilogy and special thanks go to Nicki Dennis who has overseen the project since its inception and to Alice Aldous and Matthew Turpie for their help in managing the publication process. Michael Humphreys would also like to thank his friends and family for their gracious support during this time.

Michael Humphreys
Fergus Nicol
Susan Roaf
October 2014

PART I

Foundations

1

INTRODUCTION TO ADAPTIVE BEHAVIOUR

Adaptation is at the heart of the survival of all species in the ever-changing climates of our planet. All of us, more or less consciously, are aware of adapting to weather and climate. Before we describe the beginnings of the adaptive model of thermal comfort, we explain what we mean by adaptive comfort and describe some of the experiences, both personal and research, from which the adaptive approach to thermal comfort arose.

Each of us practises a range of adaptive behaviours every day. These manifest themselves when we put on an overcoat to go outside, move into a new climate or experience extreme weather events, experiences that often awaken our thermal awareness. Toward the end of the worst winter of the twentieth century, in 1947, Michael Humphreys remembers how, as a boy of ten, he helped his mother rear hundreds of baby chicks by the Sterling solid-fuel cooker in their remote Kent farmhouse. If chicks got cold they huddled so closely together for warmth that they were in danger of crushing each other but they soon learned to put their rear ends up against the front of the stove, and then move away again when they had warmed up. This was behavioural adaptive thermal regulation in action.

He now sees the behaviour of the hill sheep near his home in rural Mid Wales, as they crowd together in places sheltered from the cold winter wind, or find a sunny place if they can. In the heat of summer both sheep and cattle seek shady places. These are all adaptive behaviours employed in thermal regulation, supplementing the body's own defences against heat and cold. It is thought that in the wild many mammals achieve perhaps two-thirds of their thermal regulation behaviourally. Edgar Folk's beautiful books about behavioural thermal regulation give many examples.[1]

People, like other animals, have always adapted too, and when and where fuel was scarce and warmth needed they largely controlled their comfort behaviourally – that is, adaptively. The 'Ice-man', whose well-preserved body appeared from a melting glacier in 1991, was wearing sophisticated hay-insulated clothing while out on his hunting expedition.[2] This was behavioural thermal adaptation by choice of suitable clothing, an example of a conscious, intelligent, adaptive response.

British Iron Age dwellings had a wood-fuelled fire burning at the centre of the dwelling and radiating heat to the occupants, who would have moved closer to the fire to get warmer and further away to get cooler. The dwelling itself was designed in response to the outdoor conditions, its walls and roof moderating the daily swings of outdoor temperature, giving shelter from the wind and rain, and providing shade in summer. The form and nature of each building reflected the location and landscape and helped to keep occupants safe from weather and climate, so that they were not wholly dependent on the body's built-in systems of thermal regulation. Where people differ from animals is in the greater conscious use of designed solutions, such as fire, clothing and house building and eventually adding mechanical technologies to their adaptive opportunities and strategies.

Humphreys presented a thought experiment in 1995 to explain how we might go about understanding the basics of adaptive thermal comfort.[3]

> The popularity of the *The Lord of the Rings* and *The Hobbit* films has recently introduced many people to hobbits. Suppose we were exploring Middle Earth and encountered some hobbits.[4] Being interested in thermal comfort, our curiosity would certainly lead us to enquire what thermal environments were best for them. Perhaps we would first enquire into the climate in those parts of Middle Earth where hobbits lived and thrived. Next we might notice that although the hobbits spent some time sitting outdoors in contented contemplation, they spent much more time in their holes. So if the opportunity arose, we would measure the thermal conditions inside a representative sample of occupied hobbit holes. We would observe which rooms they most frequented and thereby obtain a more accurate knowledge of their preferences for different micro-climates at different times of year. We might notice in what circumstances hobbits opened or closed doors and windows or stoked up the fire to control the temperature of the room they occupied and how they positioned themselves to make the best use of the heat it gave out. We would notice too what the hobbits chose to wear from their extensive stock of clothes – a choice related to, among other things, their desire for comfort. Finally, if we had learned a smattering of the hobbit tongue, we would ask whether theirs was a good hole, or whether really they might prefer one which was warmer or cooler.
>
> In the course of our enquiries, we would have learned to know a good hobbit hole when we saw one, and we would also have learned a good deal about the preferences and adaptive strategies of hobbits in their quest for comfort. So if,

some time later, we were told to expect a visit from a party of hobbits, we would know how to provide accommodation comfortable for them. Various strategies suggest themselves:

- We could construct a good traditional hobbit hole for them, appropriate to the climate, having all the usual facilities for thermal control, and well-stocked with hobbit clothes. Then we could trust the hobbits to make themselves comfortable.
- We might (unless the hobbits were on a holiday visit) provide them with individually adjustable environment-controlled workstations, capable of providing the range of the thermal environments we knew hobbits liked best. Then we could trust them to adjust the workstations to suit themselves. Whether they would aim for maximum comfort or for maximum task efficiency is a question we could ask the experts in performance studies, for maximum efficiency and maximum comfort may not always coincide.
- Alternatively, we could provide a thermal environment for the visiting party, regulated according to an algorithm which estimated the best conditions for hobbits, according to time of day and season of year, based on our observations when visiting Middle Earth.

So, either we give control to the hobbits, knowing that the hole or the hardware can provide any conditions within the range hobbits like best, or (perhaps less satisfactorily since hobbits generally prefer to be in control of their own thermal environment) we give the hobbits the precise temperatures which our experience has suggested they like best.

Our observations from Middle Earth would have been more than sufficient to enable us to specify and provide comfort conditions for our guests. Observation of the habits and preferences of the hobbits and the measurement of their environments was sufficient to enable us to understand how hobbits got comfortable, and to:

- produce a 'deemed to satisfy' design for building thermally acceptable hobbit holes;
- identify the range of environmental adjustment which a hobbit might need;
- suggest the best daily and seasonal temperature profile for spaces occupied by hobbits.

No invasive measurements or tiresome experimental routines were necessary. An enquiry about comfort and the concurrent measurement of the thermal environment yielded all that we needed to know. This simplicity makes field study work an effective, practical method for investigating adaptive thermal comfort in the huge range of climates and cultures that exist on earth.

It is interesting to notice what we did *not* need to know:

- We did not need to know anything at all about the thermal physiology of hobbits, such as the diurnal cycle of their body temperature, the metabolic heat production of their various activities, whether they could sweat or shiver or pant, or whether the Dubois relation between height, weight and skin surface-area held good for hobbits.
- We did not need to know anything about the heat exchange between hobbit skin and the hole, such as the surface heat-transfer coefficients by convection or by radiation, the mean skin temperature and at what sites it is best measured, the thermal insulation of their colourful clothing ensembles, or the vapour permeability of their clothing materials.

Such knowledge would help to explain quantitatively the thermal balance of hobbits, would give us a theoretical explanation of their comfort conditions there, and might be useful in identifying potentially dangerous environments but it would not be needed to enable us to provide comfortable apartments for our hobbits. This is not surprising if we recall that achieving thermal comfort pre-dates by thousands of years the development of thermal physiology and the theory of heat exchange.

The thought-experiment illustrates what has become known as the adaptive model of thermal comfort. This adaptive approach notices that people use numerous strategies to achieve thermal comfort. They are not inert recipients of the environment, but interact with it to optimise their conditions. Among the means that people use are:

- the choice of areas of the globe suitable for habitation;
- the choice of the building site (for example, shelter from wind, shade from trees);
- the choice of design and construction of the building (for example, shape, orientation, thermal capacity, glazed area, thermal insulation);
- the choice of spaces occupied within a building at different times of day and year;
- the choice of heating or cooling systems, whether simple or sophisticated;
- the use of controls (thermostats, switches, valves, openable windows, blinds, ceiling fans);
- the choice of clothing suitable to the climate, season, indoor temperature and social setting;
- the operation of sometimes unconscious changes of posture and activity, and of any physiological acclimatisation there may be to the season of the year;
- the change of attitude to their environment, so that they are willing to experience a wider range of sensation without protest.

The adaptive approach notices that, provided there are adequate possibilities for selection and adjustment, people will act to make themselves comfortable if they wish to do so. It follows that the room temperatures that people find comfortable ('comfort temperatures' for short) are changeable rather than fixed. Discomfort is often caused by excessive constraints being placed on these processes of choice and adjustment, rather than solely by room temperature. Thermal comfort, then, is not to be seen primarily as a matter of the physiology of heat regulation and the science of clothing, but rather as a wide-ranging and intelligent *adaptive behavioural* response to climate.

Although the approaches from thermal physiology and from human behaviour are in principle complementary there have been and remain difficulties in reconciling their findings. From this has arisen the mistaken idea that the two approaches are irreconcilable rivals.

Humphreys first wrote this at a time when the 'rational' approach to thermal comfort, based on thermal physics and physiology, and measured in climate chambers, was dominant. He was making a case for a return to the field survey as the basic tool for understanding comfort.

There was a serious problem, he thought, with rational indices. Being based on work in climate chambers, they were not good at evaluating adaptive actions in everyday living. Further, the rational indices tended to limit needlessly the range of acceptable indoor environments, sometimes excluding conditions that field studies had shown to be well accepted. Also the rational indices tended in practice to favour constant and isothermal environments,[5] and the standards based on them were most easily met by providing a centrally controlled air conditioning system. Field results showed that comfort could equally well be achieved in what are commonly called 'passive buildings', where the control of temperature is achieved largely by thoughtful climatological design, and by giving control of the thermal environment back to the occupants. Such a building typically makes careful use of thermal inertia to stabilise the temperature, of blinds to control the admission of solar radiation, and of openable windows to control ventilation.

Constant indoor conditions can be comfortable, but so can variable conditions, if the changes happen in such a way that there are sufficient opportunities for people to adapt.[6] Well-designed field surveys can then be used to discern the weaknesses and strengths of differing comfort strategies and technologies and to suggest ways in which they can be optimised. Only with the adaptive approach can all parts of the whole system – occupant behaviour, the buildings themselves and their microclimates – become part of the comfort solution.

This book is the second in our trilogy on the adaptive approach to thermal comfort. The first volume outlined the principles and practice of adaptive thermal comfort. The third volume deals with the subject of how to design a comfortable building. This, the second volume, provides an overview of the history of the subject, as remembered by the authors, and continues with an overview of adaptive comfort methods and models developed over the past decades.

Notes

1 Folk, G. E. (1969) *Introduction to environmental physiology.* Lee & Febiger, Philadephia.
2 Spindler, K. (1995) *The man in the ice.* Trans: Ewald Osers. Phoenix, London.
3 The following paragraphs are an extract from Humphreys, M. A. (1995) Thermal comfort temperatures and the habits of hobbits, in: *Standards for thermal comfort*, Eds: Nicol, F. *et al.*, E. & F. N. Spon (Chapman & Hall), London, pp. 3–13.
4 Tolkein, J. R. R. (1937) *The hobbit.* Allen & Unwin, London.
5 Olesen, B. W. and Parsons, K. C. (2002) Introduction to thermal comfort standards and to the proposed new version of EN ISO 7730, *Energy and Buildings* 34(6), 537–48.
6 Humphreys, M. A. *et al.* (2013) Updating the adaptive relationship between climate and comfort indoors; new insights and an extended database, *Building and Environment* 63, 40–55.

2

ADAPTIVE BEGINNINGS

The scientific exploration of the adaptive model of thermal comfort rests on field study research, and the field study remains the chief, but not the only, method for its development. This is because adaptive behaviour is best studied in the normal habitat – that is to say, in the buildings where men and women live and work and in the outdoor spaces they frequent. The earliest and classic thermal comfort field study by Dr Thomas Bedford of the UK Medical Research Council laid down the methodological foundations. He had several years of experience in thermal physiology and in the measurement of the thermal environment before he started his field study in the early 1930s. His report – *The warmth factor in comfort at work* – was published in 1936[1] and concerned the thermal environment of workers in light industry during the English winter when 'artificial heating was in use'.

Bedford's work is still impressive:

- It was the first large-scale systematic collection of thermal comfort field-study data.
- He took careful and extensive measurements of the thermal environment of his respondents.
- It was innovative in the application of statistical methods to such data.

He chose 12 factories for his field study: 'a colour printing works, and factories devoted to the manufacture of aircraft, aluminium pistons, radio valves, wireless receivers [radios] and components, dry-cell batteries, waxed paper cartons, paper bags, starch and polishes, boots and shoes, ladies dresses and furniture'. In each factory, he chose a number of places at which to take measurements, where he interviewed the nearby workers and measured their thermal environment. He conducted 3085 interviews. Most of his respondents he interviewed just once, others

FIGURE 2.1 Thomas Bedford, thermal comfort field study pioneer.

Source: British Occupational Hygiene Society (BOHS).

twice, and a few on three occasions. Nearly all were young women and all were engaged in very light sedentary work. No one was interviewed until they had been at their workstation for at least half an hour.

He used a structured interview to establish each respondent's feeling of warmth. First he asked: 'Do you feel comfortably warm?' If they said 'yes', he then asked: 'Are you really quite comfortable, or would you rather have the room slightly warmer or slightly cooler?' If they said 'no', he asked: 'Are you feeling too warm or too cool?' If they said 'too warm', he asked: 'Just definitely too warm, or much too warm?' If they said 'too cool', he asked: 'Just definitely too cool, or much too cool?' From the interview, he had seven categories of possible response. The labels he gave to these categories form what is now known as the Bedford Scale:

Much too warm $(7)^2$
Too warm   (6)
Comfortably warm (5)
Comfortable   (4)

Comfortably cool   (3)
Too cool   (2)
Much too cool   (1)

Bedford's environmental measurements were comprehensive. At each measurement site, he took the air temperature with a mercury-in-glass thermometer, its bulb shielded from thermal radiation by aluminium foil. He used a sling hygrometer to obtain the humidity, which he expressed both as the relative humidity and as the water vapour pressure. Air speed was estimated from the cooling time of a silvered kata thermometer and the air temperature. (The researcher needed a thermos flask of hot water to warm the kata thermometer, a soft cloth to dry its bulb, and a stopwatch to time how long it took to cool.) He calculated the mean radiant temperature from the readings of a Moll thermopile, 46 readings in different directions being required for each estimate of the mean radiant temperature. The temperatures were measured at a height of 1.2 m, the average height of the heads of his respondents at the workbench, and he also measured the speed and temperature of the air 150 mm above the floor.

Additional readings were obtained from devices then in use that purported to measure the warmth of the environment as perceived by a human being. There was Vernon's 150 mm diameter black globe thermometer, Leonard Hill's plain (unsilvered) kata thermometer, and Dufton's 'eupatheoscope' – a somewhat complicated device then used at the UK Building Research Station (BRS) to measure the warmth of a room, which Dufton called its 'equivalent temperature'. A eupatheoscope was an electrically warmed black cylinder, 0.6 m tall and 0.2 m in diameter. Equivalent temperature combined the effects of the air temperature, thermal radiation and air movement into a single number.

At each interview, Bedford used the Moll thermopile to measure the temperature of the respondent's forehead and the surface temperature of her clothed body (the average of three representative places). With a specially constructed thermocouple thermometer, he measured the temperature of the palm of the hand and of the ankle, in between the ankle joint and the shoe.

It was Bedford who developed what was for thirty years the standard kit for measuring the indoor thermal environment: a 150 mm diameter black globe thermometer, a silvered high-temperature kata thermometer, and a sling wet-and-dry-bulb ventilated hygrometer. From the readings of these instruments and by the use of special nomograms,[3] the mean radiant temperature of the room surfaces, the average air speed, and the relative humidity of the air could be estimated. The air temperature was given by the dry bulb of the sling hygrometer.

A collection of data of this kind and on this scale had not been made before, so Bedford was the pioneer of the systematic thermal comfort field study. It was an astonishingly comprehensive set of measurements to take in a field study. He does not say how long it took to complete the measurements and interviews at each measurement place, but it could not have been much less than an hour. The whole survey would have taken about 1,000 hours of experimental time (if we assume that there were three respondents for each measurement site), more than could be done by one person in a single winter. He does not say how many researchers there were in his team to complete this prodigious effort.

The labour of statistical analysis is also astonishing, when we remember that he would have been using mechanical calculators. The results from one factory he separated for analysis, because the intense local radiant heat from soldering irons warmed the hands, chest and face of the workers, and could not be adequately expressed by the mean radiant temperature. Also the temperature at floor level was much colder. This separate analysis increased the labour.

First he calculated the mean, range and standard deviation of each of eleven environmental variables, and cross-tabulated the warmth sensations (comfort votes) against the air temperature. Then he argued that the test of any device or index that purported to indicate the warmth of a room for its occupants must be how well it agreed with the comfort votes of its occupants. So he calculated the correlation coefficient (Pearson's r) and its probable error for all the rival indices. Dufton's equivalent temperature (the eupatheoscope) came out top – but only just. In second place was the Vernon globe thermometer. Contrary to expectation, neither was much better than the simple air temperature or the mean radiant temperature. Neither was Yaglou's effective temperature any better than the air temperature, and Hill's kata thermometer's cooling power was much worse.

It occurred to Bedford that his large multivariate dataset would enable him to build a new index of subjective warmth, for it would be possible to estimate from it the contribution of the various aspects of the thermal environment to the sensation of warmth of his respondents. For this he used a method called multiple regression analysis, a laborious statistical procedure that had never before been used in heating and ventilating research. The statistical method had recently been explained for researchers by R. A. Fisher, perhaps the most eminent statistician of the time.[4] Multiple regression had already been used in agricultural research. It could predict the value of one variable from a combination of two or more others, for example the yield of wheat from the rainfall, the temperature and the dose of fertiliser applied. Bedford treated the responses on the Bedford scale as numerical values, and set up a multiple regression equation to optimise the prediction of the responses from the environmental data. All the calculations had to be done by hand or by using mechanical adding machines, so it took weeks to calculate a multiple regression equation. He found that the mean radiant temperature and the air temperature were almost equal in their effect on warmth. There was a significant cooling effect from air speed, while the effect of the humidity was very slight. The resulting equation to predict the Bedford scale value, S, representing how warm people felt, was:

$$S = 0.522ta + 0.478tw + 0.0372f - 0.001474\sqrt{v}(100 - ta) - 11.16$$

(*ta* is the air temperature, *tw* is the wall temperature (the mean radiant temperature), *f* the water vapour pressure and *v* the air speed. We have left the equation in the original units: temperatures on the Fahrenheit scale, vapour pressure in mm of mercury, and the air speed in feet/minute.)

From this equation he derived a revised form of the equivalent temperature. However, and probably much to his disappointment, *S* calculated from this formidable equation did not correlate significantly better with the comfort votes than did the simple Vernon globe temperature. Nonetheless he strongly defended its use on the grounds that it *would* have performed better had there been more variation in the air movement.

Bedford could not have known quite how capricious multiple regression can be, and he was perhaps fortunate to have found such a sensible equation from his data by this method, as we will later explain. He found the optimum equivalent temperature for comfort in the factories to be 65°F (18°C), rather low by modern standards for light sedentary tasks. People wore more clothes in winter in those days.

The above account gives only the flavour of Bedford's analysis. We have not discussed his analysis of the skin temperatures or of the clothing surface temperatures, or of his tables of the correlations among all his variables. His study was remarkable for its scope, for the quality of its measurements, and for the depth of its analysis. The original report still repays study, and we recommend it to any young researcher – or any older researcher for that matter! He set a high standard for thermal comfort field research, and his work remained influential for some forty years. Many field studies of thermal comfort were conducted in the UK and abroad after the publication of his report, and used the basic pattern he pioneered. Few used as many respondents, few were as comprehensive in their measurements and few were analysed as thoroughly.

Bedford's data are no longer extant. It would be fascinating to be able to re-analyse them from an adaptive perspective. He separated the data from one of his factories because the use of soldering irons caused the thermal environment to be of a different character, and the occupants to be accustomed (we would say 'adapted') to a different thermal environment. It is a pity he did not go the extra step of separating all the factories before the analysis. Because he did not, we are left wondering if all the factories had the same mean indoor temperatures, and, if not, whether the workers had adapted to their different environments. All this Bedford could have done without adding much to the burden of calculation. Nowadays we would certainly separate the factories in the first stage of analysis, and, if there were notably different temperatures in different zones within a single factory, we would separate these too. Had Bedford done so, he could well have stumbled across the adaptive model of comfort, as Humphreys and Nicol did some thirty years later, from a not dissimilar set of data.

By the early 1960s, the focus of research was shifting away from winter heating and steady-state heat loss calculations towards modelling the dynamic response of buildings in summertime, especially during the onset of hot spells. Highly glazed buildings were overheating on sunny days and during heat waves. Before the Second World

War, few offices occupied lightweight buildings. Heavy construction and relatively small windows were the norm. With the fashion for 'modern' architecture following the war, coupled with new construction methods, typically with concrete floor plates covered from below with false ceilings, and rooms separated by lightweight partitions, the new buildings were thermally typically 'lightweight'.[5] Couple this with a fashion for 'curtain walling', consisting of large glazed areas hung from the end of the floor plate to give optimal usable space, and the result was often serious overheating on sunny days. A situation (solar overheating in England) which before the war would have sounded like a joke had inadvertently been brought about by architects and developers,[6] and led to the increasing use of air conditioning in office buildings.[7]

The Admittance Method was developed by Edward (Teddy) Danter at the BRS at Watford (Figure 2.2) to estimate temperatures in unheated buildings in summertime. It was the first procedure of its kind, and made use of the analogy that exists between the flows of electricity in a circuit and the flows of heat into and out of the fabric of a building. Danter's research team were using an early analogue computer built from resistors and capacitors to perform the calculations. From this they developed a simple model of the diurnal heat flows into the building from the occupants and the sun, and the heat flows from the building by ventilation and fabric heat loss. In this way they produced the resulting indoor temperature response of the rooms in the

FIGURE 2.2 Teddy Danter, originator of the Admittance Method for estimating peak temperatures in buildings.

Source: BRS photograph, 1978.

building. Fergus Nicol started at the BRS in 1963, working with Alec Loudon and Peter Petherbridge in Danter's team, monitoring buildings to check the predictions of the Admittance Method. A group of typical 'modern' cellular offices at the BRS was fitted out with a variety of internal and external blind systems and their internal temperatures recorded. The radiant characteristics of the blinds were measured and, together with the solar geometry, used to estimate the thermal load on the building. This thermal load, combined with an allowance for the effect of the position of the blind inside or outside the building, was part of the office's dynamic thermal load. Feeding this input into the admittance model allowed the researchers to estimate the diurnal temperature swing. From this diurnal swing, coupled with the long-term mean temperature trends, the internal temperatures could be estimated. Thus the predictions of the model could be compared with the actual measured temperatures in the offices.

Also working at the BRS at this time was Charles Webb, who was the head of the Thermal Comfort Section. He had left the BRS before the Second World War to teach physics at the University in Singapore. When Singapore fell to the Japanese he was interned, and on release he resumed his lectureship. He remained there for some years before returning to England in the 1950s, where he re-joined BRS, working first in the Overseas Division (known as the Tropical Section), and later transferring to the Physics Division.

During his years in Singapore, he had become interested in how people achieved thermal comfort in an equatorial climate, and had studied vernacular houses to see how they provided comfort for their occupants. He was familiar with Bedford's field study on thermal comfort in the English winter and thought there was a need for similar research to provide a useful thermal comfort index for the equatorial climate of Singapore. Most of Bedford's respondents had been asked on just one occasion how warm or cool they felt. With such an experimental design it was not possible to quantify any systematic differences there might be between the respondents. So rather than take thousands of people and ask them just once, Webb decided to use a small group of people and get them each to keep records of how warm or cool they felt over a period of some weeks. The records were then to be compared with the thermal environments that they had measured at the same time (air temperature, mean radiant temperature, air speed and humidity). These longitudinal data records would enable an index to be established and also the differences between people to be quantified, for not everybody is comfortable in the same thermal environment. The work was first proposed in 1940, but the project was delayed until 1946 by the war. The fieldwork was conducted in 1949–50.

Webb did an exploratory analysis of his data while in Singapore, and completed it on his return to BRS. He called the combination of variables that produced the warmth response the 'calidity', a word he coined from the Latin, *calor*, heat. He supervised the calculation of the necessary multiple regression equations, so constructing a suitable thermal comfort index for the tropics – the 'Singapore Index'.[8,9]

He found there were indeed differences among his respondents in the environments they found to be comfortable.

Webb then did similar projects in dry tropical climates: summertime in Baghdad, Iraq and the hot–dry season in Roorkee, near Delhi in north India. Again he chose to use rather few respondents, but each provided records over an extended period (see Figures 2.3, 2.4). He calculated some basic statistics but never completed a full statistical analysis of the data.[10] On transferring to the Physics Division he started similar work in the UK. The aim was to update Bedford's equivalent temperature to put it on a sounder statistical footing, using the methods developed for his Singapore Index.

In 1964 Nicol was transferred to work in the Thermal Comfort section with Webb. The Admittance Method could predict what the peak temperatures would be in hot spells, and how the temperature varied during the day, but no one knew what temperatures were acceptable. Thus the need for Webb's work on thermal comfort became evident. When Nicol joined the section, Webb was still working on a physical model of calidity. Dorothy Hume was helping to analyse his Singapore data using multiple regression analysis. The Thermal Comfort section was the first at the BRS to use a digital computer. This involved a three-hour round trip to the privately owned ICL[11] computer workshop in Greenford, for an hour of using the computer.[12]

Webb then started a thermal calibration of six young men at the BRS. One of them was Nicol. They gave their responses every half-hour on Bedford's scale and then measured their thermal environment, using the standard equipment as recommended by Thomas Bedford (a 150 mm globe thermometer, a whirling wet and dry sling hygrometer, a silvered kata thermometer, a thermos flask of hot water to warm the kata thermometer bulb, and a stopwatch). They were encouraged to take readings at work and at home and so capture the full range of their everyday thermal experience.

FIGURE 2.3 Some of Webb's respondents at the Building Research Station at Roorkee, India.

FIGURE 2.4 Webb's respondents in Baghdad.

The individual calibrations again showed significant differences among the young men. Also the Pearson correlations between their comfort-votes and their individual environmental equations were remarkably high – above 0.9 – and much higher than the value Bedford had obtained when everyone's data were pooled (about 0.5 against equivalent temperature). The square of the correlation coefficient indicates the extent to which one variable is explained by the other, so it seemed to Webb that the thermal environment could explain about 80 per cent (0.9 × 0.9) of an individual's thermal sensation. This he compared with some 25 per cent (0.5 × 0.5) for Bedford's pooled data. The young men were comfortable at temperatures much lower than Webb had found for his respondents in Singapore. So it seemed that people had somehow adjusted (adapted) to their normal thermal environments. The actual temperatures for comfort were uninteresting, Webb thought; what was important was the calidity, the way the thermal variables combined to produce comfort.

On the strength of this finding, Webb proposed a larger project. The environmental variables were to be collected automatically using the electronic data-logging techniques that were then just becoming available. Respondents would note their thermal response on the Bedford scale by pushing a pressure-switch selected from an array of seven. There was another array with which to note their skin moisture. The data would be analysed in monthly batches, keeping the data from each respondent entirely separate. Six sets of instruments were built for the pilot study, each having a miniature globe thermometer (50 mm diameter) with a thermocouple sensor at its centre, a fan-ventilated wet-and-dry thermocouple hygrometer and a sensitive hot wire anemometer. These were to be controlled centrally by a specially built unit that included a data-logger. Webb planned to expand the experiment to twelve desk-units and to use them for a whole year, to capture the year-round differences in calidity among the individuals.

Michael Humphreys was employed in 1966 by the British civil service as a Senior Scientific Officer at BRS in the Education Section, and seconded to the Environmental Physics Division. The Physics Division's research covered heating, lighting, acoustics, thermal modelling, air movement within and around buildings, meteorology and thermal comfort. The task of the Education Section was to develop and run residential courses for building science teachers from technical colleges and polytechnics. The environmental physics syllabus would include courses on the Admittance Method and on thermal comfort. Thus it was that Humphreys joined Webb's section to produce course notes and to devise practical exercises for the course members. To be capable of teaching a course on this work, he needed to understand statistics and computer programming and become familiar with thermal comfort literature, with thermal physiology, human heat exchange with the environment, and also with the psychophysics of subjective sensation.

Webb had a large filing cabinet full of photostat copies of journal articles on human thermal response, going right back to the eighteenth century, and filed by date. In his bookcase were some useful books: Bedford's 1936 report, Newburg's *Physiology of Heat Regulation and the Science of Clothing*, Burton and Edholm's *Man in a Cold Environment*, and the encyclopaedic *Temperature: Its Measurement in Science and Industry*, which included Benzinger's articles on the physiology of thermal regulation. Mastering these materials was Humphreys's first task.

To consolidate his understanding of human comfort, Humphreys built a simple mathematical model representing the heat flow from the body core to the surroundings. It was based on Burton and Edholm's work.[13] He kept the model simple so that he could 'see' the basic processes and how they related to each other. Thermal models are of two kinds. Some aim to give exact predictions. They need not be easy to understand, because precision is more important than clarity. Other models are for teaching. These need to make the processes clear, but they need not be precise. Humphreys's model was of the second kind – a teaching aid. It showed that the globe thermometer temperature ought to be an adequate description of the thermal environment at moderate temperatures and in the absence of draughts, and that clothing insulation and metabolic rate had profound influences on the temperatures required for comfort. Further, the air around the body could be treated as if it were a layer of clothing. Increasing the air movement was equivalent to removing an article of clothing.[14] The model was used for teaching on the courses, and a revised version of it is given in Part II of this volume.

Portable instrument-sets (handsets) were made for teaching purposes so that the course members could calibrate themselves thermally over two weeks. The handsets (see Figure 2.5) had an analogue read-out (a voltmeter) for the environmental variables, and the course members kept an hourly record of the variables in a thermal diary, together with their responses on the Bedford scale. These records were punched on to paper tape, and were taken off-site to analyse on the ICL computer, using a multiple regression program. The course members then studied and compared their calidity equations, so obtaining insights into their thermal comfort. During the course, Webb explained the origin and development of thermal comfort indices, and Humphreys explained the simple thermal model and how it gave insights into the way a thermal environment might be expected to affect the occupants of a building.

Not everyone in the Environmental Physics Division had a high opinion of thermal comfort research. In their opinion, it was not proper science and it was not building research. Why waste effort on something incapable of producing precise results? And if there must be comfort research, why not do proper experiments in a thermal laboratory where conditions could be controlled and balanced

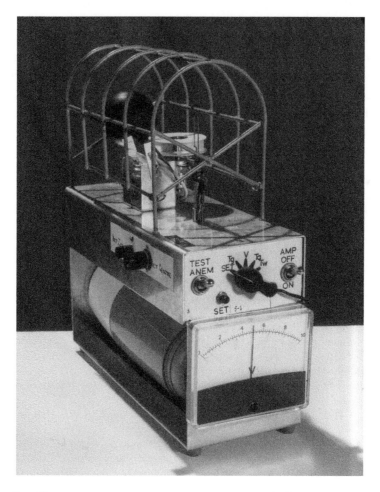

FIGURE 2.5 Portable instrument to measure air temperature, wet–bulb temperature, globe temperature and air speed, 1967. The thermos flask in the base contained water and ice, which served as the reference temperature and kept the solid-state amplifier at a constant temperature. The length of the instrument box was about 240 mm.

experimental designs used? Webb's response was that there was an honourable British tradition of empirical scientific field investigation. To understand camels, it was better to go to Arabia and study them in their native habitat than to look at them in a cage at the zoo. So it was better to look at people in their everyday life rather than in a thermal laboratory (climate chamber). Even if thermal indices were to be developed in laboratory conditions, they would need to be validated in everyday life – so why not go straight to the field study to obtain the index, as Bedford had done? Furthermore, buildings were for people, and it was not sufficient for BRS to concentrate research on bricks and mortar and heating systems; we needed to know how the buildings affected their occupants.

Fergus Nicol, now an experienced researcher, re-joined the Thermal Comfort Section in 1966 on his return from a year teaching building physics in the School of Architecture of the Kwame Nkrumah University of Science and Technology in Kumasi, Ghana. His time there had strengthened his curiosity about thermal comfort in hot climates. Now back in the UK, he spent half his time at the BRS as an experimental officer and the other half teaching building physics at the Architectural Association in London. On the closure of the Education Section of the BRS, Humphreys was transferred to the Thermal Comfort Section. So began the years of friendship and collaboration between Humphreys and Nicol.

Notes

1 Bedford, T. (1936) *The warmth factor in comfort at work.* Medical Research Council Industrial Health Board, HMSO, London.
2 The numbering is ours. Bedford numbered 1: Much too warm, through to 7: Much too cool.
3 A nomogram was a chart designed to solve an algebraic equation without doing the maths. Nomograms disappeared, along with the slide rule, with the advent of cheap computing.
4 Fisher, R. A. (1925) *Statistical methods for research workers.* Oliver & Boyd, Edinburgh and London. The book went through 14 editions, the last appearing in 1970.
5 Gray, P. G. and Corlett, T. (1952) A survey of lighting in offices. Appendix 1 of *Postwar Building Research No. 30.* HMSO, London.
6 Loudon, A. G. (1968) *Window design criteria to avoid overheating by excessive solar gains.* Building Research Station Current Paper 4/68.
7 Black, F. and Milroy, E. (1966) Experience of air conditioning in offices, *J. Inst. Heat. & Vent. Eng.,* September, 188–96.
8 Webb, C. G. (1959) An analysis of some observations of thermal comfort in an equatorial climate, *British Journal of Industrial Medicine* 16(3), 297–310.
9 Webb, C. G. (1960) Thermal discomfort in an equatorial climate, *J. Inst. Heat. & Vent. Eng.* 27, 297–303.
10 Webb, C. G. (1964) Thermal discomfort in a tropical environment, *Nature* 202(4938), 1193–4.
11 International Computing Limited, a large British computer hardware and software company.
12 The hire-charge was expensive, and often the session resulted in no progress because the punched paper tape had a hole or two in the wrong place.
13 Burton, A. C. and Edholm, O. G. (1955) *Man in a cold environment.* Arnold, London.
14 Humphreys, M.A. (1970) A simple theoretical derivation of thermal comfort conditions, *J. Inst. Heat. & Vent. Eng.* 38, 95–8.

3

THE EMERGENCE OF THE
ADAPTIVE POINT OF VIEW

Charles Webb's comfort research team[1] at the BRS was formed in response to the need to define acceptable temperatures in increasingly lightweight and highly glazed office and school buildings, a need that required a better understanding of thermal comfort. Comfort studies at the BRS in the 1930s had centred on the British winter climate, but after the Second World War the need was seen for the study of thermal comfort in summertime and in some of the hotter climates in the British Commonwealth countries around the world. In the UK, there was a concern that modern buildings might require air conditioning, and that this, together with the increasing use of central heating systems in dwellings, might place great demands on the supply of energy.

Humphreys redesigned the desk monitor units used in field studies to overcome some early problems. Twelve new units (see Plate 3.1) were assembled and connected to the central data-logger unit (see Figure 3.1). He wrote a computer algorithm to calculate Yaglou's effective temperature (a thermal index then widely used) to avoid having to read off each value from the strangely shaped ET nomogram. Beryl England, an experimental officer, wrote data reduction programs that converted the raw voltage outputs from the desk monitors into standard units. Twelve volunteers were found from the research and clerical staff in the physics building to be the subjects. One of these was M. R. Sharma, who had been one of Charles's respondents during his Indian project. He was at BRS for a year on secondment from the BRS at Roorkee. Sharma's place on the study meant that his Indian and his English responses could be compared.

Each month, England took the punched paper tapes from the data-logger to the ICL computer at Greenford for processing, because there was still no computer at BRS. She returned with printouts of the regression analyses. To avoid disputes about which variables should be entered into the regression equations used to predict the warmth sensation, she had set up the program to calculate all 127 possible

FIGURE 3.1 The central data-logger unit (the big cabinet) with some of the desk instruments and the paper tape punch (right of picture) at BRS in 1968.

Source: BRS photograph.

combinations! So each month there was a huge data output that was scrutinised by Humphreys, Nicol and Webb to identify patterns and trends.

The data arrived in the form of tables of regression statistics and these were transformed into graphs by a program Humphreys had written to print rudimentary scatter plots of the variables against each other, and of the principal variables against the Bedford scale. This was done for each respondent for each month's data.

After several months it became clear that no single combination of thermal variables could be found that best predicted the sensation of warmth. The 'best set' of variables changed from month to month, so no stable personal calidity equations could be derived from these data. No combination was consistently better than the air temperature or the globe thermometer temperature. Also, a person's mean comfort vote varied almost randomly from month to month in a manner unrelated to the room's mean temperature for that month. So it seemed that a person had no obvious pattern of thermal response, but was about as different from himself or herself in successive months as from other people in the same month (see Figure 3.2). These findings were disappointing. What could be made of such data? What patterns did it contain?

There was also an anomaly related to the regression coefficient of the Bedford scale vote on the room temperature. The regression coefficients were correlated with the standard deviations of the room temperature, and that should not have been so. The greater the standard deviation, the higher was the regression coefficient. The research team could not explain it, because a regression coefficient is theoretically independent of selection on the axis of the predictor variable. It should therefore be independent of the standard deviation of the temperature. This led them to consider more deeply the properties of correlation and regression, but there was no explanation to be found. Not until many years later was the explanation found. It is the topic of Chapter 22.

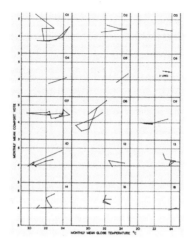

FIGURE 3.2   Facsimile of an old slide showing the month to month variation in mean warmth sensation against mean temperature for the 12 respondents, showing the absence of any systematic dependence.

Source: BRS slide.

The air movement had less effect than expected on the Bedford scale votes. Webb and Humphreys thought that this was perhaps because the response time of the hot wire anemometers was too rapid. Kata thermometers had usually been used in thermal comfort work and they had a slow response. The team spent a month constructing and calibrating their own 50 mm diameter warmed-body anemometers with a longer response time. At this stage any of the respondents who wished to be free were released, and others were recruited to take their places. Analysis later showed the air speeds recorded by the new anemometers had even less effect on the Bedford votes. The explanation lay elsewhere.

At about this time, Charles Webb reluctantly had to retire, leaving Humphreys in charge, with Nicol part-time. Together they were responsible for finishing the experimental work, completing the data analysis and writing the papers for publication. Once the experimental phase was complete and all the paper tape processed, it became obvious that the mean Bedford vote for each respondent's batch of data depended little on the mean room temperature for that batch, no matter what thermal index was used. The mean vote always had much the same magnitude, whether it came from Bedford's 1936 study, from Webb's data from Singapore, or from the summer data from Baghdad or Roorkee (see Figure 3.3). Furthermore, Sharma, when in Roorkee,[2] had found temperatures in the mid 30°Cs comfortable, but when in England and in normal English clothing he found temperatures in the low 20°Cs comfortable. He had fully adapted. It seemed that the actual room temperature mattered very little – what mattered was that the person was accustomed to it. He, and the others in the study, had *adapted* to the mean room temperature experienced during that month. It was the *departures* from the temperature they were accustomed to that were associated with discomfort. Some but not all of this adaptation could be explained by differences in clothing between summer and winter, and between England, Singapore, Baghdad and Roorkee. But the clothing insulation was not quantified in any of these projects – nor had it been in any previous thermal comfort field study.

The air speeds in the BRS offices and laboratories were found to be almost always very low, even when windows were open for summer ventilation. In these circumstances it was the natural convection from the person, rather than the air speed in the room, that governed the convective heat losses. So the measured air speed would have very little place in any equation predicting warmth in these data.

The air temperature and the globe temperature differed little from each other. So it would not matter much which of them were included in any predictive equation. The very high correlation between them also meant that it was unwise to include both in the same regression analysis, because the coefficients in multiple regression equations can become unstable when the predictor variables are strongly correlated. If both air temperature and globe temperature were included in the same equation, their relative effect on the warmth sensation varied wildly from person to person and from month to month. Bedford had been lucky with his multiple regression equation.

So it seemed for these data that analysis in terms of the globe temperature would be sufficient. This was also what we had surmised from the simple thermal model, which had suggested that, for still air and moderate humidity, the globe temperature should be an adequate index for thermal comfort. The humidity would have little effect on the rate of heat loss from the body at the temperatures encountered in the BRS offices, and the analyses showed that it had barely any effect on the warmth sensation.

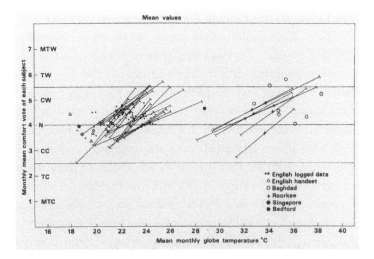

FIGURE 3.3 Facsimile of an old slide showing the relation between the English data and that from Singapore, Iraq and North India.

Source: BRS slide.

Webb's pilot study had found very high correlations – 0.9 and above – between the sensation of warmth and the thermal environment for the separate respondents, but in the main project the correlations were much lower, averaging only about 0.5. In the pilot study, the young men had experienced a considerable variety of thermal environments because they had taken measurements both at home and at work, day and night. In this study, the subjects were in their offices. A person in a BRS office during the working day did not experience such varied environments, because the building was heavyweight and the heating system quite well controlled. Indeed, the standard deviation of room temperature during the working day for a month in an office in the BRS building was typically only about 0.6 K. For a given underlying regression equation, the smaller the variance of the predictor variable the lower is the correlation coefficient. The small variation of room temperature was good for comfort, but bad for statistical analysis (see Figures 3.4 and 3.5).

So what was learned from the project? Three things stood out:

- The globe temperature was a sufficient index of the thermal environment for these data, and there was no case for using any more complex index.
- The occupant's monthly mean thermal sensation in the office was virtually independent of the monthly mean air or globe temperature. This applied from 18°C to 25°C. If the data from Baghdad and Roorkee were brought in, the upper limit extended to the mid 30°Cs (see Figure 3.3).
- But *within* each month for each respondent there was a correlation between the temperature and the warmth sensation. So thermal sensation was related to the variations of the temperature in the office during the month – it seemed that discomfort was caused by too great a departure from the currently normal temperature rather than by that temperature itself.

Items two and three are the basis of what has become known as the adaptive model of thermal comfort, although it was not so expressed at the time. Nicol and Humphreys just knew that people had adapted to their ordinary room temperatures.

Based on the project findings a joint paper was produced suggesting that an adaptive view of thermal comfort best explained the experimental findings. Unconventional methods of graphical presentation had to be devised because of the complex structure of the study and the large number of variables. This necessitated the results being presented in compact but clear diagrams. Despite continuing discussions and some concerns at the BRS, the paper was well received when presented at a London meeting of the Environmental Group, a group that brought together forward-thinking engineers, architects and researchers who shared an interest in the quality of the indoor environment. Some months later it was passed for publication, by which time it had become two papers – the first presenting the simple heat flow model and the second the experimental work and its results. They were published in August and November

FIGURE 3.4 The Physics Building at BRS. The appearance is modern but the construction traditional and heavyweight (concrete and brick).

Source: BRS photograph.

FIGURE 3.5 The Physics Building. Office interior showing internal venetian blinds, openable windows and manually controlled radiators. Michael Humphreys is at his desk.

Source: BRS photograph.

1970 in the *Journal of the Institution of Heating and Ventilating Engineers*, a British journal that normally published the BRS research in the field of heating and ventilating.[3,4]

By this time Humphreys and Nicol were aware that their viewpoint on thermal comfort had undergone a fundamental change. The focus was no longer on the physics of heat exchange and the physiology of human thermal regulation. Though these remained important components, it was human adaptive behaviour that was driving the system. By what processes did people achieve comfort in a changing environment? What hindered the processes and led to discomfort? What were the limits on these processes? As they put in the paper:

> People do not stay uncomfortable if they can help it. As well as changing their environment they can adapt to it by changes in activity, posture, clothing or physiology. If such adaptation were always complete, the comfort vote would depend little upon the temperature of the environment . . . This is perhaps a new way of looking at thermal comfort. The purpose is no longer to predict an optimum temperature. The optimum depends on clothing, activity and physique, which are rarely known in detail. What matters is the [room temperature] variability which can be permitted without causing excessive discomfort.

Nicol was no longer teaching at the Architectural Association and now divided his time between working at the BRS and working at the Human Physiology department at the Medical Research Council's Hampstead premises research unit, under Otto Edholm's direction, and working closely with Ronald Fox. The unit specialised in human thermal physiology and in particular the response to the thermal environment. Most of the research concerned extreme environments and the limits of human tolerance, but Edholm and Fox were both interested in comfort research. So Nicol began a project on thermal comfort and productivity in telephone exchanges, but had to abandon it because of an industrial dispute. He started a new project in an insurance company's office building, but the building was so solidly constructed that there was not enough temperature variation to permit statistical analysis. The tiny variations in room temperature scarcely affected the comfort votes, a circumstance that was good for adaptive comfort but bad for the research project!

Notes

1 The BRS comfort team was happy and collegiate and Charles Webb a kindly leader. Over cups of Blue Mountain coffee he would regale them with anecdotes and interesting observations. He was never able to calibrate his own thermal sensation because the malaria he contracted in Singapore had disordered his body-temperature regulation. This resulted in his thermal sensations being largely unrelated to his room conditions. His best remembered anecdote related to Dr Bedford, who had once been to India to advise on thermal comfort and had found the tropical climate most disagreeable. When it was suggested he might be more comfortable if he took off his jacket and tie, undid his collar and wore his shirt outside his trousers, he replied: 'There are limits.' Social decorum was in conflict with comfort.
2 Sharma continued thermal comfort research in Roorkee and later published an excellent paper on the application of multiple regression analysis: Sharma, M. R. and Ali, S. (1986) Tropical summer index – a study of thermal comfort in Indian subjects, *Building and Environment* 21(1), 11–24.
3 Humphreys, M. A. (1970) A simple theoretical derivation of thermal comfort conditions, *J. Inst. Heat. & Vent. Eng.* 38, 95–8.
4 Humphreys, M. A. and Nicol, J. F. (1970) An investigation into thermal comfort of office workers, *J. Inst. Heat. & Vent. Eng.* 38, 181–9.

4

SUMMERTIME OVERHEATING
IN SCHOOLS

It had become clear from studies in offices that there is no such thing as a single 'comfort temperature' at which all office workers are comfortable. The experience of comfort was the result of the complex feedback processes that comprise adaptation. An opportunity then arose to see whether the same was true of thermal comfort in schools.

Many new schools had been built since the end of the Second World War. One concern of the modern movement has always been the importance of light in architecture, and in particular the provision of daylight. Combined with a lightweight construction this had led to a growing and widespread complaint that many of the recently built highly glazed 'modern' schools were overheating on sunny days (see Figure 4.1). Humphreys was then working in John Langdon's newly formed Human Factors Section at the BRS. Langdon, collaborating with Alec Loudon, had already correlated the peak summertime classroom temperatures, calculated from the building design by using the new Admittance Method, with what the teachers said about the severity of the overheating in their classrooms. This correlation had demonstrated two things: that the predictions of the Admittance Method bore a clear relationship to summertime temperatures in classrooms in daily use, and that the retrospective comfort evaluations of the teachers were valid. If either had been untrue the correlation could not have existed.[1]

A schools comfort research project was designed including both primary and secondary schools, in which summertime temperatures in classrooms would be measured, and their effect on education and upon the comfort of the children and teachers assessed. In the 1960s, novel methods of teaching were being introduced, centred on the child rather than on the subjects being taught. Particularly in the primary schools, the classroom atmosphere was relaxed and teachers were free to teach largely as they thought best (see Plate 4.1). There were no national curricula. The research team consisted of Humphreys and Langdon with two newly recruited

FIGURE 4.1 A classroom block at a new secondary school, Hemel Hempstead, Hertfordshire. Lightweight construction, large area of glass, but with the possibility of cross-ventilation.

Source: BRS photograph.

psychology graduates, Mike Crook and Pat Denison. Research methods were discussed and a number of schools visited. Meetings were held with local government education officers to hear their experiences of summer overheating in the schools in their region, and to be shown round a selection of schools of various designs.

A similar problem had also been encountered in Scandinavia where new schoolrooms were overheating in the low-angle spring sunshine. Ib Andersen, a medical researcher, and Gunnar Lundqvist, an engineer, had been measuring temperatures in classrooms and found temperatures as high as 23°C in the spring sunshine.[2] It was difficult to understand why there should there be complaints at 23°C. Perhaps the teachers were adapted to a temperature some degrees lower? David Wyon, a young researcher who had just completed his PhD on the thermal comfort of surgeons in operating theatres at the Medical Research Council's Experimental Psychology Unit at Cambridge, had moved to the Swedish Building Research Institute to research the effect of temperature on children's learning. There they had a suite of temperature-controlled experimental classrooms where the children could be observed through walls of one-way mirror.

Humphreys went to visit Anderson, Lundqvist and Wyon in the autumn of 1969 and also visited a young Danish researcher named Ole Fanger, who had already published some impressive research. His heat exchange model used the latest climate-chamber research data from Denmark and the USA and could be expected to be far more accurate than Humphreys's educational model. They discussed heat exchange between the human body and the thermal environment and its relation to thermal comfort. Humphreys showed Fanger some of the unpublished BRS results, including those from Baghdad and Roorkee, but they did not fit Fanger's sophisticated heat flow model[3] any better than they did Humphreys's own simple one. Why might this be? Could people really adapt to achieve comfort in these hot environments? Humphreys believed they could, but Fanger thought they could not.

It seemed to Humphreys that results from laboratory tests of children performing mental tasks would be of doubtful relevance to education in the flexible ethos that then prevailed in British schools. So it was decided that the research must take place within the ebb and flow of the day-to-day lives of children at school. It was to be fieldwork again. Teachers' experiences indicated that children who were interested and well motivated would learn well – otherwise they would not. Only in extreme conditions would the thermal environment act as a limiting factor on learning, for example on days when the heating system had broken down and the children were cold despite wearing overcoats, or on a very hot summer's afternoon, when classroom tasks were apt to continue at a more leisurely pace. It was necessary to remember that the children and their teachers were human beings. Children should not be regarded as machines set to run on mental 'full-throttle' all day. Experiments that looked at how quickly and accurately children did arithmetic and various other mental tasks under test conditions, when they would almost necessarily be highly motivated, had little to say about children's education.

In addition to these considerations, Wyon's hypothesis about the effect of temperature on physiological arousal, and the postulation of differing optimal levels of arousal for different kinds of task, suggested a highly complex relationship between room temperature and task performance. Different kinds of mental work would need different room temperatures for their optimal performance, assuming the same level of clothing insulation. This has proved to be correct and since those days the effects of temperature and air quality on the performance of mental tasks have been demonstrated beyond reasonable doubt, both for children and for adults. Their practical relevance remains questionable. Task performance cannot be equated with overall productivity, and 'productivity' is a strange word to use for the long-term outcome of the whole process of educating children.

An altogether different approach was therefore adopted. It seemed likely that clothing was the principal method that the children would use to adapt themselves to their indoor thermal environment. Further, if children were successful in changing their clothing to achieve comfort it would be possible to deduce their comfort conditions from records of clothing alone. So, during a summertime pilot study, we kept a record of the children's clothing during their lessons, the numbers who were in the lightest socially acceptable clothing being counted. For boys, this was to be without a jacket or pullover,[4] while for girls it was to wear a light summer dress with no pullover. The clothing was found to depend strongly on the classroom temperature (see Figure 4.2). From these results, Humphreys developed a method of deducing comfort conditions from the clothing records. It is explained in Chapter 31 in Part II of this book.

A record of window-opening and the use of blinds was also kept. The window-opening correlated with both the classroom temperature and the outdoor temperature. So a powerful comfort-control mechanism was in action. The combined effects of changes in window opening and in clothing enabled the children to stay comfortable for most of the time despite large variations in the outdoor temperature.

In the light of this pilot study, it was decided to concentrate on clothing behaviour and leave the measurement of the children's performance to other research teams. A full-field time-lapse cine-camera system, designed, built and installed in a classroom in a local school, would be used to obtain a record of the children's clothing and the room temperature. From the colour photographs (shot into a convex mirror so as to view the entire classroom area, including the windows and blinds), it was possible to count the number of children who were wearing minimum dress. Each

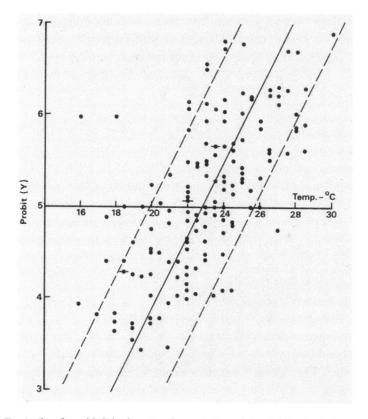

FIGURE 4.2 Facsimile of an old slide showing the variation of the children's clothing with the room temperature. Each point is the proportion in minimum clothing, expressed on a Probit scale, and represents a count of the class – usually some 30 children.

photograph also included the temperature and the time (see Plates 4.2, 4.3). Analysis of the photographs showed a coherent pattern relating the room temperature and the clothing, so the following summer (1970) the project was extended into more schools. Margaret Gidman and Barbara Buller painstakingly analysed photographs from more than 900 lessons.

The clothing data did not yield the same temperature for comfort in all the schools, and enquiry revealed that there were various social pressures acting on the children, sometimes affecting their comfort:

> different attitudes to clothing existed in the different schools. In school C it was the custom for children not to wear their blazers in the classroom, but to leave them in the cloakroom. This at once removed any 'respectability' that might attach to wearing the blazer in class, and which sometimes makes children wear blazers when they are not required for thermal comfort. . . At school B the children were encouraged to wear their full uniform when going from one room to another (the purpose was to reduce lost property). This rule would tend to cause over-dressing, and this effect is confirmed by the data. At school A there was no pressure from rules, but the staff were aware of some social pressure among the children. The girls preferred a skirt and jumper to the summer dress they were permitted to wear, and the boys were reluctant to remove their blazers. This too causes over-dressing, and the effect is evident from the data.[5]

This work in secondary schools had shown that, although the variation of clothing was a principal means of adaptation to the thermal environment, it was subject to various pressures that were sometimes not immediately apparent. These pressures meant that populations would sacrifice a certain amount of thermal comfort for what we might call 'social comfort'. Rules and conventions about clothing, clothing fashions and personal preferences could act to prevent complete adaptation. They acted as 'constraints' upon a powerful control system that had thermal comfort as its goal.

The research method also lent itself to the exploration of how clothing varied with time. English summer weather is very variable. There are often large changes of temperature from day to day, rainfall is sporadic and sunshine unpredictable. The effect of such weather on lightweight classrooms having large areas of glazing is often to cause large swings of temperature within a school day and large changes from day to day. Such conditions are ideal for behavioural clothing research, but not good for thermal comfort. The research question arose: If there were a change in the room temperature, how long would it be before the clothing 'caught up' and comfort was restored? Children rarely changed their clothing during a school day, despite the often quite large swings of room temperature. Their response was only about a tenth of that needed fully to adapt to the temperature change. The changes from one day to the next were rather greater, and those from one week to the next greater still – about half that needed for full compensation for a change in temperature.

Earlier in the project the question arose about how to use a moving average temperature to describe the variation of the children's clothing in the data collected during the summer of 1969. After much thought and many pages of pencil and paper sketches and algebraic formulae, it became clear that the mathematics of clothing change could be similar to the mathematics of radioactive decay – exponential in character:

> Because the children's clothing would not be expected to follow small or rapid temperature variations without some time delay, it seems likely that clothing will follow a moving average temperature of some kind. A possible type of moving average to try is the exponentially weighted running mean. In this, the present value of temperature is given greatest weight, and each previous value is given a weight which reduces according to how long ago it occurred. It follows that the running mean tends towards the instantaneous temperature at a rate proportional to its displacement from it. This running mean has two connected features. It smoothes the series of temperatures, and lags somewhat behind any changes that occur. Both these features seemed applicable to the variations of the clothing.[6]

Using the moving average implied a time-related structure within the scatter of the clothing data. The existence of such structure gave persuasive evidence that the variations of clothing were not just random. It was important to establish this to justify the unusual statistical model developed for analysing the clothing data. These moving averages have since proved useful in describing several kinds of adaptive behaviour and have the advantage of being very simple to use. This started an interest in the statistical analysis of mutually correlated time series – for in comfort studies the data can be regarded as several correlated time series: the outdoor temperature; the indoor temperature; the clothing insulation; and the sensation of warmth.

The team's psychologists had no success in finding effects of the environment on the children's learning in the schools despite the various procedures and tests that were tried. None showed significant effects in everyday classroom life. That did not imply that the children's learning was unaffected by room temperature – rather that the background scatter attributable to the numerous other influences on learning was so large that any effects of temperature were obscured. But this in itself suggested that room temperature was not a very important influence on learning in the everyday classroom conditions encountered. So if the aim of the research was to improve children's education, close control of classroom temperature would not be the first priority.

When experimental work in secondary schools had been completed, attention was turned to primary school children. It had not been possible to relate room temperature to the children's learning in secondary schools, so it seemed very unlikely that it would be possible in primary schools, where classroom life was more varied in its activities and teaching methods. Teachers in local primary schools said that they had noticed changes in the children's mood and behaviour that seemed related to the weather, and so perhaps to classroom temperature. The research plan was therefore to include records of the classroom temperature, the children's subjective warmth response, and their clothing. Teachers would be asked how well they thought the lessons had gone, to see if this opinion might be related to the weather or to the classroom temperature. A set of semantic differential scales was designed to capture the teachers' perceptions of the children's responsiveness to their lessons.

Young children could not be assumed to respond to room temperature in the same way as adults. Their basal metabolism is higher, they are often much more active and their activities vary more often. All this suggested that the young child's relation to the thermal environment would be less precise than that of older children or adults. Nor could it be assumed that young children would be capable of using rating scales of warmth and comfort as could older children. Langdon was familiar with the recent advances in the psychology of child development from his collaboration with Piaget, whose thinking on child development was, and remains, influential. Piaget's elegant experimental work with young children demonstrated to the satisfaction of most researchers that children were physically incapable of grasping abstract concepts until a certain age. As the child grew older the brain changed physically, and so the range of possible concepts broadened.

Only in recent years has this orthodoxy been seriously questioned. Recent work shows that young children are much more mentally capable than Piaget thought. What he had attributed to children's mental inability could often be attributed to problems in communication (misunderstandings) between a very young child and the researcher. Newer work with improved methods of communication, recently summarised by Michael Siegal, shows that young children are able to grasp more concepts than had been thought.[7]

It was necessary to discover whether these young children could use rating scales, and if they could, scales would have to be devised that were suitable for them to use. This work is described in Chapter 19. Checklists were made for the children to describe what clothes they were wearing, and Barbara Buller drew beautiful icons of the various clothing items the children wore at school. The children would tick these on their response cards (Figure 4.3). Children's fashion is not static, and in the months between the preparation of the drawings and starting the experimental work, some little girls started to wear shorts with shoulder straps (called 'hot-pants') to school. They would draw their own little icons and tick-boxes for these.

After completion of the pilot studies, which had shown that most young children were indeed capable of using our simple rating scales, the fieldwork began in the schools. Ten classes of young children were investigated, two classes in each of five schools, for

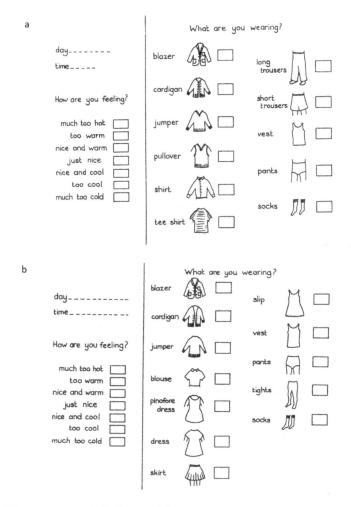

FIGURE 4.3 The response cards for boys and for girls.

Source: Barbara Buller.

two weeks in June 1971. The period was unseasonably cool, so the study revealed little about the effects of overheating on the children. The study was repeated in 1972 and again it was unseasonably cool. The weather began to warm up in the following week, and one teacher had asked to continue the experiment for a further week. So there were data showing the response to the onset of hot weather from just one class.

Each day the teachers posted us a packet containing the children's completed response cards (see Figure 4.3, Plate 4.4), together with the temperature chart from the thermograph and their assessments of how well the children had responded in each of their lessons. On opening the packets, each school could be identified by the smell – a characteristic mixture of the children's body odours, of organic compounds from the building materials and furnishings, of the smells of the cleaning products and polishes used by the caretaker, and of teaching materials such as paints and pastes. It did not occur to the team that these contaminants in the air might affect the children's attention and well-being. These packages meant that the quality of the data could be checked day by day and any problems identified in good time. One day, a boy had ticked boxes for both long trousers and short trousers. Two pairs of trousers seemed unlikely, but a visit to the school showed that it was not a mistake. On wet mornings, some boys were sent to school with long waterproof over-trousers, and some kept them on all day.

The cool summers were unfortunate from the point of view of obtaining criteria for classroom overheating, but nonetheless a lot was learned about the children's thermal behaviour, and something about school thermal design. The orientation of the classroom was important. Lightweight classrooms whose glazing faced east could maintain a fairly steady temperature during a sunny school day (9 am to 3.30 pm). The thermographs showed that the morning sun had warmed the room before the children came to school, and, as the outdoor temperature rose during the day, the sunshine moved off the façade, so keeping the indoor temperature fairly steady. Classrooms with west-facing glazing could be too chilly when the children arrived in the morning, because the room had cooled down overnight. It could become too hot in the afternoon because of the joint effect of afternoon sun on the façade and the higher outdoor temperature in the afternoon. So the new classrooms could be successful, but only if careful attention was given to their orientation, or if there could be cross-ventilation from open windows on opposite façades. One of the new schools in the sample had classrooms with three walls out of four entirely of glass – but it had copious through ventilation. It was difficult to control, but it did not overheat if enough windows were open.

It was believed that overheating was attributable to the combination of lightweight buildings and large glazed areas, and it was assumed to be confined to the newly built schools. To check this assumption, an older school had been included in the sample. Older schools were solid brick buildings of high thermal capacity, with high ceilings and tall windows giving good daylight penetration. In the older school, the indoor temperature in the classrooms during the afternoon was often below the outdoor temperature, showing the effect of the thermal inertia (thermal mass) of the structure. Also there was little temperature variation in the classrooms during a school day. The temperature variation was so slight that a correlation between the children's warmth sensations and the room temperature could not be detected. This was good for the comfort of the children, but bad from the experimental point of view.

What about the children's thermal behaviour? Did they follow the same adaptive pattern as was found from the office-workers at BRS? The mean classroom temperature was calculated for each of the classrooms during the school day for the two-week experimental periods. The mean temperatures ranged from 17.5°C to 23.1°C. The corresponding mean thermal sensation of the children was also calculated. Nearly all the mean sensations lay between 'just nice' and 'nice and warm'. There was no significant correlation between the mean classroom temperatures and the mean votes of the children – the children did not feel significantly warmer in a classroom whose mean temperature was 23.1°C than in one whose mean temperature was 17.5°C (see Figure 4.4). These young children were adapting to their mean room temperature just as had the adults in the offices at BRS. Some of this adaptation may have been helped by adult influences (mum helping to choose sensible clothes) but the overall effect was the same.

Some of this adaptation was attributable to changes in clothing. The children wore less in the warmer classrooms, but the difference was not enough to explain their adaptation. Where had the rest of the adaptation come from? Perhaps the children became less active in warmer rooms? Maybe their posture became more open?

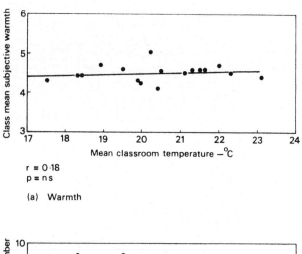

r = 0·18
p ≡ ns

(a) Warmth

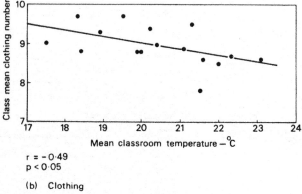

r = -0·49
p < 0·05

(b) Clothing

FIGURE 4.4 Young children's thermal responses. The upper figure shows how little their mean thermal sensation depended on the mean classroom temperature. The lower figure shows that their clothing depended on the temperature. Each point is the mean for that class for the two-week experiment.

Source: M. A. Humphreys.

Such factors will have influenced the response of the children – all will have tended to create a negative feedback and reduce the likelihood of discomfort.

Like the older children, the young children rarely changed their clothing during the school day in response to a rising classroom temperature. They did change their clothing from day to day according to the prevailing temperature, but whether this was their own decision or at their mum's suggestion was difficult to tell. So a pattern was beginning to emerge for clothing behaviour: little change during the day; more change from day to day; still more from week to week; clothing lagged behind changes in temperature; social pressures can affect choice of clothing.

The responsiveness of the children to their lessons (as assessed subjectively by the teachers) was found to be dependent both on time of day and upon the classroom temperature. The effect of the time of day was far stronger; the best time for learning seemed to be the first half of the morning. The effect of the classroom temperature on the children was complex and interesting. The behaviour of children in classrooms is multifaceted, and therefore required multivariate methods of analysis. Factor analysis (principal component analysis), and classical discriminant analysis were used. These methods were becoming available in statistical packages – but it was necessary to grasp the principles on which they worked, so that the team would not misapply them or be misled by them.

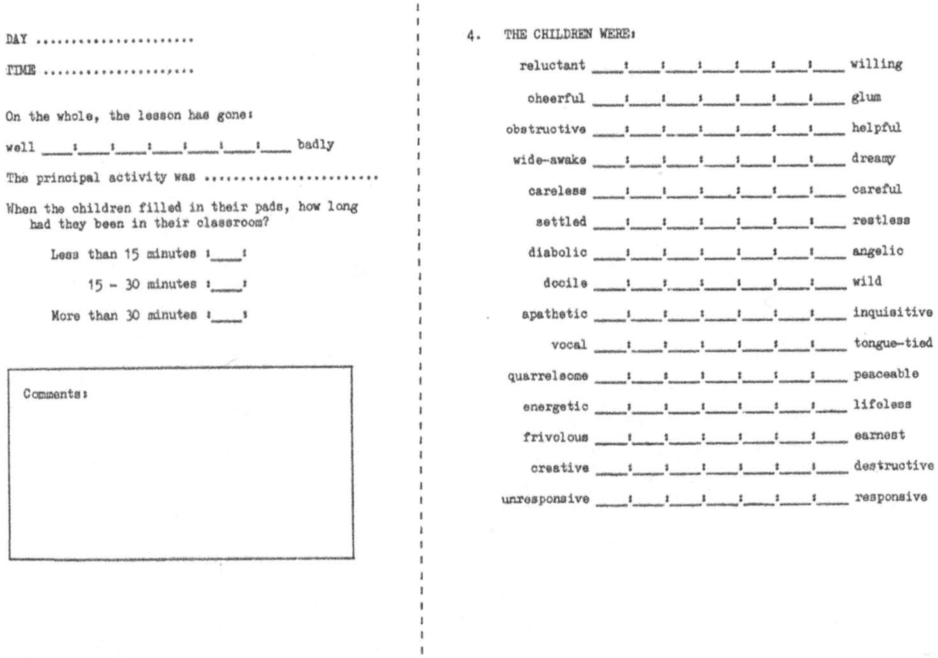

DAY ....

TIME ....

On the whole, the lesson has gone:

well ___:___:___:___:___:___:___ badly

The principal activity was ....

When the children filled in their pads, how long had they been in their classroom?

Less than 15 minutes :___:

15 – 30 minutes :___:

More than 30 minutes :___:

Comments:

4.   THE CHILDREN WERE:

reluctant ___:___:___:___:___:___:___ willing

cheerful ___:___:___:___:___:___:___ glum

obstructive ___:___:___:___:___:___:___ helpful

wide-awake ___:___:___:___:___:___:___ dreamy

careless ___:___:___:___:___:___:___ careful

settled ___:___:___:___:___:___:___ restless

diabolic ___:___:___:___:___:___:___ angelic

docile ___:___:___:___:___:___:___ wild

apathetic ___:___:___:___:___:___:___ inquisitive

vocal ___:___:___:___:___:___:___ tongue-tied

quarrelsome ___:___:___:___:___:___:___ peaceable

energetic ___:___:___:___:___:___:___ lifeless

frivolous ___:___:___:___:___:___:___ earnest

creative ___:___:___:___:___:___:___ destructive

unresponsive ___:___:___:___:___:___:___ responsive

FIGURE 4.5 Facsimile of an old slide of the cyclostyled response sheet used by the teachers.

Source: M. A. Humphreys.

Principal component analysis of the teachers' response sheet (see Figure 4.5) revealed two main components. The first was related to how industrious the children were, and the second to how active they were. On the whole, it seemed that the best temperature for learning was between 19°C and 21°C – the centre of the experimental range. Again this would be expected on the adaptive hypothesis. The children and their teachers seemed to have adapted to give the best result at the temperatures that they most commonly experienced.

The contribution made by this work in primary and secondary schools to the advance of the adaptive approach was invaluable. The design of the classroom, the operation of the windows and the change of clothing in response to the changing room temperature together provided sufficient means of adaptation so that the children were comfortable over a wide range of outdoor temperatures. Together they formed a powerful feedback control mechanism.

The work also uncovered the importance for comfort of minimising the constraints that were placed on the choice of clothing, and showed that the speed of adaptation in response to changing temperatures could be quantified. Clothing changed little within any day, more from day to day, and still more from week to week. It followed that the building should be designed so that within a day the classroom temperature should be stable, and that drifts of temperature from day to day should be kept small. However, quite large longer-term temperature drifts over a period of weeks could be accommodated without discomfort.

As to setting maximum temperatures for classrooms, the work could be seen as a failure – but, on the adaptive assumption, a fixed upper temperature limit would not be expected – adaptation is complex and simple answers do not exist. The temperatures that are acceptable depend on the adaptive state of the population, which in turn depends on the recent thermal history of the population.

Notes

1 Langdon, F. J. and Loudon, A. G. (1970) Discomfort in schools from overheating in summer, *J. Inst. Heat. & Vent. Eng.* 37, 265–79.
2 Lundqvist, G. R. (1973) Thermal field measurements in schools, in: *Thermal comfort and moderate heat stress, Proceedings of the CIB Commission W45 (Human Requirements) Symposium.* Eds: Langdon, F. J. et al., HMSO, London.
3 Fanger was writing his classic book at that time: Fanger, P. O. (1970) *Thermal comfort.* Danish Technical Press, Copenhagen.
4 Names for garments in English differ from country to country. A pullover, jumper or sweater is a knitted long-sleeved garment to keep the upper body warm. Traditionally it was of wool, but nowadays may be of acrylic or other man-made fibre.
5 Humphreys, M. A. (1973) Classroom temperature, clothing and thermal comfort – a study of secondary school children in summertime, *J. Inst. Heat. & Vent. Eng.* 41, 191–202.
6 Humphreys, M. A. (1973) Clothing and thermal comfort of secondary school children in summertime, in: *Thermal comfort and moderate heat stress,* Eds: Langdon, F. J. et al., HMSO, London.
7 Siegal, M. (2008) *Marvellous minds: the discovery of what children know.* Oxford University Press, Oxford.

5

THE FIRST INTERNATIONAL CONFERENCE ON THERMAL COMFORT

John Langdon had been quick to grasp the differences between the results of field studies and the predictions of the physiological models, and it was his suggestion that an international conference on thermal comfort be convened where this and other matters could be openly debated. The BRS lecture theatre and library complex had just been built, and would be an excellent venue for an international conference. Since the formation of the International Building Council for Research and Innovation (CIB), building research had become more international in outlook, and the conference was announced as a meeting of the CIB Human Requirements Commission. The date was fixed for September 1972 – shortly after the completion of the experimental work on the young children, but before its analysis. It was the first international conference devoted solely to thermal comfort and moderate heat stress.

The response was outstanding and the conference memorable. Delegates came from twenty countries and included established researchers into thermal comfort and heat stress, some from behind the Iron Curtain. Some names were known only from their research publications – A. Pharo Gagge and Ralph G. Nevins from the United States, Baruch Givoni from Israel. There were younger researchers too – Milos Jokl from (then) Czechoslovakia, K. Ibamoto from Japan, Ole Fanger from Denmark, David Wyon from Sweden, Don McIntyre and Ian Griffiths from the UK, who all became well-known figures in human thermal research.

Humphreys and Nicol gave two papers. One explained how to derive comfort information from observing the clothing behaviour of school children (see Chapter 31),[1] and the other, based on reflections on the assembled data, looked at thermal comfort as a feedback system[2] whose goal was to secure comfort – the basis of what has become known as the adaptive model. This paper drew on a wide range of research. It showed that window-opening was related to indoor and outdoor temperature, and that posture changed with room temperature in such a way that at higher temperatures the area for heat loss was increased. Air speed was correlated with the room temperature. In hot

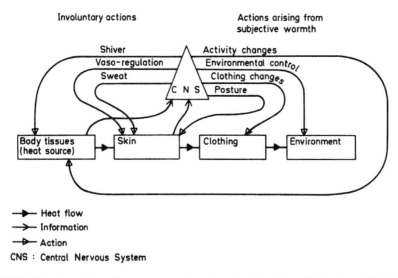

FIGURE 5.1 The thermal regulatory system showing physiological and behavioural feedback
loops, together with the path of heat flow from the body core to the environment.

Source: BRS Current Paper 4/75.

rooms the air speed was higher because of open windows and fans – a behavioural rather
than a physical relationship. People take off clothes when they are warm and put them
on when they are cold. So in field studies light clothing is often associated with warmth
and heavy clothing with coldness. This makes no sense in terms of heat-flow physics,
but perfect sense if people are using clothing as a means of controlling thermal comfort.
Nicol drew a beautifully constructed flow chart to show this system of physiological and
behavioural feedbacks and their relation to the flow of heat from the body core to the
room environment (see Figure 5.1). The diagram carefully distinguished between heat

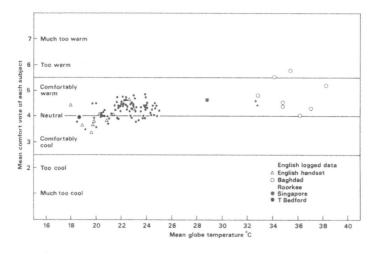

FIGURE 5.2 Facsimile of the slide that caused the rumpus.

Source: BRS slide.

flow, information flow and the consequent actions. It set out the logic of the control system at work at the heart of their adaptive model.

The power of this control system was illustrated by a slide (see Figure 5.2) that included the new BRS English data and Webb's data from Singapore, Baghdad and Roorkee, showing that the control system could give comfort at room temperatures as low as 16°C or as high as 38°C. The paper was a concise statement of their understanding of adaptive thermal comfort at that time.

This was the first public presentation of the adaptive model and the last paper of the conference. As well as presenting the evidence for the several adaptive processes, Nicol showed the slide with the data from England, Singapore, Baghdad and Roorkee all together on the same figure (see Figure 5.2). The slide was received with surprise and incredulity. The discussion was lively. We reproduce the relevant part:

> **Dr P. O. Fanger**: . . . the results referring to the hot climates are new to me. Are they published anywhere? What were the air velocities? Is there any information about their clothing?

> **Mr Nicol**: In all these field studies the clothing was decided by the people themselves. In Baghdad, the subjects were office workers. In Roorkee they were workers in the Building Research Institute. We do have information about the air velocities. In Roorkee the subjects spent more time adjusting their environment than one would expect to spend in Northern Europe. They used fans and opened and closed windows as part of their daily routine. For the present purpose it is enough to note that they are regulating the thermal environment according to their own thermal sensation. The effect of the air velocity was probably equivalent to not more than two or three degrees. We hope to publish a more detailed analysis of the Baghdad and Roorkee data in the future . . .

> **Mr I. Lotersztain**: I was very surprised by the large differences in the thermal environments found to be comfortable. I cannot see an explanation for it. I wonder if the meaning of the Bedford scale is perhaps quite different in Roorkee and in England.

> **Mr Humphreys**: One ought not to suggest that these differences in comfort temperatures are difficult to account for in terms of the physics of heat exchange. Leithead and Lind quote experiments dating back to the eighteenth century in which much more extreme environments are found to be habitable. The clothing and air velocity changes are adequate to explain them, without recourse to much physiological or psychological adaptation, except perhaps an increase in sweat-rate. The point of interest to us is that over this wide range of environments people can make themselves comfortable.

> **Mr M. Hoffman**: The results are given comparing the mean globe temperature. I think you should consider separately the other thermal measurements: air

temperature, relative humidity, and radiation. I expect that in India the difference between the wall temperature and the air temperature is much greater than it is in England.

Mr Nicol: The buildings in Roorkee are quite adequate. The globe temperature was similar to the air temperature, less than a degree different on average, as also was found in England. Certainly there is nothing there to produce such a large difference in preferred temperature.

Mr Humphreys: Perhaps we should point out that full environmental measurements were taken. You could redraw this diagram with, for example, air temperature as a horizontal axis, or effective temperature. It makes little change to the general appearance of the diagram.

Mr D. L. Thornley: I too would like to express my concern, because this diagram conflicts with so much that we have learnt throughout the rest of the symposium. There must be something amiss with it. It could be because of diurnal variation, or because of the humidity, which is more noticeable at these high temperatures. However, the design temperature for the air conditioning systems in Baghdad is 27°C. So I really don't see how people can be voting 'comfortably warm' at about 33°C.

Mr Nicol: People adapt to such conditions by a change of clothing, of custom, and of attitude to the environment. These appear to account for the differences. Our purpose in showing the diagram is to show that these variations can and do occur.

Dr D. P. Wyon: The point being made is that this vast difference in comfort temperatures can be achieved by the behavioural means spoken of in the paper. However, there would be a difference between the state of the people voting 'comfortable' at the different temperatures, because they would have achieved their comfort in different ways. High air velocities would prevent much paperwork from being done. Another available strategy is reducing body heat production to a minimum and maximising surface area. Probably this would decrease the amount of work being done. It is important to distinguish between comfort conditions and conditions for performing mental work. I think the amount of work these people were doing in the different situations would not have been the same.

Mr Humphreys: The Building Research Station at Roorkee produces some quite good work. But I agree that there could be differences.

Mr I. D. Griffiths: At the risk of throwing a spanner in the works, what do you think is the possibility of a contrast explanation for these results? For example 'comfortably cool' simply means 'cooler than we have been for the last ten weeks'?

Mr Humphreys: You would have to be cautious, in view of Dr Fanger's evidence. If you take people from hot environments and bring them quickly to Denmark, you find no difference in their preferred temperatures. No large contrast effect is found.

Mr Griffiths: But then they are in Denmark: presumably they have expectations about temperatures in that country and it is these expectations they will use as their criteria.

Mr P. Jay: Mr Nicol might have drawn evidence of social effects in heating from two BRS surveys of local authority heating. It was shown that where tenants had to pay for their own heating they used much less energy than when they had to pay a fixed sum in a monthly rental. The extremes were that, in a two-bedroomed local authority flat with oil-fired central heating included in the rent, 300 therms per annum were used. With electric floor heating, paid for by the tenant, it could be as low as 30 therms per annum. Both tenants said they were satisfied with their heating. When this was published, just as we have been hearing today, some specialists said that although these people said they were satisfied, they could not really have been. I think one should be careful, when presented with such facts, to remember that they are facts which need to be explained. It is no answer to say that they cannot be true because they conflict with our theory. I think this is the danger that some of us are in.[3]

Nicol and Humphreys were grateful to Peter Jay for this appeal to look at the evidence rather than depend on received opinion. Neither of them enjoyed conflict nor thought of research as a battle to be won or lost. The discussion had raised several points that remain to this day matters of research and debate:

- Does the meaning of a thermal sensation scale change with its social and climatic context?
- What is the effect of attitude to (or expectation of) the environment?
- Does physiological acclimatisation affect thermal comfort?
- Is there a conflict between climate chamber findings and the results of field research?

But the discussion, in focussing on just one slide from their presentation, had missed the main thrust of the paper – that thermal comfort could best be viewed as a self-regulating adaptive system. After the conference, Nicol completed and published a full analysis of the data from Roorkee and Baghdad, thus concluding an analysis that Sharma and Webb had started, and Gidman had continued.[4] The analysis quantified the role of air movement in improving comfort and reducing skin moisture.

The discrepancy between the field data from hot climates and the predictions of the physiological models was now on public record.

Notes

1 Humphreys, M. A. (1973) Clothing and thermal comfort of secondary school children in summertime, in: *Thermal comfort and moderate heat stress*, Eds: Langdon, F. J. *et al.*, HMSO, London.

2 Nicol, J. F. and Humphreys, M. A. (1972) Thermal comfort as part of a self-regulating system. Proc. Symposium *Thermal comfort and moderate heat stress*: CIB Commission W45, Eds: Langdon, F. J. *et al.*, HMSO, London.

3 Extract from pp. 272–3 in: Eds: Langdon, F. J. *et al.* (1973) *Thermal comfort and moderate heat stress*, HMSO, London.

4 Nicol, J. F. (1974) An analysis of some observations of thermal comfort in Roorkee, India and Baghdad, Iraq, *Annals of Human Biology* 1(4), 411–26.

6

ADAPTING TO ENFORCED CHANGES OF TEMPERATURE

In the winter of 1972–73 there was a world fuel crisis. The oil-producing nations in the Middle East had come to an agreement to limit oil production in order to increase the price of crude oil. The economic effect was worldwide and severe. The price of crude oil became much higher and there were shortages. In the UK, petrol became expensive and hard to find. Oil for heating was also expensive and in short supply. Added to this there was industrial unrest in the coal industry, starving the electricity generating stations of their supply of coal.

Government offices were instructed to reset their indoor temperature to 17°C in order to economise on fuel. By the end of February 1973, the district heating system at BRS had been re-balanced and all the radiators had been adjusted so that the maximum room temperature would be 17°C when all the radiator valves in the room were fully open and the windows shut. This was some five degrees cooler than had become customary in the offices in wintertime.

Here was a ready-made experiment in thermal adaptation – and one that would have been impossible to get permission to do. How had people responded to the temperature change? What adaptive actions did they take? Were they adapting successfully to the new lower temperature? In the middle of March, a questionnaire was sent to all members of staff right across the BRS, with a carefully worded covering letter. Seventy per cent responded and the results were interesting.

Staff members were asked how they had felt just after the temperature reduction. Sixty per cent had found their room 'too cool' or 'much too cool'. The mean subjective warmth on Bedford's scale was 2.8, just below 'comfortably cool'. This was a very low value compared with results from the previous data-logging project, where the norm had been 4.2, just warmer than 'neither warm nor cool'. Apart from feeling generally chilled, the main complaints were of cold feet, hands, fingers and legs – which would be expected to result from vasoconstriction.

Five per cent of staff considered that the new temperature made their work easier while 56 per cent considered that it made their work more difficult. The sensation of cold, they reported, made sustained intellectual tasks difficult. Cold hands reduced their efficiency in writing, drawing, typing, and in operating desk machines and in using instruments. A few (3 per cent) said additional clothing hindered their work, and some said their machines or work materials functioned poorly at this lower temperature.

They were asked to report on how they had tried to improve their comfort. Their replies indicated that most commonly they increased their personal insulation (extra clothes, thicker clothes, rugs or blankets, fur-lined boots). Next was the use of extra heat (sit against a radiator; have frequent hot drinks; use a hot water bottle).[1] Then there were attempts to reduce heat losses from rooms by taping over cracks round the windows, keeping doors and windows shut and lowering the blinds. Some increased their activity by walking around the office and walking even further, rubbing their hands and stamping their feet. People also admitted privately to urinating more often, which results from the coldness increasing muscle tension in the bladder and abdomen in response to cold, thus reducing the capacity of the bladder. The increased muscle tension also slightly raises the metabolic rate, generating more heat within the body. Thus people had adapted by using a range of strategies rather than by any single adjustment.

Staff were asked how they felt at the time they were completing the questionnaire, about a month after room temperatures had been reset to 17°C. By mid-March the mean warmth sensation had risen from 2.8 to 3.7 (just below 4: neither warm nor cool), which showed substantial but incomplete adaptation. They were still below their reported desired value of 4.6, about half way between 'neither warm nor cool' and 'comfortably warm'. This was an unusually high desired value, perhaps because many of them were feeling rather cool when the question was put, and people who are feeling cold when questioned tend to say they would like to feel rather warmer than neutral, and vice-versa. Their adaptive learning processes had reduced but not entirely removed their discomfort.[2]

The questionnaire had revealed massive cold discomfort during February 1973, just after the downward step-change of temperature. Sixty per cent were suffering cold discomfort. This was not surprising, because to be comfortable at 17°C a sedentary person would need more thermal insulation than provided by the indoor clothing than was normal in UK offices in the 1970s, or indeed at any time since the nineteenth century. BRS offices were normally at about 22°C before the change, and so the occupants were adapted to that temperature. They could have become adapted to an office temperature of 20°C by adopting ordinary but slightly warmer indoor clothing, but any lower temperature would require a change from what was considered normal indoor clothing. Such social changes can and do take place, but they need time. People had not felt free to wear clothing that would have been thermally sufficient. There are strong social constraints, and these are internalised. People do not often wish to transgress such social norms, whatever they may be.

At a subsequent meeting in London, called to discuss a Government proposal to make permanent the 17°C upper limit for heating in government offices, Humphreys for the BRS argued against the proposal on social grounds, pointing out that it would require the equivalent of winter outdoor clothing to be worn all day in all government offices. Those arguing for the proposal referred to a report they had obtained from BRS that said 17°C would cause no problems. It turned out to be the report written by Humphreys himself. This incident demonstrated one of the dangers with the adaptive model; that people take it to mean that any thermal environment whatsoever is acceptable because people can always adapt to it. They fail to consider the powerful social norms that constrain and guide people's behaviour. The limit was eventually set at 19°C, still rather low, and not very practical to enforce.

But is 17°C in any absolute sense too cold for office work? Humphreys wrote much of this book in his upstairs study in his old stone house in Wales, where there is no central heating, and in winter he uses a pair of low-output electric heaters, placed beside his feet in the kneehole of his desk. In a well-insulated but light and flexible outdoor jacket he is perfectly warm and comfortable at temperatures above 14°C. If they dropped lower than that, he would need to add insulated trousers of some kind, perhaps ski-pants, to his ensemble, but this would require some explaining to casual callers at the front door. Even thermal comfort researchers are subject to social constraints on their clothing.

There are social constraints in hot weather too. During the hot summer of 1976, the BRS Enquiry Desk fielded a phone call from a London Bank. The conversation went something like this:

'We are dying in here – it is 30°C!'

Humphreys glanced at a thermometer on his desk. It registered 30°C.

'It's 30°C in here too; it's not a problem.'

'How come?'

'I'm wearing shorts and a short-sleeve shirt.'

'Enough said!'

With that he rang off, and it was assumed he told his staff that they were permitted to take off their jackets and ties. Conventions can be very strong.

Notes

1 A hot water bottle is a flexible flat rubber bottle that can be filled with hot water. Its normal use is to warm a bed in wintertime. Here it was being placed inside the clothing to provide extra warmth.
2 Nicol observed while at a conference in New Zealand in 1992 that a similar restraint on office temperatures applied locally at that time, a drought having caused problems with the hydroelectric supply. A couple of months after the imposition of the temperature limit, he asked a friend in conversation how she felt about the temperature drop. She said it was hard at first but that now she was not even sure whether it was still in force, she had become so adapted to the new conditions.

7

THE FIRST META-ANALYSIS

After the international conference had ended, Langdon, Nicol and Humphreys edited the proceedings for publication[1] but there was still a need to respond more fully to the objections to the adaptive model that had been raised at the conference. Perhaps the disbelief that had been encountered was partly because so many of the results came from BRS's own datasets. A single point on the presented graph represented Bedford's data and all of the other data-points came from studies that had been conducted or initiated by Webb, so the next logical step was to collect together the results of all the field studies that were available from the whole world. If these fitted the same pattern, then it would be a powerful confirmation of the findings from the BRS data and hence of the adaptive model. Papers referred to in all the known published field studies were obtained, then their references checked in turn, until no more could be found. It was surprising how many published field studies were found:

The data (Table 7.1) represented over 200,000 'comfort votes', from a wide variety of countries, climates and seasons, and spanned nearly forty years. Humphreys analysed these data in spare hours at work and at home because it was not part of any official research project, and so the BRS mainframe computer could not be used. A clever new statistical calculator was available for use. It could do some basic statistical tests, such as the Chi-squared test for goodness of fit, Student's t-test for differences between mean values, and the F-test to compare variances. It could also calculate univariate regression equations. Probit regressions were beyond it, so these had to be fitted by eye. Fitting Probit regressions by Finney's method requires iterative adjustments to weighting coefficients, and would have taken far too long using the calculator. And often the requisite details were absent from the publications. With such limited resources the wide-ranging analysis took about two years to complete.

From most of the published studies, the mean indoor temperature experienced by the respondents during the period of the survey could be extracted, along with the

TABLE 7.1 Published field studies.

Year	Author	Place
1936	Bedford	UK (light industry, winter)
1938	Sa	Brazil (Rio de Janeiro)
1938	Newton	UK (a.c. offices)
1940	McConnell	USA (a.c. offices)
1947	Rowley	USA (a.c. offices)
1952	Ellis	Aboard warships in the tropics
1952	Rao	India (Calcutta)
1952	Mookerjee	India (North, summer)
1953	Ellis	Singapore, on land
1953	Mookerjee	India (dry tropics)
1954	Black	UK offices
1955	Malhotra	India, tropical
1955	Ambler	Nigeria (tropical)
1955	Hickish	UK factories (summer)
1957	Angus	UK lecture room (winter)
1959	Webb	Singapore
1962	Hindmarsh	Australia (Sydney) offices (year-round)
1963	Goromosov	USSR dwellings (summer)
1963	Wyndham	Northern Australia (manual workers)
1965	Ambler	North India
1966	Black	UK (a.c. offices)
1966	Grandjean	Switzerland (offices, winter)
1967	SIB (anon)	Sweden (classroom teachers)
1967	Ballantyne	Papua (Caucasians, tropics)
1968	Wyon	UK hospitals (operating theatres)
1968	Grandjean	Switzerland (offices, a.c. and n.v., summer)
1969	Auliciems	UK school children (winter)
1970	Humphreys and Nicol	UK offices (year-round)
1971	Pepler	USA school teachers (a.c. and n.v.)
1972	Pepler	USA school children (a.c. and n.v.)
1972–73	Davies	UK school children (summer)
1973	Auliciems	UK school children (summer)
1973	Humphreys	UK school children (summer)
1973	Wanner	Switzerland (a.c. offices)
1974	Nicol	India and Iraq (offices, summer)

Note: a.c. means air conditioned, n.v. means naturally ventilated.

optimum temperature for comfort. These were expressed either as the air temperature or as the globe thermometer temperature. Where both could be extracted, they were always very nearly the same (as had been found in the BRS office surveys), so it was sufficient just to say 'temperature'.

If people were well adapted to their normal indoor environment, as the adaptive model would predict they should be, the optimum temperature for comfort would be strongly correlated with the mean temperature they had experienced during the survey period, if that were sufficiently long. Figure 7.1 shows that this was indeed so.

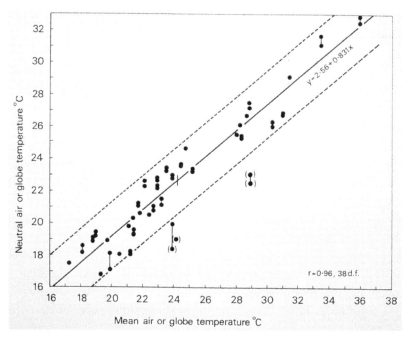

FIGURE 7.1 Comparing the mean temperatures in the accommodation with the temperatures that people preferred or found to be thermally neutral. Each point represents a separate body of data.

Source: Building Research Establishment (BRE) CP75/76.

The range of neutral or comfortable temperatures was too wide to be explained by the predicted mean vote (PMV) equation. Humphreys wrote at the time:

> The temperatures at which the respondents in the various studies were thermally neutral can be derived [from the data] . . . For adults the neutral temperatures range from 17°C to 33°C. The latter, which . . . is from Nicol's analysis of data from Baghdad, would have been equivalent to about 30°C if the air velocity had been only slight. The observed range of neutral temperatures is therefore effectively 13°C. Three factors, apart from the thermal environment, can combine to produce different neutral temperatures. They are the level of activity, the thermal insulation of the clothing and the physiological state which is considered by the respondents to be 'neutral'. For elderly people such as Fox's respondents [neutral at 17°C] the metabolic heat generation would have been rather low, both because of their age and because of the relative inactivity which usually accompanies it. For people in hot climates, such as Nicol's respondents, the metabolic heat generation also tends to be reduced, because the heat inhibits unnecessary physical exertion. It therefore seems probable that the two groups would have had very similar levels of physical activity. There was certainly a difference in clothing

insulation between these two groups of people, but it is difficult to imagine that it would have amounted to more than 1 Clo Unit (. . . approximately equivalent to a suit worn with normal underwear). Physical models agree in attributing a difference of 5°C or 6°C in preferred temperature to this source for sedentary people in thermal comfort. Since this is only about half the observed difference it follows that the groups have described as 'comfortable' or 'neither cool nor warm' different physiological states. This conclusion does not rest upon Nicol's result alone. Of the forty neutral temperatures . . . those from fourteen independent studies differ by 6°C or more from that of Fox's respondents. It is therefore necessary to conclude that acclimatisation has affected the temperature required for thermal neutrality. Fanger's results, which suggested that acclimatisation did not affect thermal comfort requirements, may therefore require some qualification.

This was a challenge to the physiological heat exchange modellers to explain why there was such a discrepancy between the predictions of the models (the BRS teaching model included) and the findings of field surveys. How might the physiological models be adjusted to bring them into line with the findings from the worldwide field data?

The dependence of the mean comfort vote upon the mean room temperature is shown in Figure 7.2. Only a slight dependence on the mean temperature was found, in agreement with the BRS datasets. There were two outlying points, and both could be explained by applying adaptive theory. They came from surveys where the mean temperature during the survey had not been typical of the average temperature experienced by the people surveyed, and so they would not be expected to be adapted to it. One outlier was a study of Swedish schoolteachers in spring.[2] The researchers had deliberately chosen a single sunny spring day for the survey, when the classrooms would be hotter than usual for the time of year. Being adapted to a cooler temperature, the teachers found 24°C to be too warm. For comparison the researchers had repeated the survey on a day when the outdoor temperature was average for the time of year. The mean classroom temperature on this day was 21°C. The teachers were comfortably adapted to this temperature. The other was from Goromosov's summertime study of flats (apartments) in southern USSR.[3] Most of the observations had been taken during the heat of the day. The indoor temperature at that time of day averaged 29°C, much above the average conditions experienced by the occupants during the whole day.

These analyses of all the available worldwide data confirmed for us that thermal comfort could be treated as a self-regulating adaptive process, a model whose inner feedback loops caused the neutral temperatures to be close to the mean temperatures experienced by the respondents. The data formed coherent patterns, enabling predictions to be made even without the help of a physiological model of human thermal regulation.

Humphreys sent a copy of the pre-publication draft to the researchers whose work had been included and who could be contacted. This was to ensure they were

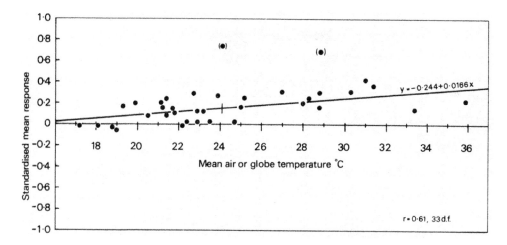

FIGURE 7.2 The mean thermal responses for the various bodies of data plotted against the mean temperatures. The mean thermal responses have been normalised because not all studies used seven-point scales. +1 is the maximum possible response and −1 the minimum possible.

Source: BRE CP75/76.

content with the use made of their work and the information extracted from it. It also unpremeditatedly ensured that this unpublished typescript became known to researchers right across the world, leading to its frequent citation instead of the later published versions. Dr Andris Auliciems and Professor R. K. MacPherson, both in Australia, independently sent comments and suggested that a further analysis in terms of the climate might be interesting.

This analysis of all the available data confirmed what had been surmised from the BRS data, that the data from Roorkee and Baghdad were not anomalous. They fitted into the worldwide pattern perfectly, so validating the BRS results.

Humphreys presented the paper at an international conference in Prague in September 1975, and then published it in the *Journal of the Institution of Heating and Ventilating Engineers*, where the two previous papers by Nicol and Humphreys had appeared. The paper, which includes much more than has been mentioned here, is one of the foundation documents for the adaptive model and continues to be cited.[4]

Notes

1 Langdon, F. J. *et al.* (1973) *Thermal comfort and moderate heat stress.* HMSO, London.
2 SIB (Anon) (1967) *Teachers' opinions of classroom climate – a questionnaire survey.* Statens Institut for Byggnadsforskning, Report No. 31, Stockholm.
3 Goromosov, M. S. (1965) *The microclimate in dwellings.* State Publishing House for Medical Literature, Moscow. (In English: BRS Library Communication No 1325, 1965.)
4 Humphreys, M. A. (1975) *Field studies of thermal comfort compared and applied.* Department of the Environment, Building Research Establishment, CP 76/75. (Reissued in: *J. Inst. Heat. & Vent. Eng.* 44, 5–27, 1976, and in *Physiological Requirements on the Microclimate.* Symposium, Prague, 1975).

8

CLOTHING OUTDOORS
AND DURING SLEEP

In the early 1970s, people were becoming concerned that some tall buildings were causing unpleasant and even dangerously strong gusts of wind at street level in towns in the UK.[1] There was also a growing awareness that the presence of buildings altered the urban thermal microclimate, as did large areas of pavement and urban parklands with copious vegetation. The Human Factors Section at the BRS was asked to look at thermal comfort outdoors.[2]

Their previous studies of children's clothing had shed light on the processes by which they adjusted (adapted) to their thermal environment and achieved thermal comfort. Could the same principles be applied to people trying to achieve thermal comfort outdoors? It was decided to look at the way people dress outdoors for some everyday activities and at how they change their dress in response to changes in the weather.

Data were collected in the summer of 1973 from people shopping in the main streets of two local towns (Hemel Hempstead and St Albans) and also from people at leisure in rural parkland (Whipsnade Zoological Park) in the summers of 1973 and 1974. The microclimate (air temperature, wet-bulb temperature, wind speed, globe temperature) was measured, and by observation the clothing of the people who passed by was classified as 'heavy', 'medium' or 'light'. Men and women were counted separately. If asked what they were doing the researchers would explain briefly and courteously, but in a manner designed to close the conversation, generally with the desired effect. Some 41,000 clothing observations were collected over the two summers (1365 counts of 30 people who were passing by).

Since we are tracing the story of the adaptive model of comfort, the adaptive use of clothing is the focus of interest rather than the outdoor microclimate. The general pattern of clothing behaviour was the same as for school children. At 16°C hardly anyone wore the 'light' clothing. At 28°C almost everyone did (see Figure 8.1).

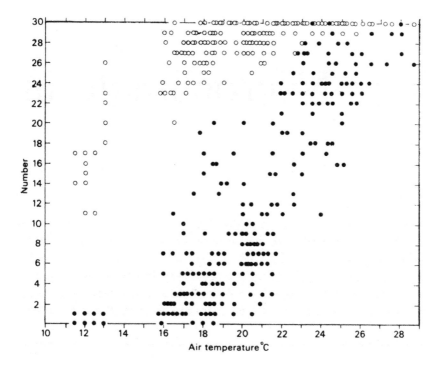

FIGURE 8.1 Scatter plot of clothing and outdoor air temperature. (Women shopping in Hemel Hempstead.) The solid points are the numbers, in each batch of 30 women who passed by, dressed in 'light' clothing; each open point is the number in light clothing and the number in medium clothing added together.

Source: BRE slide.

When the temperature was higher people compensated by reducing their clothing. In the towns women did so more than men, for in those days for an Englishman to appear in public in his shirt sleeves was still not thought polite, particularly in town, and especially if his braces (suspenders) could be seen. So men tended to keep their jackets on, sacrificing some thermal comfort for 'social comfort', conforming to an unwritten social code.

This was not so in the zoo-park, where the ethos was more relaxed and people were in holiday mode. There the social code was different, and men felt free to shed their jackets. They adapted just as well as did the women and children, being equally responsive to the air temperature in their choice of clothing.

Again it was found that people did not change clothing in response to fairly short-term changes in the thermal environment, such as those caused by changes in cloud cover on a bright day with drifting clouds. But if people were in the sun or shade for longer periods, or if the wind was persistent, they did change their clothing to allow for these changes in the local weather. For example, the shops in St Albans were only on the sunny side of the street, and the shoppers' clothing was correspondingly lighter on sunny days. In Hemel Hempstead there were shops on both sides of the main street, and as people went from shop to shop they might move from shade to sun. Their clothing did not respond to these short-term changes. Their clothing therefore depended not only on the present microclimate, but also on previous microclimates, there being a tendency not to change the clothing in response to a changeable microclimate. (In statistical language the clothing showed a strong serial correlation.) So changes in the microclimate were a potential source of thermal discomfort, because the clothing changes would not keep up with them. All these findings fitted with the model of thermal comfort as a self-regulating system operating in the presence of various constraints – the adaptive model. The study of outdoor clothing was published a few years later.[3]

One of the lasting effects of the oil crisis of 1972–73 was that the government asked that research should explore energy conservation. Much energy was used to heat dwellings, and we were asked to shift our attention from schools and offices to homes. Among the matters to be resolved were the limits on bedroom temperature for health and comfort. To advance our understanding of clothing behaviour, it was decided to look at thermal comfort and bedding. In those days most British bedrooms were unheated, and people compensated for seasonal changes in the bedroom temperature by altering the number of blankets covering them. (A blanket is a woollen fabric layer perhaps 5–8 mm thick and usually tucked in around the edge of the mattress.) In cool weather, the blankets were topped with an eiderdown – a light feather-filled quilt. Volunteers from the staff at BRS were recruited along with their families and friends for a preliminary study starting in July 1975 and continuing until wintertime. They completed a brief questionnaire each morning about bedclothing, quality of sleep and thermal discomfort. Clockwork thermographs in white-painted metal boxes recorded temperatures in the bedrooms of 21 adults and several children.

If the bedroom temperature was in the range 13°C to 17°C people typically had a sheet, three blankets and an eiderdown to cover them. By 19°C most had shed the eiderdown, and then the number of blankets reduced until at around 26°C just a sheet was used as covering. The quality of sleep began to deteriorate at around 24°C–26°C as the incidence of warmth discomfort increased. There was no sign of cold bedrooms adversely affecting sleep (see Figure 8.2).

The onset of heat discomfort, particularly evident for the people in double beds, coincided with the limit of what could be achieved by reducing the number of blankets. At that room temperature, only a sheet is used for covering, and with a typical English highly insulating mattress no further adjustment is easily possible. The insulation downwards cannot be reduced without using a different type of mattress – such as a canvas camp-bed.

The particular upper limit for bedroom temperature is related to the material culture of the society, and in our case attributable to limitations of the type of bedding that was normal in the UK. To overcome the cultural barrier required ingenuity

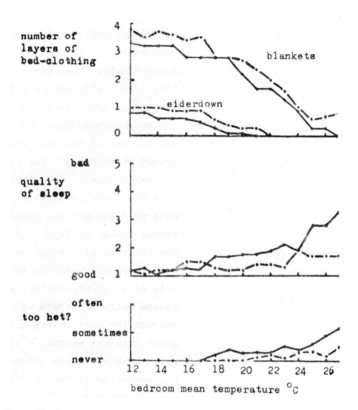

FIGURE 8.2 Effect of bedroom temperature on bedclothing, quality of sleep and thermal discomfort. The solid lines are from people in double beds and the broken lines for people in single beds.

Source: Facsimile of an original slide.

and a willingness to behave differently. In that hot summer, one couple in our sample slept outdoors on a verandah – a successful adaptive strategy, but one that defied normal English convention. We also noticed that people did not change their bedclothing immediately in response to the onset of hot weather. They took a few nights to 'catch up'. So to predict the likelihood of discomfort it is necessary to know the room temperature for the previous nights as well as for the night in question.

The envisaged full-scale study did not take place – other projects were considered more urgent. The exploratory study was briefly reported in a paper that drew together our experience of time and clothing changes and set out the derivation of exponential time series to quantify it.[4] The various projects that had included some investigation of people's clothing behaviour confirmed and extended what had been learned from the study of the school children: clothing change is a powerful adaptive action, but takes time to complete, and can be subject to cultural constraints that place limits on it and lead people to 'trade' comfort for conformity.

Notes

1 Penwarden, A. D. and Wise, A. F. E. (1975) *Wind environment around buildings.* Building Research Establishment Report, Department of the Environment. HMSO, London.
2 It is worth noting here that, at this time, the BRS was a government establishment and could be directly required by the government to address a particular issue of concern. Nowadays this is managed less directly through the funding system. In addition there were far fewer university-based research groups.
3 Humphreys, M.A. (1977) Clothing and the outdoor microclimate in summer. *Building & Environment* 12, 137–42.
4 Humphreys, M. A. (1979) The influence of season and ambient temperature on human clothing behavior, in: *Indoor climate*, Eds: Fanger, P. O. and Valbjørn, O., Danish Building Research, Copenhagen, pp. 699–713.

9

META-ANALYSIS 2

Relating climate to indoor comfort

The exploration of the worldwide field studies data was continued throughout 1976, looking for any evidence of the influence of the outdoor climate on thermal comfort indoors, as suggested by Professor R. K. MacPherson and Andris Auliciems. Humphreys was now heading the Human Factors Section, and had the assistance of fellow researcher Margaret Gidman and permission to use the BRS computer. The requisite climate data were extracted from world meteorological tables, except for the BRS surveys, where concurrent climatic data were available from BRS's own meteorological station.

One problem was that there was no immediate place for the climate in the BRS adaptive model. It was the *indoor* temperature that people adjusted to their requirements, while at the same time adjusting their clothing to suit. However, there were ways in which the climate might be indirectly influential. They were to do with minimising discomfort on entering and leaving the building, coupled with social considerations, and were potentially quite complex. Humphreys wrote at the time:

> In a pleasant climate, thermal discomfort on entering or leaving a building can be minimised by keeping small the difference of temperature between indoors and outdoors. This may be achieved by permitting the indoor temperature to vary in sympathy with the prevailing outdoor temperature. One would therefore expect the preferred temperature obtained from a thermal comfort survey to bear some relation to the outdoor seasonal temperature, provided this were within the range found to be pleasant.

> Such a link between indoor and outdoor temperature would cease to be advantageous if the outdoor temperature were unpleasantly hot or cold. To link the indoor temperature directly to the outdoor temperature would then merely ensure that people were uncomfortable indoors as well as outdoors. It

is better in such climates to regard the indoors as a haven from the rigours of the outdoor environment. In these circumstances one might expect even an inverse relation between the outdoor temperature and the preferred indoor temperature. Such an effect has been noted by Goromosov,[1] who reported that in winter the preferred indoor temperature is higher in the north of Russia than in the south.

There would be uncertainty about the preferred indoor temperature if the external conditions were neither pleasant nor extreme. An example of this uncertainty occurs at the beginning of the heating season in the United Kingdom. One can choose either to have no heating and wear warm clothes, or to turn the heating on and wear lighter clothes. The two strategies give different comfortable temperatures. The choice will partly depend upon what others have chosen to do, and partly upon the cost and trouble of operating the heating appliance.

Taken together, the above considerations suggest that the preferred indoor temperature would depend upon the outdoor temperature in the manner shown in Figure 9.1. For the 'free-running' condition (no energy being consumed by heating or cooling appliances), one would expect a substantial dependence upon the prevailing outdoor temperature (zone A, Figure 9.1). This zone will be limited in extent, and centred upon the most pleasant outdoor temperatures. For heated or cooled buildings (zone B, Figure 9.1) one would expect a fairly broad and substantially horizontal band, with some tendency to depend upon the outdoor temperature in the pleasant zone.[2]

Even with the help of this tentative model it was hard to know where to begin the analysis, as it was not known what aspects of the climate might be influential. Apart from the outdoor temperature for the time of the survey, it was possible that there would be effects from the humidity, or from how hot the summer was, or how cold the winter. Maybe the hours of sunshine would be important too. Numerous hand-drawn scatter plots of the comfort temperatures against the available climate parameters were drafted to explore the question. It was also suspected that economic factors could be influential, because if money were no object one might run buildings warmer or cooler than normal, and so become adapted to different temperatures. (At this time the energy crisis had subsided and the environmental consequences of increased energy use had not yet been widely recognised.)

After some exploratory work, it was found useful to split the data into two groups: those where there was neither heating nor cooling at the time of the survey (the free-running mode) and the rest. These might be air conditioned (heating and cooling available) or just heated. The two groups were analysed separately. There were no data from air-conditioned buildings in hot climates, which was a regrettable limitation

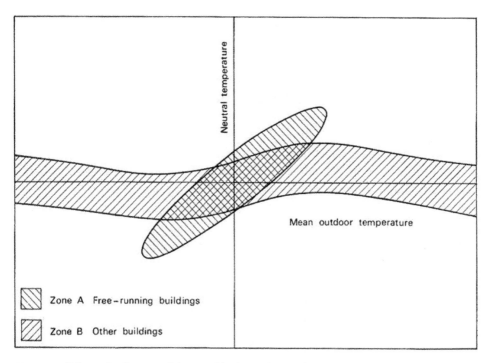

FIGURE 9.1 Schematic diagram of the possible dependence of comfortable indoor temperatures on the climate.

Source: BRE CP 53/78.

of the data. It was not that there was an absence of air-conditioned buildings in those climates, just an absence of thermal comfort studies in them.

Analysis of the free-running mode was the simpler task. Graphical analysis showed a linear relation between the outdoor temperatures (monthly mean values from the tables) and the indoor comfort temperature. The best temperature-metric was the mean outdoor temperature rather than the mean daily maximum or the mean daily minimum. No reliable influence from any other climate variable was found despite a thorough search. The correlation coefficient was an astonishing 0.97, which meant that some 94 per cent of the variation in comfort temperature could be explained in terms of the outdoor temperature alone. This was an unexpected result and raised the question of whether there was a physical relation somehow underlying it, or contributing to it in some way, for the model had not led Humphreys to expect a strong relationship.

Understanding and quantifying the effect of the climate on indoor comfort requirements when buildings were heated or cooled was more difficult. It appeared that many buildings were using heating and cooling when there was no climatic need to do so. At the same outdoor temperatures there were other buildings operating well without heating or cooling. But granted that the heating and cooling plant is in operation, whether necessary or not, what might govern the thermostat set-point? Heat exchange theory was of little help here. In very light dress people would feel thermally neutral without sweating at about 28°C or even in the low 30°Cs with some air movement, while there was no obvious theoretical lower limit. There was an example of comfort at 6°C from a comfort survey during an expedition to the Antarctic.[3,4] Good quality polar clothing ensured comfort even at these low indoor temperatures. So what might govern the choice of indoor temperature? It was likely to be dominated by social convention and by economic factors. If fuel is scarce and expensive, societies might choose cool winter interiors and wear warm clothes, as seen during the oil crisis. If clothing were expensive, societies might choose few clothes and warm interiors. But perhaps if fuel were expensive, clothes would be too. If fuel were cheap, people might choose warmer winter temperatures and cooler summer temperatures.

It seemed sensible to expect some link between indoor temperature set-points and the weather. If you were wearing light clothing outdoors in summer, there would be no sense in going indoors and needing to put on heavy clothing because the rooms were cold. But there was anecdotal evidence that this was not unusual. A BRS colleague just returned from tropical parts had been advised by the hotel receptionist: 'When you come down to dinner, Sir, let me suggest you to put on lots of clothes. We have air conditioning in the dining room and it is very cold.'

So, insofar as anything was expected, it would be a broadly horizontal band within which the comfort temperatures would lie, but having a 'kink' around the most pleasant outdoor conditions. Polynomials were used as a convenient tool to explore the data, but it would not be appropriate to use a polynomial finally to express the expected shape of the data. Polynomials tend towards plus or minus infinity at the extremes – not at all what one would expect of a comfort temperature. So a suitable algebraic function was chosen that would have the right general shape (like the shape of zone B in Figure 9.1).

Again the monthly mean outdoor temperature was found to be the best predictor of the temperature for comfort indoors. The correlation coefficient was 0.72, which meant that slightly over 50 per cent of the variation of the comfort temperature could be explained just from knowing the outdoor monthly mean temperature.

Two other variables improved the prediction. The average daily maximum temperature of the hottest month significantly improved the predictive power, as did the continent from which the data came. American data had (on average) warmer winter comfort temperatures than did European data. But it did not seem wise to include either of these two variables in the final predictive equation. Both seemed plausible, but searching among multivariate equations for possible contributing variables runs the risk of accepting features that later turn out to have been chance effects, and high order polynomials had already been found that improved the predictions in ways that seemed unlikely to be genuine. So the statistics of the 'kinky' curve were calculated for buildings that were heated in winter or cooled in summer using only the outdoor mean temperature as the predictor variable. The comment at the time was this: '[the curve] indicates the temperatures most likely to be satisfactory. An indoor temperature which varies with the season, besides being more economical in the use of fuel, is seen to be more satisfactory for the comfort of the occupants.'

This time Humphreys published the results in an international journal.[5] The data and the final curves are shown in Figure 9.2.

The exceptionally strong relation found between the outdoor temperature and the indoor comfort temperature for the 'free-running' mode was still puzzling. Was there some physical basis for this that had been overlooked? When a building is neither being heated nor cooled the indoor temperature is affected by the outdoor temperature. The warmer it is outdoors the warmer it is likely to be indoors too. The occupants would be expected to adapt to these gradually changing indoor temperatures, and so the temperature they find 'neither warm nor cool' would change correspondingly. The adaptation process would have two complementary aspects:

- People would reduce their clothing as the room got warmer.
- People would keep the room from getting too hot by opening the windows and by using blinds.

So in these 'free-running' conditions the comfort temperature indoors might be expected to be quite highly correlated with the outdoor temperature. This combination of building physics and successful human adaptation could explain the very high correlation.

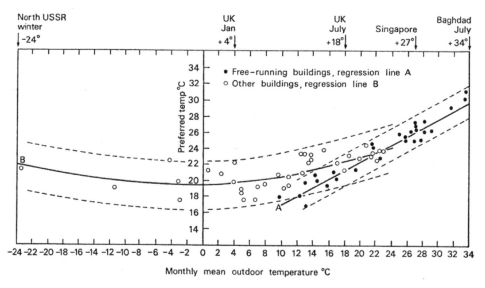

FIGURE 9.2 Scatter diagram for neutral temperatures against the monthly mean outdoor temperatures.

Source: BRE CP 53/78.

The next step was to examine the relationship between the mean indoor temperature and the mean monthly outdoor temperature. Sure enough there was a very strong relation between them – a correlation coefficient of 0.96. So it seemed that the *fundamental* adaptive relationship was between the indoor temperature and the person – the person adapting the clothing (and maybe posture and activity) to suit the indoor temperature, and adapting the building by manipulating windows, shading devices and, in heated or cooled buildings, the heating/cooling system. So the building itself should be seen as part of the adaptive model.

With this new insight, Humphreys put together a paper that combined the previous two papers that had used the worldwide data, and included comparisons of mean indoor temperature and comfort temperature at different outdoor temperatures. If, in a particular survey, the indoor temperature was lower than expected from the climate and mode of operation, so too was the comfort temperature. So the fundamental adaptation was indeed with the indoor temperature rather than the outdoor temperature. The relationships between comfort temperature and climate were an interesting consequence of this. A quote from that paper[6] summarises our understanding of the adaptive model by 1978:

> A building and its occupants can be regarded, from the point of view of thermal performance, as a single self-regulating system whose objective is to achieve and maintain comfortable conditions for the occupants. The regulating processes include:
>
> a) the careful design of the building for the climate, either by the deliberate application of modern building technology or by adherence to types of structure traditional to the locality;
> b) the provision, and use if necessary, of heating or cooling plant;
> c) the provision and use of blinds or shutters to regulate the admission of solar radiation;
> d) the use of doors, windows and other vents, or the provision of fans, to control ventilation;
> e) the choice of clothing appropriate to the prevailing temperatures and activities;
> f) the choice of activity level appropriate to the warmth of the environment, for example, resting during the heat of the day;
> g) the physiological changes associated with acclimatization to heat or cold by the occupants.
>
> The indoor temperature and the preferred temperature indicate the response of the regulatory system to the external disturbance caused by the weather. Success can be judged in terms of the thermal comfort achieved by the occupants, and this is indicated by comparing the temperature occurring within the occupied building with the temperatures preferred by the occupants.

In September 1978, Humphreys resigned from the Scientific Civil Service to follow his vocation in Christian ministry, after making the difficult choice to become a student again, this time reading theology. As George Caird has commented: 'The most difficult choices in life are not between the good and the evil, but between the good and the best.'[7]

In 1973, Nicol left the Medical Research Council to start a socialist bookshop in North London. Continuing the human factors work at the Building Research Establishment (BRE) fell to David Hunt, a very capable young researcher who was then leading the BRS project on thermal comfort and energy use in dwellings, but a year later he too left to train for Christian ministry. The division head wondered what was the matter with the office, as: 'Everyone I put in it leaves to go into the church.' Despite repeated advertisements, the post remained unfilled for several years.

Notes

1 Goromosov, M. S. (1963) *The microclimate in dwellings.* State Publishing House for Medical Research, Moscow. (English translation: BRS Library Communication No 1325, 1965.)
2 Humphreys, M. A. (1978) Outdoor temperatures and comfort indoors, *Building Research and Practice (J. CIB)* 6(2), 92–105 (the quoted text is on p. 92).
3 Goldsmith, R. (1960) Use of clothing records to demonstrate acclimatisation to cold in man, *Applied Physiology* 15, 776–80.
4 Palmai, G. (1962) Thermal comfort and acclimatisation to cold in a subantarctic environment, *The Medical Journal of Australia* January, 9–12.
5 Humphreys, M. A. (1978) Outdoor temperatures and comfort indoors, *Building Research and Practice* 6(2), 92–105.
6 Humphreys, M. A. (1981) The dependence of comfortable temperature upon indoor and outdoor climate, in: *Bioengineering, thermal physiology and comfort,* Eds: Cena K. and Clark J. A., Elsevier, Amsterdam, pp. 229–50.
7 Caird, G. B. (1963) *The gospel of St. Luke,* The Pelican Gospel Commentaries, A490, Penguin, London, p. 141.

10

THE ORIGIN OF THE OXFORD
THERMAL COMFORT UNIT

In 1991, some twelve years after Humphreys had left his work on thermal comfort, Dr Susan Roaf, from the Oxford School of Architecture, contacted him with an invitation to give a lecture on the subject. He declined the invitation, being out of touch with the progress of thermal comfort research during the intervening years, but agreed she could visit him to discuss her research and its relation to field study results. Humphreys had begun reading about global warming.[1] He knew that the adoption of an adaptive model for comfort could significantly reduce energy use in buildings. This would reduce the amount of carbon dioxide entering the atmosphere, and so help prevent or alleviate global warming. The prospect at once raised the importance of adaptive comfort theory from being a scientific model having implications for comfort and economy, to being a topic of ethical significance for humankind.

Roaf discussed with him her research on the thermal behaviour of courtyard houses with wind-catchers in the hot, dry climate of Iran.[2] She had investigated the thermal performance of these buildings and the thermal comfort of their occupants, and had found people were comfortable at temperatures that were impossibly high according to any of the standard comfort models. Humphreys' work on the results of field studies across the world did, however, agree with her findings. In particular, Nicol's analysis of data from Iraq and North India showed comfort at temperatures similar to those she had found in Iran. Far from being superseded, the work Humphreys and Nicol had done so many years before was now crucially important in the context of energy use and global warming. She persuaded Humphreys to give the lecture after all.

His contact with thermal comfort work since 1978 was slight. Don McIntyre had published a magisterial book on indoor climate in 1980,[3] and had given Humphreys a copy, but that was eleven years before the impending talk. He knew that Ian Griffiths, having moved to the University of Surrey, was doing thermal comfort field surveys in Europe. Ian had turned to field studies to avoid the psychological abstraction of the laboratory setting.

A skim through the various journals that sometimes published thermal comfort research showed that climate chamber work continued to be the principal research method, ever more closely defining conditions for comfort in the laboratory setting. PMV, Fanger's Predicted Mean Vote, had become an International Standard (ISO 7730) in 1984. It had not occurred to the BRS team that a method of predicting thermal sensation could become a Standard. Several new field studies had been undertaken, but nothing that changed the overall picture. The BRS meta-analysis on field studies worldwide had been updated in Australia by Andris Auliciems, who had included some good new surveys and excluded some of the weaker ones.[4] He had developed the suggestion that indoor temperatures be profitably varied in sympathy with the seasons. People would still be comfortable and up to 50 per cent of annual energy used for heating and cooling could be saved. He had coined the term 'thermobile' for a thermostat whose setting could be adjusted according to the prevailing outdoor temperature.[5] ASHRAE (the American Society of Heating, Refrigerating and Air Conditioning Engineers) Transactions reported a lively clash between Auliciems and Fanger. Fanger reproved Auliciems for ignoring decades of climate-chamber work. Auliciems responded that Fanger's work ignored decades of field work. So there was a stand-off between the two approaches. Richard de Dear, a young Australian researcher working intermittently with Auliciems, but who had also spent some time with Fanger in Denmark, was doing high-quality fieldwork in various climatic regions of Australia. He was keeping records of clothing insulation and metabolic rate as well as making highly accurate measurements of the thermal environment. In the UK the case for an adaptive approach had continued to be argued by Ian Cooper.[6]

In the intervening decade, PMV had become almost ubiquitous despite the limitations that had been pointed out so many years previously. It was now assumed to apply across the world to all peoples in all circumstances. Auliciems and de Dear were making a stand for the adaptive behavioural approach against the general acceptance of PMV. Field studies were also being done in the San Francisco Bay region of California by Gail Schiller (later Gail Brager) and her colleagues in the school of Architecture of the University of California. Under the influence of de Dear, Brager and others, ASHRAE had begun to fund thermal comfort field studies in various climates. A connection was developing between the Australian and the Californian researchers via ASHRAE Meetings.

Using the original slides kept safe by Margaret Gidman, who still worked at the BRE, Humphreys gave his first academic lecture after an absence of thirteen years. Nicol agreed to come to the lecture to help him out if necessary. The lecture was well received and was published as a chapter in a book,[7] and led to a paper on adaptive comfort at the 1992 World Renewable Energy Congress[8] in Reading, UK. It signalled the start of the Oxford Thermal Comfort Unit through which both Humphreys and Nicol would be drawn back into the heart of the thermal comfort debate.

By 1992 the Human Factors Section at the BRE had resumed thermal comfort field studies after a lapse of a decade. Dr Gary Raw was Humphreys' successor as head of Human Factors and Health. He and Nigel Oseland, an able young psychologist, were investigating thermal comfort in small newly built dwellings,[9] and Nigel was working towards a PhD.[10] BRE invited Humphreys to conduct a literature review covering the developments in thermal comfort research since 1978, and this enabled him to become up to date. The review found that PMV was well established as the most commonly used way of quantifying thermal comfort and that a great number of climate chamber studies had been conducted during the period. The alternative to PMV was Gagge's Standard Effective Temperature (SET), its final formulation in 1986 being a result of his lifelong research into human thermal environments.[11]

An interesting field study had been published by Fishman and Pimbert,[12] using methods and equipment similar to those of that first BRS data-logging project that had led to the first formulation of the adaptive model in 1968–69. In addition to the variables the BRS had measured, every week they assessed the clothing insulation of each of their respondents, and they also estimated a likely overall metabolic rate for office work. This meant they could compare the subjective warmth of their respondents with the predictions of PMV. They found the agreement to be satisfactory. Their statistical procedure, however, differed from that of Humphreys and Nicol as they had not separated their data into monthly batches, and so any seasonal adaptation to the indoor temperature was obscured. They concluded that a fixed temperature of 22°C all year round would be best, which was close to their average indoor temperature of 23°C. It would have been interesting to re-analyse their data from an adaptive standpoint.

The literature search had found some thirty field studies published between 1978 and 1992. The subsequent report reflected on these studies:[13]

> There are a number of reasons for this continuing effort. First is the need to validate in the field the results obtained from laboratory research. Thus ASHRAE is now taking the field study seriously, and has recently invited tenders for research contracts for field studies in various climatic regions. Secondly there is the observation that the temperatures suggested by the Fanger equation differ markedly from those traditionally occurring in good buildings in hot climates and hitherto found to be acceptable to the occupants. That is to say, there is a perception that there is a human adaptation to climate which is not included in the Fanger equation . . .

> It is possible from each of these [field] studies to calculate the mean temperature experienced by the subjects, and the temperature which they found to be

'neutral' or 'comfortable'. These mean temperatures and the comfort/neutral temperatures are found to be strongly correlated, confirming the finding for field studies conducted prior to 1977. A preliminary inspection of these new studies . . . suggests that the model developed in the late 1970s at BRE will be consolidated by these results, and to some extent extended, since the new studies include data from air-conditioned buildings in hot climates, which was an important gap in the earlier surveys.

There had been some methodological changes:

In developed countries data-logging and computer analysis have become the norm, making the collection and subsequent analysis of survey data less laborious. This has also led to more comprehensive measurements of the thermal environment. Subjective responses may also be obtained by presenting the scales on a computer screen, and it is now common practice to use two types of scale: the ASHRAE or the Bedford scale for warmth, and the 'McIntyre' or some similar scale for thermal preference. The use of two types of scale side by side has confirmed what was previously suspected – that people in cold and temperate climates prefer on the whole to feel slightly warmer than 'neutral' while people in hot climates prefer to feel slightly cooler than 'neutral'. There are probably enough data now to quantify this effect. If [the effect] stands up to statistical analysis it might contribute to the reconciliation of field study results with predictions from the Fanger comfort equation or other physiological models developed from climate chamber research.

It was now possible to compare field results with the predictions of the PMV model:

Because some recent field studies of thermal comfort have included estimates of clothing insulation and of metabolic rates, it has been possible to compare their results confidently with the predictions of the Fanger equation. These comparisons show a clear difference between the predictions and the practical results from field studies.[14]

And there was a clear future for thermal comfort field studies:

There is a developing consensus that the extra relevance of field studies outweighs their lack of rigour when compared with climate chamber studies. To get the maximum benefit from field studies of thermal comfort, it will be necessary to develop an agreed core methodology, so that results from various research teams can be readily compared. . . . So it is hoped that a data-bank of field-study results will became available from a wide range of climates and cultures. This would be a valuable guide for the thermal design of buildings.

There seemed to be a problem with the ASHRAE scale in some surveys:

> There is evidence that the ASHRAE scale, now the most commonly used of the scales, can yield a bi-modal distribution of response when used in field studies.[15,16] This seems to be attributable to the category 'warm', which some respondents take to be a pleasant condition, while its context in the scale indicates an unpleasantly warm sensation.

These comments indicate the progress made in the adaptive model from 1978 to 1992 by the numerous field studies that had been published, with the prospect of many more now that ASHRAE was beginning to commission field surveys. With BRE and ASHRAE both acquiring fresh data rapid progress could be expected.

One word was missed in the literature review that was later to become important – *expectation*. The BRS team had normally used it to express a simple physical difference between the thermal environment that a person *expected* to find and that which they actually found. For example, people might go to the office *expecting* the room temperature to be 20°C as usual, only to find that for some reason it was 24°C. Being dressed for the expected temperature of 20°C, people would find 24°C too hot. That the temperature differed from their *expectation* led to people being too warm. Auliciems was using 'expectation' in its psychological and cultural sense, as in the phrase: 'The thermal environment in this building falls short of my expectation.'

With this sense of the word, one can speak of people having 'high' or 'low' expectations of the environment. To fall short of 'expectation' is likely to produce dissatisfaction. The two meanings are quite different. The first draws attention to a physical difference, while the second implies a quality judgement. Auliciems drew a flow chart showing the adaptive processes. It is not altogether clear how the chart is to be understood, because it does not distinguish between heat flow, flow of information, and action. But it does include 'thermal expectation' and relates it directly to satisfaction (Figure 10.1). The two senses of expectation are both important in the adaptive model, but they need to be distinguished to avoid confusion.

Nicol returned to thermal comfort research and joined the staff of Oxford Polytechnic (soon to become Oxford Brookes University). He was leading the newly formed Thermal Comfort Unit (TCU) in the School of Architecture. Humphreys was invited to join the TCU, being released from his normal ministerial duties for a day or two a week. A rough division of labour emerged. Nicol was team leader and in charge of field-research projects. Humphreys would help with conceptual modelling of the adaptive processes and with questions of statistical analysis while Roaf would work on the practical implications of adaptive comfort for sustainable building design and construction.

Nicol was also working part time with Mike Thompson at the University of East London, teaching on the Masters course Thompson had developed there. As a contribution to re-establishing the adaptive approach, Nicol wrote a short handbook for researchers which laid out the bones of the adaptive approach and suggested

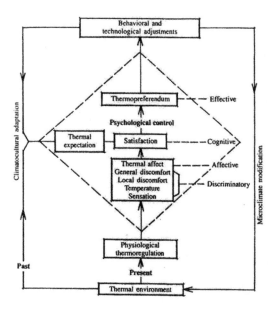

FIGURE 10.1 Auliciems's adaptive flow chart from 1983.[17] Notice the inclusion of 'thermal expectation' as a contributor to 'satisfaction'.

how one might best conduct thermal comfort surveys in the field.[18] The book was cheaply produced by the University and became a common reference for Masters or PhD students undertaking field studies: it was the foundation on which the first volume of this trilogy was based nearly twenty years later.

The TCU team saw that there were a number of things to be done if the adaptive model were to advance:

- More field studies were needed. These should increase the range of climates for which there were adequate data, with the ultimate aim of a complete climatological coverage. This would fill out the relation between the climate and the temperatures people liked indoors. These field studies would need to be well designed, well conducted and well analysed if they were to carry conviction.
- The reason for the systematic differences between the field results and the predictions of PMV needed to be quantified and understood. Field studies to quantify the discrepancies would need to measure all six main factors (air temperature, mean radiant temperature, air speed, humidity, clothing insulation, metabolic rate) with sufficient accuracy. The result of this endeavour would lead to controversy, as it would question the universal validity of PMV.
- The ways people adapt to their thermal environment needed further investigation, with the aim of understanding their adaptive actions, quantifying them and identifying their limits. So surveys would need to record window-opening behaviour, clothing behaviour and other adaptive actions. These actions take place in time sequences, so longitudinal studies would be needed if they were to be modelled and quantified.

- There was a need to stimulate the interest of more researchers in the adaptive approach, as it was obvious that these research objectives could only be accomplished if there were a worldwide network of researchers and interested practitioners (architects and building engineers). Such a network could be formed through conferences and 'teach-ins' dedicated to developing the adaptive approach and exploring its application.

Nicol and Roaf had been to the PLEA (Passive and Low Energy Architecture) Conference in New Zealand in 1992, where they had met Gail Brager. They found that she and de Dear were considering a broadly similar programme for advancing the adaptive approach. This meeting led to fruitful collaboration and friendship during the subsequent years, and contributed to the formation of an international network of researchers working on field studies from an adaptive perspective.

Notes

1 Gribbin, J. (1990) *Hothouse earth*. Bantam Press, London.
2 Roaf, S. C. (1988). *The windcatchers of Yazd,* PhD thesis, Oxford Polytechnic.
3 McIntyre, D. A. (1980) *Indoor climate*. Applied Science Publishers, London.
4 Auliciems, A. (1981) Towards a psychophysiological model of thermal perception, *International Journal of Biometeorology* 25, 109–22.
5 Auliciems, A. (1986) Air conditioning in Australia III: thermobile controls, *Architectural Science Review* 33, 43–8.
6 Cooper, I. (1982) Comfort and energy conservation: a need for reconciliation? *Energy and Buildings* 5(2), 83–7.
7 Humphreys, M. A. (1992) Thermal comfort in the context of energy conservation, in: *Energy efficient building*, Eds: Roaf, S. and Hancock, M., Blackwell, Oxford, pp. 3–13.
8 Humphreys, M. A. (1992) Thermal comfort requirements, climate and energy, in: *Renewable energy, technology and the environment*, Ed.: Sayigh, A. A. M., Pergamon Press, Oxford, pp. 1725–34.
9 Oseland, N. A. and Raw, G. (1990) Thermal comfort in starter homes in UK, *BRE note PD 186/90: Proceedings of Environmental Design Research Association 22nd Annual Conference on Healthy Environments.*
10 Oseland, N. A. (1997) *Thermal comfort: a comparison of observed occupant requirements with those predicted and specified in standards.* Unpublished PhD thesis, Cranfield University, UK.
11 Gagge, A. P. *et al.* (1986) A standard predictive index of human response to the thermal environment, *ASHRAE Transactions* 92(2b), 709–31.
12 Fishman, D. S. and Pimbert, S. L. (1982) The thermal environment in offices, *Energy in Buildings* 5(2), 109–16.
13 Oseland, N. A. and Humphreys, M. A. (1994) *Trends in thermal comfort research*. Building Research Establishment Report No. 266, BRE Watford, Herts.
14 Humphreys, M. A. (1992) Thermal comfort requirements, climate and energy, in: *Renewable energy, technology and the environment*, Ed.: Sayigh, A. A. M., Pergamon, Oxford, pp. 1725–34.
15 Grivel, F. and Barth, M. (1981) Thermal comfort in office spaces: predictions and observations, in: *Building energy management*, Eds: de Fernandes, O. *et al.*, Pergamon, Oxford, pp. 681–93.
16 Oseland and Raw (1990), Op. cit.
17 Auliciems, A. (1983) Psycho-physiological criteria for global thermal zones of building design, *International Journal of Biometeorology* 26(Supplement), 69–86.
18 Nicol, F. (1993) *Thermal comfort: a handbook for field studies toward an adaptive model*. University of East London, London.

11

FIELDWORK IN PAKISTAN

In 1993, Roaf, Nicol and Humphreys were joined by Ollie Sykes, a young recruit to the unit, and colleague Mary Hancock for a thermal comfort fieldwork project in Pakistan. The Oxford School of Architecture had contacts there, and some students spent part of their course studying the thermal design and performance of the ancient traditional buildings of Peshawar. This was an initiative set up by Roaf with Yasmeen Lari, a Pakistani architect who had studied architecture at Oxford. The students were recording buildings in danger of demolition – not especially studying their thermal performance. It was instructive to study such traditional buildings because recent buildings had often been designed with scant regard for the climate, and needed a lot of energy to maintain the indoor temperatures recommended by international standards. This demand for energy was putting a serious strain on electricity genera-tion and on the economy of the country, so there was a need to establish realistic thermal comfort standards – standards that would provide comfortable conditions without requiring excessive amounts of energy – and we had been asked to help. A collaborative project between the Oxford School of Architecture and ENERCON, the Pakistan government's energy conservation centre, had been agreed. Gul Najam Jamy and Aruf Allaudin of ENERCON, whom Roaf and Nicol had met in Islamabad on their return from the 1992 Passive and Low Energy Architecture conference, were the principal ENERCON collaborators. Jamy, who had studied for his PhD in Japan, was particularly interested in our findings, touching as they did on his PhD topic.

Funding for a small project was obtained from the UK Government's Overseas Development Agency to help ENERCON develop realistic indoor temperature standards, as a contribution to new building regulations envisaged by the Pakistan Government. (They were at the time using American standards for air-conditioned spaces.)

The Brookes team flew in carrying 25 dataloggers and the associated instrumentation. In a hotel room in Islamabad, 25 handsets – hand-held units that would record
air temperature, globe temperature, humidity and air movement – were assembled.
Sykes had flown out earlier with a prototype and, with Jamy's help, got local trad-
ers to fabricate the instrument-boxes, and made other preparations for the research
team. The assembly of the handsets involved considerable cutting and filing to make
everything fit together properly (see Plate 11.1).

Pakistan's size and topography leads to diversity of climate, and the country can
be divided into as many as 16 climatic zones. These can be reduced to five major
zones.[1] Jamy had selected a town in each of these five (Karachi, Multan, Peshawar,
Quetta and Saidu-Sherif/Mingora) where he had contacts, who in turn found five
people who were willing to serve as subjects in the project. Each was to be given a
handset and a pad of response sheets. One sheet was to be completed every waking
hour for a week, insofar as might be practical. The response sheets were printed in
Urdu. Aware of the occasionally strange behaviour of the ASHRAE scale and of
possible problems arising from its translation into Urdu, the team decided to use a
seven-point semantic differential scale of warmth, labelled 'too hot' at one end and
'too cold' at the other, with the centre region labelled 'comfortable'. Unfortunately
some respondents treated the semantic differential as a three-point scale, using only
either end of the scale and the centre point (see Figure 11.1).

The Brookes research team then dispersed to the five chosen regions (see Figure 11.2).
Humphreys travelled to Mingora/Saidu-Sherif, in the mountainous region of the

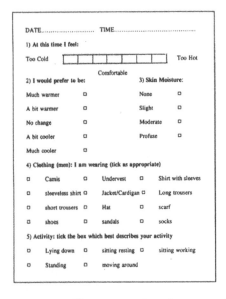
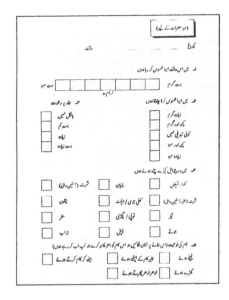

FIGURE 11.1 The response sheet for men – a) English and b) Urdu versions.

Source: J. F. Nicol.

FIGURE 11.2 The climatic regions of Pakistan. The locations of the surveys are shown within the major climatic regions (I–V).

Source: after Raja 1996.

Northwest Frontier province. Sykes went to Karachi on the warm, humid coastal region of the lower Indus plain, where he worked in collaboration with the School of Architecture through Kauser Bashir, the head of the school. Nicol travelled to Multan in the heart of the hot desert in the south of the Upper Indus Plain, and Hancock to Peshawar in the northern Upper Indus Plain, a region with a composite climate. Roaf went to Quetta in the cold, high desert of the Baluchistan Plateau on the Afghan border. Figure 11.3 shows the monthly mean temperatures for each of the sites. Notice that the temperature curves are almost parallel, except for Karachi where the influence of the sea ensures a smaller annual temperature range.

The researchers were each working with a local team recruited by Jamy, and all had a story to tell. Sykes recounted how his student respondents had found it necessary to have a letter from the university in the turbulent city of Karachi to explain why they were carrying a box with wires and electronic equipment, in case they were questioned by the police. Roaf got into serious conversations with the owner of the hotel where she was staying about improvements which could be made to the building.

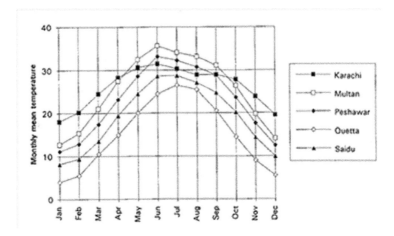

FIGURE 11.3 30-year monthly mean temperatures in each of the survey sites.

The fullest report came from Humphreys who recalls:

I took a flight to Peshawar and thence travelled by minibus through the mountains to Mingora – an alarming experience. I was unfamiliar with the style of driving and I had difficulty understanding and being understood. And my five sets of instruments had been lashed to the roof in a cardboard box despite my attempted protest. Would the equipment survive? The man sitting beside me was good to me: 'How may I help you? Islam teaches us to be kind to strangers.'

At Mingora bus station, he put me and my equipment in a taxi to my destination, a bank in the town centre. I arrived moments before they were due to close for the day, much to my relief and theirs, as I had been unable to contact them.

I spent many hours in the bank over the next week (see Plate 11.2), because all five of my respondents either worked there or were the sons or daughters of those who did. Being immersed in a culture that was strange to me was instructive and fascinating. I was surrounded by kindness and curiosity wherever I went. The curiosity led to some deep conversations when they discovered that I was the Christian equivalent of an Imam.

I had bought some ordinary local summer clothing (a lightweight cotton-polyester shalwar-kamis with cotton undervest, a chadar (a simple rectangular cloak for the cooler times of day), and a Pathan hat) (see Plate 11.3). Now I experienced the comfort of these light and flowing garments in the heat. My European sandals and my socks were considered winter-wear, and I was gently teased for wearing them. Dressed like this, although still unacclimatised to the heat, I found 30°C pleasant and comfortable.

I wished to understand the thermal routines of my respondents, and needed to visit them at home for this purpose. But how could I decently talk with my unmarried female respondents in their homes? The solution was to make me an honorary uncle, which was a privilege. I could then speak with a young woman in the presence of her father. I was interested to learn about their fluid intake, diet, comfort during sleep, and the choice of different living spaces at different times of day. The visits turned into family occasions, with everyone discussing and explaining to me their daily routines. Fluid intake was high – I estimated it to be some 5 litres a day – and the diet was light and varied, with a balance of meat, vegetables, fruit and local bread.

Most people slept indoors on string beds, the web of string being covered by a thin padded cotton 'mattress' (uncompressed thickness about 6–10 mm) so the thermal insulation downwards was much less than that of an English mattress. They usually slept in the clothes they had worn during the day, and changed into a fresh set after washing or showering on rising in the morning. During the night they would use light bed-coverings as needed. Some had a pedestal fan running at night. This served two purposes – to provide some cooling from the air movement, and to drive mosquitoes away. One or two people had western-style beds with foam mattresses, and because of the higher insulation of the mattress they needed to use an air-conditioner in the bedroom. Both the western bedding and the air conditioner seemed to have some social status attached to them.

These visits convinced me that much, probably most, of their thermal adaptation was behavioural in character, and that this behaviour was culturally embedded, often to the extent that our respondents were unaware of it. It was 'second nature' to them. Thermal comfort was a bit like a cybernetic 'black box'.[2] The overarching input to the box was the climate. Within the box, and often hidden from view, were the various adaptive behaviours that together produced the output from the black box, which, if all was well, would be thermal comfort (see Volume 1, pp. 30–33, 84, 158).

The respondents collected their data for a week, providing some hundred sets of comfort responses each. At the end of the week, the Brookes team re-assembled from the five climatic regions, this time in Peshawar, so that they could reflect on their separate experiences of data-collection. It was hot, and for some their first experience of a 'hot wind'. The afternoon air temperature outdoors was so high (46°C) that any breeze had a heating rather than a cooling effect on the body.

Nicol and Humphreys visited the British Council in Peshawar to explain the work. The receptionist/guard in his booth at the entrance had no ceiling fan. They were led through the library, where people were sitting reading. Ceiling fans were running. In the director's office, an air conditioner was running and he was wearing a thick tweed suit and a tie. The room was cold: an interesting interplay between hospitality, status and energy-use.

The Oxford Brookes students put on a public exhibition of their beautiful models and drawings, and there was a meeting with local architects and engineers to explain the project and its implications for the design and operation of buildings. The visiting engineers and architects all came in business suits while the Brookes team were all in Pakistani dress. This seemed to have arisen from reciprocal shows of courtesy and respect. The clothing was conveying social and cultural meanings, which might conflict with their thermal function.

Then it was back to Islamabad to download the data and store the equipment ready for a return visit the following winter (see Plate 11.4), and back to the UK after a day off in the cool of the hills (see Plate 11.5).

The procedures were much the same on the return visit in December/January 1993–94. The aim was to get a picture of winter thermal comfort in the same places. Few buildings had any heat, apart from those in Quetta, because the period of the year when heating would have been desirable was so short. Subsequent analysis showed that heating systems began to be used when the mean outdoor temperature fell below 10°C (remarkably close to the result subsequently found for Europe), and only in Quetta did the mean outdoor temperature drop substantially below this threshold (see Figure 11.3).

The mean summer indoor globe temperature experienced by our respondents was 29.4°C in Saidu and 31.4°C in Karachi. In the winter, they were 12.7°C in Saidu and 24.6°C in Karachi, with the other three sites being around 19°C. The most obvious behavioural difference was in the clothing. The material was thicker in winter, and people wore several layers. Plates 11.6 and 11.7 from Saidu illustrate the differences. Similar differences were found in other locations, except for Karachi where 'winter' temperatures were kept relatively high by the influence of the Indian Ocean.

Humphreys reports:

> The manager of my hotel in Mingora kindly gave me a winter-weight shalwar-kamis and a matching Pathan hat. Wearing these with a winter under-vest, a pullover, and a traditional fully lined winter-waistcoat, I attended a social evening in someone's home. We were sitting on mats on the floor and conversing. I was comfortably warm. There was no heating. I measured the globe temperature: 16°C. I saw no sign from their posture that anyone was feeling cold. People at the offices I visited were comfortable at around this temperature too.

However, our respondents did not fully adapt to a mean room temperature of 13°C in their unheated homes. It was too cold for them. People reading and studying in rooms at this temperature, as some of my Mingora respondents were, would need much more clothing insulation if they were to be comfortable – nearly twice as much as they had. 13°C would have been fine for them if they had been engaged in slightly more active occupations, such as light household tasks. (When clothing insulation is high, the temperature for comfort is strongly dependent on activity – see Chapter 20.)

Research in Pakistan was an enriching experience, especially for those of us without experience of the tropics, increasing our awareness of the importance for comfort of the local culture. Back in Oxford, Nicol and Sykes analysed the data from all five regions, summer and winter. The result was a substantial report confirming that, in a variety of climates, the comfort temperature changes with the climate, that clothing is an important factor in the changes of comfort temperature, and that air movement from wind or the use of fans also has a substantial effect on the comfort-temperature.[3] Humphreys, both at Mingora and Peshawar, had taken the opportunity to conduct small cross-sectional surveys of a wider group of people working at the bank.[4]

What had been learned from the project about fieldwork that could advance our understanding of the adaptive approach to thermal comfort?

- *Year-round surveys are needed*: It is not sufficient for an adaptive comfort project to look only at a seasonal extreme. It is quite possible that people would not or could not adapt fully to such extremes. For example, it may not be worth the cost of installing and running winter heating in Mingora for the short period when it would be desirable, or summer cooling just for brief hot periods. Many people would not be able to afford to install and run such equipment. So future surveys would when possible last for the entire year. The spring and autumn are of particular interest because at these seasons outdoor temperatures are changing most rapidly, and so the processes of adaptation become more evident.
- *Longitudinal or transverse surveys*: There were problems with using the longitudinal experimental design (a few respondents each giving many responses) when the data-collection period was as short as a week. The extraction of a comfort temperature becomes uncertain or even impossible by the normal analytical methods (regression analysis; Probit analysis) if the respondents are making significant adaptive adjustments in response to their changing thermal environment. The presence of adaptation is inconsistent with the assumptions underlying these statistical methods. In this circumstance, a conventional analysis can give very misleading conclusions. So to extract comfort temperatures for our respondents we used a method devised by Ian Griffiths.[5] We have

explained this method in our first volume, and it is given a fuller discussion in Chapter 28 of this volume.

- *Different cultures and climates have different adaptive limits*: Each climatic region of Pakistan had its own band of indoor temperature within which the respondents could quite easily adapt and so become comfortable. But the bands were different in the different climatic regions. This is probably because people's adaptive strategies are learned responses, and part of their regional culture. So a temperature much outside the band that was normal in that region would be likely to cause discomfort. Understanding the regional culture was essential for establishing the limits of adaptation. These limits, of course, are not absolute, but to adapt to conditions beyond them would entail learning new behaviours. Such cultural change can take a long time – perhaps months or years.
- *Semantic differential scaling for subjective warmth*: Semantic differentials are useful tools for psychological scaling, but the problems we had encountered from the absence of category-labels outweighed their theoretical advantage. In future, we would revert to using the ASHRAE scale or the Bedford scale. We were aware that there could be problems with these scales too, but they have the advantage of being widely used, making comparison across surveys easier.
- *Other adaptive actions*: We had calculated that about two-thirds of the adaptation could be ascribed to changes in clothing, but other adaptive actions were also important. Our discussions with our respondents revealed a considerable variety in the adaptive actions that they made – opening and closing windows and doors, adjusting the ceiling fan speed, opening and closing shutters, moving between sun and shade, and occupying different rooms at different times of the day and in different seasons. It would be desirable to keep a record of such actions in future surveys.

The experiences in Pakistan are reported at some length because, for Humphreys and Nicol, the project was a return to practical thermal comfort fieldwork after a lapse of more than a decade. For Roaf, Hancock and Sykes it was their first experience of the systematic collection of thermal comfort data. The Pakistan experience influenced the protocols which were adopted for subsequent surveys conducted by the Oxford Thermal Comfort Unit, and of other projects in which the Unit collaborated.

One aspect of the original Pakistan project with which Jamy in particular was unhappy was the small number of different subjects involved. Funding for a wider, more inclusive study was obtained from the renamed Department for International Development and a plan for the project agreed. Jamy agreed to act as overall coordinator in Pakistan if we could gather the instrumentation and design the experiment.

Dr Iftikhar Raja had joined the unit shortly after the completion of the first Pakistan project. As well as working on the UK projects, with the help of Jamy and Nicol he arranged for a year-round survey in Pakistan, his home country. The experimental design was the repeated transverse survey approach, in which surveys are carried out using the same people at monthly intervals throughout the year.[6] This design permits the observation of seasonal changes in comfort temperature and the extent of discomfort. The unit has since used this experimental design to great effect in other research projects. The regions where the surveys were to be carried out were the same, apart from the replacement of Peshawar by Islamabad (in the same climatic region), which had the advantage that Jamy, a resident of Islamabad, could himself do the survey.

Jamy contacted a person in each region to act as researcher and each was asked to locate five workplaces where surveys could be conducted. Some of the researchers were already known to the UK team from the previous surveys. In all, 31 workplaces were identified, some occupying more than one building. Raja and Nicol travelled to Pakistan to visit each region and workplace, to give the local researchers a set of instruments, and to train them in their use. The instruments were locally sourced except for vane anemometers brought from the UK.

A total of 65 transverse surveys were conducted between April 1995 and July 1996. Each of the 846 subjects in the 34 buildings was visited once a month and was asked for their comfort vote on the 7-point Bedford scale, their preference on the 5-point Nicol scale, and for their skin moisture on a 4-point scale (none–slight–moderate–profuse). A note was taken of their clothing, their activity and their use of controls (doors, windows, ventilators, lights, heaters, fans or air conditioning). The globe temperature was measured using a small pocket thermometer with a painted table-tennis ball mounted over the sensor. The relative humidity and air temperature were measured using a combined humidity and temperature indicator.

The fieldwork in the second Pakistan survey was not without incident, and Raja reminded us of some of them:

> In Multan (Muzaffargarh), while visiting various offices, the instruments were left in the car. After an hour or so when we (Fergus and I) returned to the car we found that the thermometers had burst from the heat. (. . . we had used alcohol thermometers.) In Saidu Sharif [Mingora] our representative had brought all the respondents into one room and filled in the comfort forms with a single set of climatic variable readings. This was realized while entering the data of first two months back in Oxford, and corrective measures were taken. A problem with a hand-held vane wind speed measuring set was holding it in the right direction while the fans (ceiling or pedestal) were running.

Unrest in Karachi meant that for the safety of the researchers results were not collected for seven of the months.

The first thing that was extracted from the data was the prevalence of discomfort (from heat or from cold) among the subjects in each individual survey. Discomfort was taken to mean comfort votes which were either 'too warm' or 'much too warm', 'too cool' or 'much too cool' on the Bedford scale. The proportion of subjects reporting discomfort was calculated for each monthly survey in a particular building. The results are shown in Figure 11.4 and suggest that the incidence of discomfort for these office workers was low if the indoor temperature was between about 20°C and 30°C. Within this range, they made themselves comfortable. In the colder climate of Saidu-Sharif, they could be comfortable at lower indoor temperatures (though interestingly not in Quetta where there was heating) and in the hotter climates of Multan they could make themselves comfortable at higher indoor temperatures.

The data were accurate enough for our purposes, but did not allow rigorous comparison with PMV. In particular the clothing, chiefly the Shalwar-Kameez, a long-sleeved shirt with loose-fitting trousers, was not then among the clothing types whose thermal insulation had been measured, and, in any case, as noted above, it can

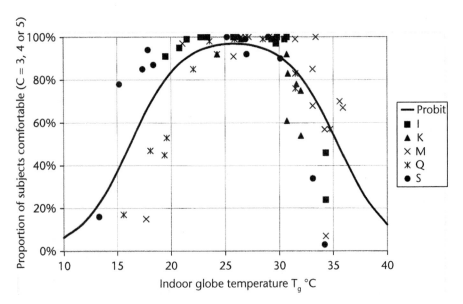

FIGURE 11.4 The proportion of subjects comfortable, recorded in each month in each town as a function of mean indoor globe temperature. The bell-curve was deduced from the pair of Probit regression equations for warmth discomfort and cold discomfort.

Source: J. F. Nicol.

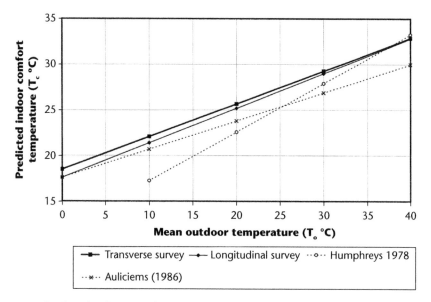

FIGURE 11.5 Predicted indoor comfort temperatures and the outdoor monthly mean temperature. The 'longitudinal' is the first survey and 'transverse' is the second (year-round) survey.

Source: after Nicol *et al.* 1999.

come in a variety of fabric thicknesses and can be worn with a number of additional garments.

The relation between the outdoor temperature and the neutral indoor temperature agreed well with the result of the earlier winter-and-summer field study (see Figure 11.5).[7]

The close agreement between the two surveys in the relation between the outdoor temperature and the temperatures for comfort indoors was reassuring. The results are given in Nicol *et al.*, 1999, cited above. Both surveys have since been more fully analysed, the analysis including the use of clothing and other thermal controls. For an analysis of the use of controls in the second survey, the reader is referred to Rijal *et al.*, 2008.[8]

Notes

1 Raja, I. A. (1996) *Solar energy resources of Pakistan.* Oxford Brookes University, Oxford.
2 Humphreys, M. A. (1994) Field studies and climate chamber experiments in thermal comfort research, in: *Thermal comfort: past, present and future,* Eds: Oseland, N. A. and Humphreys, M. A., Building Research Establishment Report, Watford, pp. 52–72.
3 Nicol, F. *et al.* (1994) *A survey of thermal comfort in Pakistan: towards new indoor temperature standards.* Final Report, School of Architecture, Oxford Brookes University: Oxford.
4 Humphreys, M. A. (1994) An adaptive approach to the thermal comfort of office workers in North West Pakistan, *Renewable Energy* 5(5–8), 985–92.

5 Griffiths, I. D. (1990) *Thermal comfort in buildings with passive solar features.* Report EN3S-090-UK, The Commission of the European Communities: Brussels.
6 See Volume 1, section 9.2.1, p. 114.
7 Nicol, J. F. *et al.* (1999) Climatic variations in comfort temperatures: the Pakistan projects, *Energy and Buildings* 30(3), 261–79.
8 Rijal, H. *et al.* (2008) Development of adaptive algorithms for the operation of windows, fans and doors to predict thermal comfort and energy use in Pakistani buildings, *ASHRAE Transactions* 114(2), 555–73.

12

RAISING AWARENESS OF THE ADAPTIVE APPROACH

When the Oxford Thermal Comfort Unit was in its infancy, few were aware of the adaptive approach to thermal comfort. There was a need for more researchers to become involved if progress were to be made. So the unit held an informal 'teach in' at Oxford on the adaptive approach, to bring those who attended up to speed on adaptive concepts and research. Attendance was good, and included Nick Baker from the Martin Centre at Cambridge, who soon afterwards contributed to the development of the adaptive model, conducting comfort surveys in the UK, France and Greece as part of the EU PASCOOL project. He saw that the success of a building was likely to be related to the 'adaptive opportunity' it afforded its occupants. A building with openable windows, adjustable blinds and ceiling fans, all of which the occupants could control, would provide multiple opportunities for adaptation, and summertime discomfort would be less likely. 'Adaptive opportunity' has since become an important concept in the adaptive approach.[1]

About this time, an informal network, the rather awkwardly named UK Thermal Comfort Interest Group, was set up to keep researchers and research groups in contact with each other. It met annually for some years in one or other of the affiliated universities, and gave those who came a chance to keep abreast of developments. Researchers and students could present their research to an interested and informed audience for critical discussion.

When thermal comfort fieldwork restarted at the BRE in the early 1990s, Nigel Oseland and Gary Raw suggested that BRE should again host an international conference on thermal comfort. It took place in June 1993 in the lecture theatre where Nicol and Humphreys had first presented the adaptive model all those years before. Some who had been present on that occasion returned for this conference.[2] There were new faces too: people who had more recently become interested in thermal comfort research and its application, and who went on to make important contributions to the development of the adaptive approach. Such included Gail Brager from UC Berkeley and Richard de Dear from Sydney, who were to become so influential through their work on the ASHRAE 55 adaptive comfort standard. The conference was wide-ranging, but here we comment only on matters that concern the adaptive approach.

Fanger was present and in a keynote paper explained how to use the PMV/ PPD equation, as set out in ISO 7730. He touched on the existence of discrepancies between the predictions of the equation and the findings in recent field studies and put them down to poor input data, such as overlooking the thermal insulation provided by an office chair, or ascribing an incorrect metabolic rate for office work: 'Poor input data will provide a poor prediction.' More colloquially, this is the 'garbage in – garbage out' effect. There was truth in his critique – it is no easy thing to ascribe a precise thermal insulation to a clothing ensemble or a precise metabolic rate to an activity. It is hard to do even in the laboratory, and almost impossible in the rough and tumble of a field survey.

He went on to make a fundamental objection to the adaptive approach:

> There may occur some physiological adaptation which would require that occupants experience cool or warm discomfort, probably for weeks. Nicol has the intention to develop an 'adaptive model' for thermal comfort. The idea is that people gradually should adapt to the temperatures that happen to occur in free-running buildings. The application of this idea provides an important energy conservation potential. This may work very well in dwellings for people who desire to save [money] and are also ready to suffer a certain discomfort during the adaptation process. . . But the idea of adaptation is in contradiction to the basic rule of ergonomics: that the machine should be adapted to the human. In contrast to this it is Nicol's idea that the human should adapt to the machine (the building). This principle, especially the physiological adaptation is probably less likely to be acceptable in office buildings.[3]

This was a serious objection to the very idea of an adaptive model. It would have substance if physiological adaptation were the principal process by which adaptation proceeds, for physiological adaptation does indeed entail exposure to thermal stress that can be uncomfortable. But the adaptive actions occurring in normal life are more concerned with such things as choosing suitable clothes and adjusting the room temperature or the air speed to meet one's requirements. When this is so, the adaptive model is not 'in contradiction to the basic rule of ergonomics'. It meets another ergonomic requirement: that a machine (in this case the building) should be controllable by its user – which all too often is not so in centrally air-conditioned buildings.

With hindsight it became apparent that Fanger's main objection to the adaptive model arose from a different use of the word 'adaptation'. We had used it in its everyday wide-ranging sense of accommodating oneself to one's circumstances, and

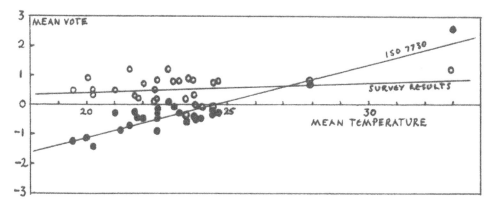

FIGURE 12.1 Facsimile of the original slide: Temperature-dependent discrepancies between actual and predicted mean warmth. Actual mean comfort vote (open points) and predicted mean vote (filled points) for field surveys published between 1978 and 1992.

Source: M. A. Humphreys.

one's circumstances to oneself. He was thinking of what we would have called *physiological acclimatisation*. The misunderstanding was unfortunate. Different uses of the same word, whether in the humanities or in the sciences, can lead to disagreements that are notoriously hard to resolve. The protagonists misunderstand each other, as happened in this case for a number of years.

In a paper that used the findings of the numerous field studies published since 1978, Humphreys pointed out that the discrepancies between the actual mean vote in a survey and the predicted mean vote were systematically related to the mean indoor temperature (see Figure 12.1).[4] Such an outcome was impossible to attribute to the 'garbage in – garbage out' effect, if by that we mean only random errors of measurement and assessment. There could, however, be temperature-dependent systematic errors in the methods of assessing the clothing insulation from garment lists, and metabolic rates from activity lists, as we discuss below. Or it could be that the PMV equation had structural approximations that caused the discrepancies.

The paper also contained an extensive discussion of the role of feedback in achieving thermal comfort. This was, we believe, the first paper to include such a discussion and to extend the adaptive model to include non-thermal aspects of the environment:

> Characteristically, people seek to be comfortable, and take actions to secure thermal comfort. The motivation to do so is powerful. People are not often passive recipients of the provided environment. They will modify the environment, or move to a more suitable one. They will modify their activity, their posture or their clothing to make themselves comfortable. In the longer term, they will seek a better climate, modify their heating or cooling systems, devise

new systems, or even engage in thermal comfort research. Such feedback processes will profoundly affect the relationship between a person and the thermal environment.

Most actions people take to secure thermal comfort can be classified under four headings:

- modify the internal heat generation;
- modify the rate of body heat loss;
- modify the thermal environment;
- select a different thermal environment.

Any mismatch between the person's actual thermal state and their desired one is likely to initiate one of these adjustments or adaptations. The desired state need not be constant, and in practice is not constant. The perception of mismatch between the currently desired state and the actual state is the feedback in the person's thermal comfort control system.

Such a view of comfort can usefully be extended to cover non-thermal aspects of the environment. So, for example, a person coming home after a day at work may take off a coat, turn on some lights, put on some background music, open a window, turn up the heating, and then sit down with the paper while the kettle boils. If the settings prove unsatisfactory, they will be altered according to the feedback received through the senses. The desires and needs may also alter as time goes by, requiring different settings and choices to achieve satisfaction. There will be a hierarchy of responses, the more troubling discrepancies capturing the attention and taking priority. We should therefore expect comfort to be sought by making successive attempts to satisfy a set of continually varying desires and needs. In this way a person is in dynamic equilibrium with the environment. If the various facets of the environment are not independently controllable, the person may have to 'trade off' one kind of comfort against another: opening a window to cool the air may let in too much noise, and so a compromise must be found.

Humphreys pointed out that this feedback would cause a strong correlation between the mean temperature a group experienced and the temperature the group would prefer, and proceeded to discuss the effect of constraints on the adaptive processes:

The state in which a feedback control system settles – that is, its condition of equilibrium – depends on the constraints which are imposed upon it. The room temperature which results from the interplay of the building, the occupants, the heating system and the climate may [be] . . . thought of as the 'answer' or 'solution' or 'result' arising from the complex feedback processes which are taking place. There are various factors which may operate as constraints or boundary conditions, and govern the point at which the system will settle. Thus the comfortable room temperature will depend on such factors as the wealth of the occupants, the climate, the design of the building, the cost of fuel, the cost of clothes, the requirements of other people, the socially correct dress, the controllability of the heating system, the normal dress for the time of year, whether the person must stay at a fixed location, whether the activity is fixed or may be varied, whether a uniform or special clothing is required, and so on.

The list is virtually endless, and one might despair of any possibility of prediction. Put another way, there may be a good relation between what people on average have and what they prefer, but is it possible to predict either of these apart from the other? If not, we are in much the same position as with the heat exchange equations – we can predict the temperature for comfort if we know the clothing and the activity, but how can we predict these without knowing the temperature? The procedure would be circular.

Mathematically, the room temperature for comfort (t_c) will depend upon several constraints:

$(c_1, c_2, \ldots c_n)$ hence:

$$t_c = f(c_1, c_2, c_3 \ldots c_n) \;^5$$

The thermal discomfort (TD) will depend on the difference between the comfort temperature (t_c) and the actual temperature (t):

$$TD = f(t_c - t)$$

These constraints may be gathered under four main headings:

- constraints due to the climate;
- constraints due to poverty;
- constraints due to social custom;
- constraints due to task or occupation.

. . . In practice a good degree of predictive power can be obtained from taking into account fairly few constraints. Chief of these is the climate, which has a pervasive influence on human living, influencing the architecture and many other aspects of the people's culture. It might therefore be expected that the climate would have a dominant influence on the indoor temperatures at which people were comfortable, dominating the 'solution' of the complicated feedback processes.

He went on to point to the evidence for this, and to compare the adaptive relation to a 'black box':

The relationship can be regarded as a cybernetic 'black box'. The 'input signal' is the climate. The 'output' is the comfort temperature. The invisible components within the box include the design and construction of the building, all the human heat exchange variables and all the various sensory feedback loops. The feedback overall is negative, for the variation in the comfort temperature is only about half that of the mean outdoor temperature. A 'black box' having these characteristics behaves in a stable fashion, so the relationship is likely to be solid and general. Ordinarily there is no need to know the detailed characteristic of any component within a 'black box', for the feedback largely governs the output. So it turns out that we can estimate the comfort temperature without knowing anything at all about the human heat exchange equation. However, if for some reason the feedback were inhibited from operating, such knowledge would be needed.[6]

Richard de Dear, in a perceptive and carefully reasoned paper, then presented the results from various recently conducted field studies where the six factors required for calculating PMV had been measured according to the best up-to-date protocols.[7] These field studies were from his and Auliciems' own work in the various regions and climates of Australia, from his work in Singapore, from John Busch's work in

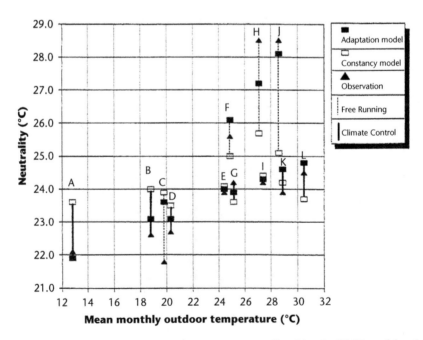

FIGURE 12.2 Discrepancies between neutral temperatures predicted by the PMV model and those actually found in field experiments. Points A and B are from San Francisco, C, D, E, F, G, and K are from Australia, H and I from Singapore, J and L from Bangkok.

Source: de Dear 1994.

Bangkok, and from the work of Gail Brager and her co-workers in the San Francisco Bay area of California. He was able to show beyond reasonable doubt that there were real differences between the predicted temperatures for thermal neutrality and those actually found, especially in naturally ventilated buildings in warm climates. The quality of the measurements and the large size of some of the discrepancies made the differences nearly impossible to deny.

De Dear's work on the adaptive model built on that developed by Auliciems (see Chapter 11). The discrepancies between the neutral temperature obtained from PMV and the empirical result of the field study he attributed to the differing 'expectations' people had of the thermal environment. For example, people would have different expectations of a naturally ventilated building in summertime than they would of a centrally air-conditioned building in the same weather. This difference of expectation allows the possibility of different temperatures for comfort.

That different expectations lead to different temperature preferences is a hypothesis that is now widely accepted although it is quite difficult to demonstrate. Nicol and Humphreys considered that, while 'expectation' of this kind might have a role to play, the discrepancies were chiefly attributable to physical factors. Values for clothing insulation are obtained from a list of garments, and do not usually allow for different weights of cloth. In warm environments, clothing is usually cut from thinner cloth, and is often worn to allow air movement within it. So values of thermal insulation obtained from a checklist of garments would be likely to over-estimate the thermal insulation of the clothing as actually worn in warm environments. Values for metabolic rates are obtained from tables of activities, and overlook the fact that in warm environments people tend to avoid needless exertion. So the same task may have a lower metabolic rate when performed in a warm environment. (There is a need for accurate non-intrusive ways of measuring clothing insulation and metabolic rates in daily life – but these are yet to be developed.) Also there could be structural faults in the PMV equation. So, in our opinion, it was premature to ascribe the discrepancies entirely to 'expectation'. The expectation model presumes a psychological cause for the discrepancies, while we believed that they could largely be assigned to temperature-dependent errors in the assessment of metabolic rate and of clothing insulation, to approximations in the physics of the PMV model and to errors in its logic.

The source of the discrepancies, and even their existence, is not of much importance purely from the point of view of the adaptive model, for the model does not depend on either the existence of such discrepancies or upon their source. But from a scientific point of view it was unsatisfactory to be getting demonstrably different answers from careful field research and equally careful laboratory research. We return to this topic in a later chapter.

At the BRE conference, the adaptive view was forcefully presented by these two papers. This time it was received with careful appraisal rather than with the incredulity of the 1972 conference. But the adaptive approach was rather overwhelmed by the sheer number of papers on other aspects of thermal comfort. Perhaps because of this, Susan Roaf thought of holding an international conference specifically for adaptive thermal comfort, particularly with regard to its implications for thermal comfort codes and standards. So she booked the Cumberland Lodge conference centre in Windsor Great Park for a few days at the end of August 1994. This UK conference centre is residential, so the participants could meet one another socially, talk together about ideas and explain their research proposals informally and at length.

This first Windsor Conference was inspirational, with some forty participants, and was notable for the diversity of cultures and climates represented: architects, engineers and researchers from all round the world. They spoke of the thermal comfort standards used in their own countries and of the associated research. Several speakers focused on the needs, desires and behaviour of the occupants of buildings rather than on the technicalities of heating and cooling systems. It quickly became apparent that the internationally recognised but western-based standards (ASHRAE Standard 55, ISO 7730) could be ill suited to local needs, either because of the climate or because of cultural and economic circumstance. But how could new standards be formulated to allow for such diversity?

Jamy and Nicol both gave papers on our field work in Pakistan and its implications for buildings and for standards there.[8,9] There were also papers from Australia, Bangladesh, the Czech Republic, Greece, India, Indonesia, Vietnam, Japan and the USA. In a keynote paper, Humphreys explored thermal comfort as if trying to establish the preferred thermal habitat of hobbits (see Chapter 1).[10] We would first measure the climate of the regions where they were found, and then the thermal environment inside their dwellings. If we understood their language we would go on to ask them if they were comfortable. We would not need to understand their thermal physiology, but could simply observe their habitat and their adaptive behaviours.[11] Comfort conditions could therefore be formulated without the help of heat exchange equations – although our scientific curiosity would of course wish to know about clothing insulation, metabolic rate, air movements, humidity and the body's thermal regulatory system and so on. The implication for thermal comfort standards was not hard to see. Buildings should provide a sufficient range of adjustment so that the occupants could make themselves comfortable.

It was at this conference that Humphreys and Nicol first suggested using an exponentially weighted running mean of the outdoor temperatures, rather than the historic values from meteorological tables, as a basis for the prediction or control of indoor temperature for thermal comfort.[12] Such a running mean could form the basis of a practical control algorithm for Auliciems' 'thermobile'. Running means had previously been used to quantify the speed with which change of clothing occurred (see Chapters 4 and 31). The proceedings of the conference were published complete with all the discussions – painstakingly transcribed by Sykes.[13]

It is impossible to tell of the expansion of the awareness of the adaptive approach without a brief mention of subsequent Windsor Conferences. The conference has become an influential forum for the discussion of adaptive thermal comfort and low energy architecture. The second Windsor Conference (*Thermal comfort standards for the twenty-first century*) was held in 2001 and attracted delegates from all branches of the thermal comfort world (see Plate 12.1). There was particular interest in the work of Gail Brager and Richard de Dear in developing a meta-analysis of data from all over the world[14] to shape the thinking behind the new adaptive standard to be included in ASHRAE Standard 55-2004. Selected papers from the conference were developed as a special issue of the journal *Energy and Buildings*.[15] This special issue became an important resource for researchers and students, and established the place of the adaptive approach in the academic world.

The third Windsor Conference in 2004 focussed on the developing discipline of the Post Occupancy Evaluation of Buildings (sometimes referred to as Building Performance Evaluation). It is the process of evaluating buildings in a systematic and rigorous manner after they have been built and occupied for some time. Inherent in the evaluation is the need to ask the inhabitants whether the building provides comfortable conditions. This necessarily includes asking about thermal comfort, which is integral to the way the building works by providing adaptive opportunities. A selection of papers appeared in a special issue of *Building Research and Information*.[16]

In the same year the Low Energy Architecture Research Unit (LEARN) at London Metropolitan University obtained a grant for the formation of the Network for Comfort and Energy Use in Buildings (NCEUB).[17] The grant enabled the Network to employ a part-time organiser for three years to organise conferences and meetings and to build a website. The network started with about 300 members: about one third of them from outside the UK and about two thirds of them from an academic background. Membership has now grown to over 500. Its website makes available all the papers from the Windsor Conferences and has become an important venue for the international exchange of information on thermal comfort research.

Since 2004 the Windsor Conferences have become a regular event every two years, usually held in April and lasting for three days:

2006: *Comfort and energy use in buildings – getting them right* (Selected papers in *Energy and Buildings* 39(7), July 2007).

2008: *Air conditioning and the low carbon challenge* (Selected papers in *Building Research and Information* 37(4), 2009 *Cooling in a low carbon world*).

2010: *Adapting to change – new thinking on comfort* (Selected papers in *Building Research and Information* 39(2), 2011 *Adaptive comfort*).

2012: *The changing context of comfort in an unpredictable world* (Selected papers in *Architectural Science Review* 56(1) *The wicked problem of designing for comfort in a rapidly changing world,* and in *Building Research and Information* 41(3), 2013 *Adaptive comfort*).

2014: *The cost of comfort in a changing world.*

The papers of the 2014 conference (see Plate 12.2) are, as usual, downloadable from the NCEUB[18] where the proceedings are available as individual pages or as a 1400 pages e-book.[19] They are also available from the special conference web-site.[20] At the time of writing, special issues of *Building Research and Information* and *Architectural Science Review*, based on the conference papers, are being prepared.

Those who have come together at the Windsor Conferences represent a wide range of interests, from theoretical and applied physiology to building and conditioning-system design, and move in a number of different professional, political and commercial circles. This has meant that the adaptive approach to thermal comfort, which has provided a foundation for all the conferences, has become known among building thermal modellers, among those concerned with renewable energy, among those working in passive, low energy and low carbon architecture, and among the various bodies responsible for formulating national and international standards and guidelines for thermal comfort. It is through such networks, meetings and friendships that regional and international understanding and appreciation of the adaptive approach grew to become an influential force in the design and servicing of buildings, and in the understanding of their ability to provide for the comfort of their occupants.

Notes

1 Standeven, M. A. and Baker, N. V. (1995) Comfort conditions in PASCOOL surveys, in: *Standards for thermal comfort*, Eds: Nicol F. *et al.*, E. & F. N. Spon (Chapman & Hall), London, pp. 161–8.

2 Oseland, N. A. and Humphreys, M. A., Eds, (1994) *Thermal comfort: past, present and future.* Building Research Establishment Report, Watford.

3 Fanger, P. O. (1994) How to apply models predicting thermal sensation and discomfort in practice, in: *Thermal comfort: past, present and future*, Eds: Oseland, N. A. and Humphreys, M. A., Building Research Establishment Report, Watford, pp. 11–4 (both quotations are from p. 12).

4 Humphreys, M. A. (1994) Field studies and climate chamber experiments in thermal comfort research, in: *Thermal comfort: past, present and future*, Eds: Oseland, N. A. and Humphreys, M. A., Building Research Establishment Report, Watford, pp. 52–72.

5 Functional notation may be strange to some of our readers. The expression means that the comfort temperature is a function of (depends on) the constraints that apply.

6 The extracts are from pp. 60–3 of the report.

7 de Dear, R. J. (1994) Outdoor climatic influences on indoor thermal comfort, in: *Thermal comfort: past, present and future*, Eds: Oseland, N. A. and Humphreys, M. A., Building Research Establishment Report, Watford, pp. 106–32.

8 Jamy, G. N. (1995) Towards new indoor comfort temperature standards for Pakistani buildings, in: *Standards for thermal comfort*, Eds: Nicol F. *et al.*, E. & F. N. Spon (Chapman & Hall), London, pp. 14–21.

9 Nicol, J. F. (1995) Thermal comfort and temperature standards in Pakistan, in: *Standards for thermal comfort*, Eds: Nicol F. *et al.*, E. & F. N. Spon (Chapman & Hall), London, pp. 149–56.

10 The idea goes back to a coffee-time conversation with Charles Webb on how one might establish thermal comfort conditions for Martians.

11 Humphreys, M. A. (1995) Thermal comfort temperatures and the habits of hobbits, in: *Standards for thermal comfort*, Eds: Nicol F. *et al.*, E. & F. N. Spon (Chapman & Hall), London, pp. 3–13.

12 Humphreys, M. A. and Nicol, J. F. (1995) An adaptive guideline for office temperatures, in: *Standards for thermal comfort*, Eds: Nicol F. *et al.*, E. & F. N. Spon (Chapman & Hall), London, pp. 190–95.

13 Nicol F. *et al.*, Eds, (1995) *Standards for thermal comfort*, E. & F. N. Spon (Chapman & Hall), London.

14 See Chapter 14.

15 Nicol, J. F. and Parsons, K., Guest editors: *Energy and Buildings* (2002) 34(6).

16 Nicol, F. and Roaf, S., Guest editors: *Building Research and Information* (BRI) (2005) 33(4).

17 www.nceub.org.uk.

18 http://nceub.org.uk//W2014/webpage/W2014_index.html.

19 Nicol, F. *et al.* (2014), Eds: *Proceedings of the 2014 Windsor Conference,* NCEUB, London (nceub.org.uk) ISBN 978-0-9928957-0-9.

20 See: www.windsorconference.com.

13

BEGINNING FIELDWORK AT OXFORD BROOKES UNIVERSITY

Once Nicol and Roaf had established the Thermal Comfort Unit at Oxford Brookes University, its research on adaptive thermal comfort in the UK began. The first project was an exploratory longitudinal survey of thermal comfort in some naturally ventilated buildings in Oxford at the end of the summer, to see how the indoor comfort temperature tracked the outdoor temperature as temperatures fell with the onset of autumn. It was for this project that Dr Iftikhar Raja joined the group in 1994. Raja had studied the solar resources of his native Pakistan and was to publish his work in a book published by Oxford Brookes.[1] A further aim of the project was to develop a robust methodology for future work. The study looked at the use of windows and blinds to control the room temperature, and all six main thermal variables were recorded.

Analysis of the survey results found that the comfort temperature indoors tracked the exponentially weighted running mean of the outdoor temperature as previously proposed,[2] and suggested that the level of thermal mass in the building was reflected in the rate at which the subjects adapted to the indoor temperature. There was also an unexpected result: one of the buildings had a central corridor with offices on either side. Offices on one side of the corridor faced north; on the other side, they faced south. Based on their understanding of the adaptive model, it was expected that people would be adapted to the mean temperature they experienced in their own office during the survey period. But people in the warmer south facing offices felt warmer than those in the cooler north facing offices. They seemed to have adapted to the mean temperature of the offices on their floor, rather than simply to the mean temperature in their own office. This observation raised an interesting question that has not yet received adequate research: Precisely what in a person's thermal experience determines the room temperature to which he or she is currently adapted?

In addition to the survey-based research, Raja and Nicol undertook an analysis of the implications of posture for heat loss from the body and therefore the role it might play in adapting to the indoor temperature. The method depended on estimating the proportion of the subject's body that was in contact with another part and therefore unlikely to be an avenue for heat loss into the environment. They estimated that the effective surface area of the body could be reduced by as much as one-third by postural changes.[3] The results from the survey and the analysis were published in a report.[4]

From the experience of this project and the first Pakistan project, a robust survey methodology was developed, and was tested in the second Pakistan survey (Chapter 11). This methodology has several features. Year-round surveys were preferred and each building would be visited each month of the year to conduct a transverse survey of thermal comfort. At each visit, the thermal environment of each respondent would be measured with a handset similar to that developed for use in Pakistan, and a comfort questionnaire administered. Clothing and activity would be noted, and also the use of thermal controls: windows, blinds and fans. A subset of volunteers would provide longitudinal data using a brief response sheet up to four times a day, while a miniature datalogger recorded the temperature at their desk throughout the period of their participation, which could vary from a week to some months. Outdoor temperatures were to be obtained hourly from data-loggers at each building and cross-checked with data from the nearest meteorological station. The Pakistan surveys had also suggested that using subjects from a variety of climates was an advantage.

Starting in 1996, this methodology was applied to surveys in Aberdeen (North-East Scotland) and in Oxfordshire (Southern England), in air-conditioned and naturally ventilated office buildings.[5] The unit was awarded an EnREI research fellowship by the Department of the Environment in recognition of the potential value of the work. This enabled them to employ Kate McCartney, an able and energetic building science graduate, to supervise the surveys. McCartney wrote to a number of employers in the Oxford area and asked them if they would take part in the survey, and nine replied in the affirmative. Gary Clark undertook to oversee the Aberdeen end of the survey and recruited six buildings for the surveys. Three air-conditioned and three naturally ventilated buildings were selected in the Aberdeen area, and two air-conditioned and seven naturally ventilated buildings in the Oxford area. In each building, volunteers were recruited to take part in the monthly surveys. Each was also asked if they would be willing to take part in the longitudinal survey which involved filling in a brief form four times a day, and these volunteers were provided with an environmental monitor unit to place on their desk. Monthly surveys were conducted in each building throughout the year, and 4997 sets of subjective and environmental data were collected from 897 subjects. The 219 respondents who volunteered to give responses up to four times a day provided in total 35,974 sets of data. It was made clear that they could stop as soon as they found it burdensome.

Some lasted just a week, while others happily continued to provide data for months. Outdoor air temperatures were continuously recorded at each building for the duration of the experiment. The surveys continued from March 1996 until September 1997. The start date varied somewhat between buildings and most of the data were collected between May 1996 and July 1997.

The survey was not problem-free. At one of the offices in Aberdeen, when researchers arrived one month they found the building empty and our equipment gone. One of the Oxford buildings was lost part way through the project, because they were dealing with confidential data and became anxious about security.

In some of the Oxford buildings the fingertip temperature of respondents was also measured, by asking them to hold a thermistor between the thumb and forefinger until its reading stabilised after a minute or so. The readings demonstrated the body's thermal control mechanism in operation. The fingertip temperature of a person in comfort did not remain steady, but tended either to the core temperature or to the room temperature, as vasodilation or vasoconstriction occurred. Intermediate temperatures were relatively rare.[6] People were in dynamic rather than static equilibrium with their environment.

One of the aims of the project was to track the changes in thermal comfort during the onset and continuation of a hot spell, to see how quickly and how well people adapted to the changed weather. Unfortunately, there was no hot weather that year either in Oxford or in Aberdeen (UK weather is notoriously uncertain). It was nevertheless possible to quantify the speed of people's response to the changing weather during the year, relating the exponential running mean of the outdoor air temperature to the day-on-day changes in the indoor comfort temperatures of our respondents. The result, though hardly definitive, was encouraging. It had been assumed beforehand, based on the judgement of the team, that the relevant time-constant (alpha) would have a value of about 0.85.[7] The experimental work gave the best fit if alpha was set at 0.8,[8] a figure that was confirmed by later research across Europe (Chapter 16).

One aspect of this study was the collection of large quantities of data about the use of various controls as part of the adaptive process. The team were able, on the basis of the data collected, to begin to develop an approach to the use of controls[9] and to challenge the building thermal simulators to develop ways to simulate this aspect of occupant behaviour.[10] This work was developed further some years later by Humphreys, Rijal and others. The variety of the available controls and adaptive opportunities among the buildings surveyed made it possible to show that the overall adaptive opportunity was not simply the sum of the individual opportunities.[11]

Doctoral students studying at other British universities began to consult Nicol and Humphreys about conducting and analysing thermal comfort surveys in their home countries. In this way, substantial surveys were conducted in Indonesia by Tri Karyono and in Iran by Shahin Heidari. Similar student contacts led to thermal comfort field studies in Nigeria (Mike Adebamowo) and in Zambia (Albert Malama).[12] Advice was also sought on a project in Tunisia (Cheb Bouden) where year-round surveys were conducted in a number of cities.[13] At the BRE, Nigel Oseland followed up his work in starter homes with a large survey in air-conditioned and naturally ventilated offices in the UK.[14] He was also able to show that people at the office and at home provided significantly different comfort assessments although the thermal conditions and the clothing were the same, suggesting that the warmth assessments depended to some extent on the social context,[15] but it is very difficult to ensure that the metabolic rate is exactly the same at home and at the office.

So with all these surveys, together with those commissioned by ASHRAE, the quantity of data expanded, as did the variety of climates and cultures from which they came. Particularly welcome was the influx of data from South East Asia, Australia, Latin America and Africa, a continent unrepresented in the 1997 database.[16–18] The growing number of surveys in previously under-represented regions, cultures and climates showed how broad is the range of the experience of comfort. At Brookes, we added the statistics from each new survey into a database of summary statistics – a database that now has over 700 entries. The number of surveys published in recent years has been such that we have been unable to keep the database up to date.

At this point we summarise the rapid progress of the adaptive model from 1991 until about 1997. In 1991, a mere handful of researchers (Andris Auliciems, Richard de Dear, Gail Schiller, Ian Cooper) had adopted an adaptive perspective on thermal comfort. By 1997, there was a worldwide informal network of researchers and practitioners working on adaptive comfort. They were from Africa, Asia, Australia, Europe, North America and South America. The quantity of data from new surveys was large and covered an increasing diversity of climates. We note the following conceptual advances during this period:

- *Social influences*: The original Nicol/Humphreys 1972 adaptive self-regulating model of comfort applied to the individual. Work on clothing in the 1970s had shown the potential effect of social pressures and norms on adaptive cloth-ing behaviour. The social nature of thermal comfort had been more forcefully brought out during our work in Pakistan. Adaptive actions were embedded in the culture of the people. This became much more obvious when working in an unfamiliar culture – working in familiar cultures can lead to blind spots.
- *Climate*: Because the climate has a pervasive influence on the culture of a people and on the design of their buildings, the comfort of an individual must be set in its social, cultural and climatic context.
- *Adaptive opportunity*: The concept of adaptive opportunity linked adaptive ther-mal comfort to the design of the building. The concept had begun to influence the way adaptive researchers were thinking of comfort, and also led to the

formation of stronger links between thermal comfort researchers and those concerned with passive and low energy architecture.

- *Psychological adaptation*: The idea that one's response to the thermal environment could be influenced by one's expectations (in the evaluative sense of the word) had led to the postulation of psychological as well as behavioural and physiological adaptation.

- *PMV's applicability*: That PMV could not explain the range of conditions people found comfortable had become clear from our 1975 worldwide database. During the period 1991–96, the limitations of PMV became increasingly evident, particularly for predicting thermal comfort conditions for people in naturally ventilated buildings in warm climates. The discrepancies were now being *quantified* by surveys that took careful note not only of the thermal environment but also of people's clothing insulation and metabolic rate.

For the period after about 1996, our account of the progress of the adaptive model becomes thematic rather than sequential. This is because there were several almost independent themes during this period, and because the increasing number of researchers makes the personal story of the authors but a small part of the progress of the adaptive approach.

Notes

1 Raja, I. (1996) *Solar energy resources of Pakistan*. Oxford Brookes University, Oxford.

2 Humphreys M. A. and Nicol J. F. (1995) An adaptive guideline for UK office temperatures, in: *Standards for thermal comfort*, Eds: Nicol, F. *et al.*, E. & F. N. Spon (Chapman & Hall), London, pp. 190–95. For such running means see also Volume 1 (p. 38) and Chapter 31 of this volume.

3 Raja, I. A. and Nicol, J. F. (1997) A technique for postural recording and analysis for thermal comfort research, *Applied Ergonomics* 28(3), 221–5.

4 Nicol, F. and Raja, I. (1996) *Thermal comfort, time and posture: exploratory studies in the nature of adaptive thermal comfort*. School of Architecture, Oxford Brookes University, Oxford.

5 Nicol, F. and McCartney, K. (1997) Modelling temperature and human behaviour in buildings – field studies 1996–97, *Proceedings of BEPAC/EPSRC Miniconference: Sustainable Building*, Abingdon.

6 Humphreys, M. A. *et al.* (1999) An analysis of some observations of finger temperature and thermal comfort of office workers, *Indoor Air*, Edinburgh. See also Chapter 23 of this volume.

7 Humphreys, M.A. and Nicol, J. F. (1995) An adaptive guideline for UK office temperatures, in: *Standards for thermal comfort*, Eds: Nicol, F. *et al.*, E. & F. N. Spon (Chapman & Hall), London, pp. 190–95.

8 McCartney, K. J. and Nicol, J. F. (2002) Developing an adaptive control algorithm for Europe, *Energy and Buildings* 34(6), 623–35.

9 Raja, I. A. *et al.* (1998) Natural ventilated buildings: use of controls for changing indoor climate, *Renewable Energy* 15, 391–4.

10 Nicol, J. F. (2001) Characterising occupant behaviour in buildings: towards a stochastic model of occupant use of windows, lights, blinds heaters and fans, *Proceedings of the Seventh International IBPSA Conference*, Rio, Vol. 2, International Building Performance Simulation Association, pp. 1073–8.

11 Nicol, J. F. and McCartney, K. J. (1999) Assessing adaptive opportunities in buildings, *Engineering in the 21st Century – the changing world*, Chartered Institution of Building Services Engineers, London, pp. 219–29.

12 Malama, A. *et al.* (1998) An investigation of the thermal comfort adaptive model in a tropical upland climate, *ASHRAE Technical Data Bulletin* 14(1), 102–11. (See also: *ASHRAE Transactions*, 104(1).)

13 Bouden, C. *et al.* (1998) A thermal comfort survey in Tunisia, EPIC 98, Lyon, France, *Proceedings ACTES*, pp. 491–6.

14 Oseland, N. O. (1998) Acceptable temperature ranges in naturally ventilated and air-conditioned offices, *ASHRAE Technical Data Bulletin* 14(1), 50–62. (See also: *ASHRAE Transactions*, 104(1).)

15 Oseland N. A. (1995) Predicted and reported thermal sensation in climate chambers, offices and homes, *Energy and Buildings* 23(2), 105–15.

16 Akande, O. K. and Adebamowo, M. A. (2010) Indoor thermal comfort for residential buildings in hot–dry climate of Nigeria, *Proceedings of the 6th Windsor Conference,* NCEUB, pp. 86–96.

17 Sangowawa, T. *et al.* (2008) Cooling, comfort and low-energy in a warm humid climate: the experience of Lagos, Nigeria, *Proceedings of the 5th Windsor Conference*, NCEUB, pp. 35–49.

18 Adebamowo M. A. (2006) Thermal comfort for naturally ventilated houses in Lagos Metropolis, *Proceedings of the 4th Windsor Conference,* NCEUB, pp. 71–84.

14

PMV AND THE RESULTS OF FIELD STUDIES

There are a number of concerns with the PMV/PPD thermal comfort model in relation to the findings of field research. They date back to its publication in 1970:[1]

- PMV could not account for the data from the hot dry summers in Roorkee, India, or in Baghdad, Iraq,[2] where acclimatised respondents reported feeling comfortable (neither warm nor cool) at temperatures above 30°C. According to PMV, people could not be comfortable at these high temperatures. Physiological factors must account for some of the discrepancies and we surmised that acclimatised people could be comfortable at sweat rates rather higher than those defined as comfortable for the PMV equation. Perhaps what mattered was not the sweat rate itself but the thermal stress on the body. When a person is acclimatised to heat, the body's temperature threshold for sweating is reduced, and moderate sweating is not stressful. In a hot, dry climate a high sweat rate can pass unnoticed because the skin stays dry.
- It was difficult to see how PMV could be used in practice – that is to say, in the circumstances of daily life. To predict the comfort-vote, the PMV model needs to know what people are wearing and what they are doing. According to the adaptive model, the clothing and activity depend on the thermal environment. The application of PMV is circular.
- Humphreys and Nicol were not persuaded of the need for complex multivariate indices of comfort for common indoor environments. Their general experience was that adding extra variables in addition to the globe temperature often *reduced* the predictive power of an index. We felt this was likely to apply to PMV too.
- Early in Fanger's 1970 book, the fundamental conditions for comfort were given as a range of skin temperatures and a range of sweat rates, the comfort ranges

varying with the activity level. But later in the book, when PMV is added into the heat balance equation for thermal neutrality, the departure from this neutrality (the comfort vote) is expressed in terms of a 'hypothetical heat load'. A 'hypothetical heat load' (W/m^2) is dimensionally different from a range of skin temperatures (K). The relation between the two depends on the clothing insulation. They are not equivalent.

A heat exchange model like PMV is theoretically compatible with an adaptive approach to thermal comfort, for heat exchange between a person and the environment is an integral component of the adaptive model, as we explained in Chapter 5. However, a disagreement between the theoretical predictions of PMV and the empirical findings of field studies did require explanation. Such an explanation could be used to improve the heat exchange model, or the field study methods, or both. The existence of discrepancies led some field researchers to distrust PMV and some laboratory researchers to distrust the field results. This was unsatisfactory.

In 1988, Doherty and Arens had compared the predictions of the PMV equation with a database of thermal comfort assessments and physiological measurements that they assembled from numerous laboratory studies.[3] They found that PMV was unbiased at about 28°C for resting, naked people. At this temperature, a sedentary naked person is on average thermally neutral. At higher temperatures, PMV overestimated how warm they felt. At lower temperatures, it overestimated how cold people felt. The errors were progressive. Doherty and Arens expected that outside the band 26°C–30°C the discrepancy would exceed 0.5 scale units for naked people. They also found that PMV predicted poorly at higher levels of activity. So it would be unwise to assume that PMV was entirely satisfactory as a predictor of thermal sensation even in laboratory conditions. This finding was very similar to what we had found from the field studies.[4] Thus both laboratory experiments and field studies seemed to agree about the general pattern of the biases in the PMV equation.

One test of the adequacy of an index is how well it correlates with people's thermal perceptions (comfort votes) in everyday life. This aspect is distinct from the question of bias, for it would be possible to have an index that contained bias (that is, it wrongly predicted the average thermal sensation) that correlated very well with the comfort votes. Correlation does not consider bias at all, but rather measures the degree to which an index is *associated* with the comfort vote (a perfect association gives a correlation coefficient of unity, while no association at all gives a correlation of zero). Back in 1936, Bedford had used the correlation coefficient to compare the various indices then in use.[5] Yaglou's Effective Temperature, although it included humidity and air movement as well as air temperature, correlated no better with people's perceptions of warmth than did the simple air temperature, so there seemed to be no advantage in using the more complicated index. Some other indices not now in use performed considerably worse than did the simple air temperature.

There are three kinds of reason why the correlation of a complex index with respondents' comfort votes might be no higher, or even be lower, than with a simple index:

1   If the air temperature and the mean radiant temperature are almost the same, the air is practically still, and the humidity is no problem, a simple measurement of the air temperature or the globe temperature is sufficient. This is common in everyday moderate thermal environments. It is no criticism of the complex index; it could become useful in less common thermal environments.

2   If an index is incorrectly formulated – that is to say, it does not give the correct relative weights to those aspects of the environment it includes – then it is likely to perform poorly. This is the reason for the eventual replacement in 1986 of Yaglou's Effective Temperature, ET, by Gagge's New Effective Temperature, ET*. It had been found that, in practice, Yaglou's version gave much too much weight to variation in humidity. If the humidity in the room varies in a quasi-random fashion, and it is incorrectly weighted in the formulation of the index, then it will increase the random scatter of the estimate of warmth, and thus lower its correlation with the thermal sensation. So it is important to get the formulation of the index correct. This need explains the years of research effort, both in the field and in the laboratory, dedicated to producing ever-better indices.

3   If a correctly formulated index includes variables that affect the sensation of warmth but cannot be measured accurately, then the error introduced by these variables will outweigh the benefit of including them. The inclusion of the clothing insulation and the metabolic rate in an index can, for this reason, introduce more scatter than they eliminate. This seems to be true of PMV when it is used to assess the subjective warmth of individuals in ordinary indoor environments. Again, this is no criticism of the index itself – rather of its usefulness in assessing everyday thermal environments.

Field studies with measurements sufficiently complete to enable the estimation of PMV and Standard Effective Temperature (SET) made it possible for the first time to correlate these indices with the actual responses of people in everyday life. Humphreys reported on these correlations in 1994,[6] drawing on the then recent surveys by John Busch and by Gail Schiller.[7,8]

The New Effective Temperature ET* is in effect the globe temperature adjusted for humidity. In Busch's data it performs no better on average than the air temperature. SET, which adds to ET* the effect of air speed, clothing insulation and metabolic rate, performed much worse.[9] If we may assume that SET attributes correct weight to each of its component variables, its poor performance must be attributed to uncertainties in measuring the air speed, estimating the clothing insulation, or the metabolic rate. The uncertainties have more than outweighed the benefit of including the extra information.

R-squared is the proportion of the variance of the comfort vote that is explained by the index. The effectiveness of the index has therefore been about halved by including the extra variables, as seen in the last column of Table 14.1 (0.10 compared with 0.19).

Schiller's surveys in offices in San Francisco gave another example (see Table 14.2). The R-squared values are low, as is common in buildings where the temperature variation during the working day is small. Again there seems to be no overall advantage – rather perhaps the reverse – in using an index that includes the clothing insulation and the metabolic rate, as do both PMV and TSENS. (TSENS is the predicted thermal sensation within Gagge's SET model.)

These examples illustrate by means of practical examples the loss of accuracy of an index attributable to increased complexity. A different kind of criticism of PMV, one that is theoretically based rather than empirically demonstrated, was presented at the joint ASHRAE–CIBSE conference held in Harrogate, UK, in 1996.[10] It examined the effect of an oddity within the PMV model. To move from the steady-state, heat-balance equation to obtain PMV, Fanger had used what he called a 'hypothetical heat load'. This was the extra heat that would need to be generated in order to compensate for the actual off-balance heat loss to the environment. PMV was a function of the hypothetical heat load. Now, a steady-state model cannot contain an off-balance heat load, hypothetical or not. Thermal balance must be restored somehow, either by an increase in the metabolic heat generation or by a decrease in skin temperature by vasoconstriction. Within the PMV equation itself, if balance is restored by vasoconstriction – the body's first line of defence against cold – the same PMV predicted different mean skin temperatures for heavy and light clothing. This is not credible if skin temperature is the criterion for comfort. This anomaly within

TABLE 14.1 Correlations (R-squared values) of the comfort votes with various indices of warmth in Busch's survey of office workers in Bangkok.

	Air conditioned	Naturally ventilated	Average[1]
Air temperature	0.19	0.19	0.19
New Effective Temperature (ET*)	0.20	0.18	0.19
Standard Effective Temperature (SET)	0.08	0.12	0.10

[1]Correlations should be averaged using Fisher's z transformation, but here a simple average is good enough.

TABLE 14.2 Correlations (R-squared values) of the comfort votes with some indices of warmth in Schiller's survey of office workers in San Francisco.

	Summer	Winter	Average
New Effective Temperature (ET*)	0.09	0.09	0.09
PMV	0.11	0.05	0.08
Thermal sensation (TSENS) (Gagge)	0.04	0.10	0.07

PMV has far-reaching consequences for the tables of PMV, for if the criticism is true it renders the values in the tables incorrect except at 0.6 clo. The errors become serious for ensembles with very low thermal insulation, and for heavily clothed people.

Ken Parsons and Linda Webb, of the Human Thermal Environment Laboratory at Loughborough University, UK, set up a climatic chamber study to investigate the matter in collaboration with the Oxford Brookes Thermal Comfort Unit.[11] Actual comfort votes were to be compared with the predictions from the PMV equation for different levels of clothing insulation, in a balanced experimental trial. A PMV value of −1 (slightly cool) was aimed at for people wearing 1.2 clo and 0.15 clo. The two groups would have had equal actual mean votes had PMV been correct. The measured conditions in the chamber were not quite as planned, and were −0.7 and −0.8 PMV respectively. So the lightly clad group should have felt very slightly cooler. In fact, as predicted, those in 0.15 clo felt *warmer* than those in 1.2 clo (p = 0.0016, paired t-test, single tail). The value may be calculated from the data in table 5 of the paper. The results provided empirical evidence of the theoretical limitation on PMV. However, the actual mean comfort votes of both groups were higher than expected, and this made the outcome less clear-cut. It is not easy, even in a climate laboratory, by applying PMV, successfully to obtain a specified thermal sensation from a group of respondents.

The advent in 1997 of the ASHRAE RP-884 database of thermal comfort surveys, compiled by Richard de Dear, Gail Brager and Donna Cooper, put a huge amount of data into the public domain for the first time.[12] The database consisted of some forty files, each giving the complete line-by-line data from a separate thermal comfort survey, each line representing an interview with a respondent. Because each line in the database gives a comfort vote together with values of the six principal variables, a value of PMV can be obtained for every line of data – and there were more than 20,000 lines in the database. Correlations between the comfort vote and the various indices are shown in column 2 in Table 14.3. The correlations are directly comparable because they all rest on the same body of data.

The overwhelming evidence from this extensive new database showed that PMV was a worse predictor of the comfort vote in everyday life than a simple index such

TABLE 14.3 Correlations (R-squared values) of the comfort votes with various indices included in the ASHRAE RP-884 database and the SCATs database.

	ASHRAE RP-884 data	*SCATs data*
Air temperature	0.264	0.124
Operative temperature	0.265	0.111
New Effective Temperature (ET*)	0.257	0.102
Standard Effective Temperature (SET)	0.185	0.040
PMV	0.213	0.062
	N=20,468	N=4,068

as the air temperature or the operative temperature, which is virtually the globe temperature. The same conclusion is evident from the database of the SCATs project, a large European research programme, of which more later. The result is shown as column 3 in Table 14.3.

The databases had used the best available ways of estimating the metabolic rate from the recent activities of the respondent, and the clothing insulation from a list of the garments worn at the time of interview. So the 'garbage in – garbage out' explanation of this poor performance had less power to convince. An index which cannot be adequately quantified by using the best existing methods, and is related less strongly to thermal sensation than is the air temperature, is of limited practical use.

PMV and SET had been vying for supremacy. Both used all six principal variables to produce an index of warmth, though they were conceptually rather different. The ASHRAE RP-884 database enabled an interesting comparison of the two. Each row in the database gave a calculated value of PMV and of SET from the environmental data in that row. Exactly the same measurement errors would therefore be present in both calculated values. So a scatter diagram of PMV and SET would show how different or how similar the two indices were in their predictions of subjective warmth in the database (see Figure 14.1), eliminating any effect of measurement errors. The result is discouraging. Had the indices been in agreement there would be no scatter – just a single line or perhaps a curve relating the two. So it is necessary to conclude that SET or PMV, or both of them, contain substantial faults in their formulation. The extent of the difference is serious. A value of PMV does not give a precise corresponding value for SET. It may be anywhere within a range of several degrees.

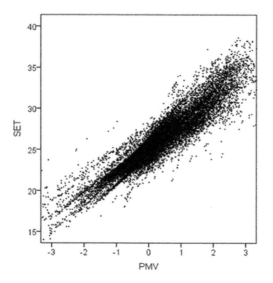

FIGURE 14.1 Comparing SET and PMV in the ASHRAE RP-884 database.

Source: M. A. Humphreys.

All these investigations were further undermining our confidence in the usefulness of complex indices in general and PMV in particular, but had no effect on its
widespread use. Some advocated controlling the thermal environment in a room to
within 0.2 units of PMV. That is impossible, for in everyday life neither the clothing nor the metabolic rate can be 'known' to the building's thermal control system;
the thermostat does not know what a person is wearing or what they are doing. To
keep PMV within 0.2 scale units must mean in practice that the control system is set
so that the combined effect of the environmental variables would keep PMV within
0.2 scale units, had the clothing and metabolic rate remained at their assumed values.
This corresponds to about 0.6 K in operative temperature if the air movement is
constant – still not easy to do, and probably needlessly stringent.

Most standards based on PMV give examples of acceptable temperature, based on
standard values of humidity and air velocity and the assumption that air temperature
and radiant temperature are equal. The standard values of clothing insulation may vary
between summer and winter. For instance in the European Standard EN15251[13] the
clothing insulation for the heating season is 1.0 clo and in the cooling season it is 0.5 clo.
Insofar as it takes account of seasonal clothing changes, PMV is being used in an adaptive
way. The danger is that the adaptation is viewed mechanistically. In real offices, clothing
changes continuously and in modern offices will rarely be 1 clo even in winter; in tropical and equatorial lands the division between winter and summer will be meaningless.
The danger is that the temperature values given as examples in the standards *become the
standard values*. Although PMV includes humidity and air velocity the standard values
are given as values of operative temperature. PMV seems to have become redundant.

The criticisms of PMV that had been made by various field researchers over the
years were having an effect, for in response Fanger and Jørn Toftum introduced an
adaptive component into the PMV equation. They first presented it at the opening
session of the 2001 Windsor Conference, and again at the Indoor Air conference
in Monterey, California, in 2002.[14,15] An adaptive adjustment to PMV was to be
applied to people in non-air-conditioned buildings in warm climates.

The adjustment had two parts. The first was an adjustment to allow for the 'expectation'
of the occupants. They argued that in regions of the world where air conditioning was
uncommon, people would have lower expectations of their thermal environments, and
consequently they would tolerate higher indoor temperatures. The expectancy factor
ranged from 0.5 (few air-conditioned buildings, warm weather all year round) to 1.0
(air-conditioned buildings common, brief warm periods during the summer). This
adjustment took up de Dear and Brager's meaning of 'expectation', though they would
not wish to speak of high or low expectations, but of *different* expectations.

The second factor concerned the estimate of the metabolic rate. The value
obtained from the usual lists of activity should, they argued, be reduced in warm
interiors, to allow for the possibility that the same kind of activity would be performed with more economy of movement in a warm climate.

So PMV, now PMV_e, incorporated two adaptive factors. Table 14.4 shows the process of arriving at the PMV_e for four cities in warm climates.

An 'expectancy factor' is hard to quantify for any particular geographical location, being a professional judgement rather than a measurable variable. The factor for reduction of metabolic rate with increased temperature, 0.067 for every scale unit of PMV above neutral, does not rest on experimental evidence. Fanger and Toftum say in the Monterey version of their paper: 'Both the decrement of metabolic rate per unit PMV and the assignment of expectancy factors for the four cities were based on professional judgment.'

The scientific credentials of the extended PMV are weak, but it shows an acceptance that adaptation can and does affect the human response to the thermal environment, surely a step in the right direction.

If PMV is subject to various biases, it is desirable to quantify them. The ASHRAE RP-884 database can be used to uncover the pattern of the various biases. A difficulty is that PMV is the prediction of the mean response of a *group of people who are all having the same clothing insulation and metabolic rate*. This never happens in an office building in daily life, and still less at home. Obtaining a mean PMV for a diverse group having different clothing and different activities and various thermal environments, and comparing it with the empirical group–mean subjective warmth, is therefore not a totally rigorous test of PMV. How then can a rigorous test be devised?

PMV could logically apply to a group of size one – that is, to an individual. A comparison between the PMV and the actual vote for an individual would of course be of very low precision, but would be an unbiased estimate of any discrepancy. If a sufficient number of these unbiased estimates were pooled, the pattern of discrepancies could be revealed. A thorough analysis was undertaken of the individual discrepancies in the ASHRAE RP-884 database to see how they depended on the several variables in PMV. This analysis was presented at the Windsor Conference in 2001.[16] It showed that, while PMV was unbiased when all the data were pooled, there were biases on each of the principal variables, and that the biases were large enough to matter (see Figure 14.2).

TABLE 14.4 Comparison of observed thermal sensation votes and predictions made using the new extension of the PMV model (Fanger and Toftum's data).

	Expectancy factor	Unadjusted PMV	PMV adjusted to proper activity	PMV_e adjusted for expectation	Observed mean vote
Bangkok	0.6	2.1	2.0	1.2	1.3
Singapore	0.7	1.3	1.2	0.8	0.7
Athens	0.7	1.4	1.0	0.7	0.7
Brisbane	0.9	0.9	0.9	0.8	0.8

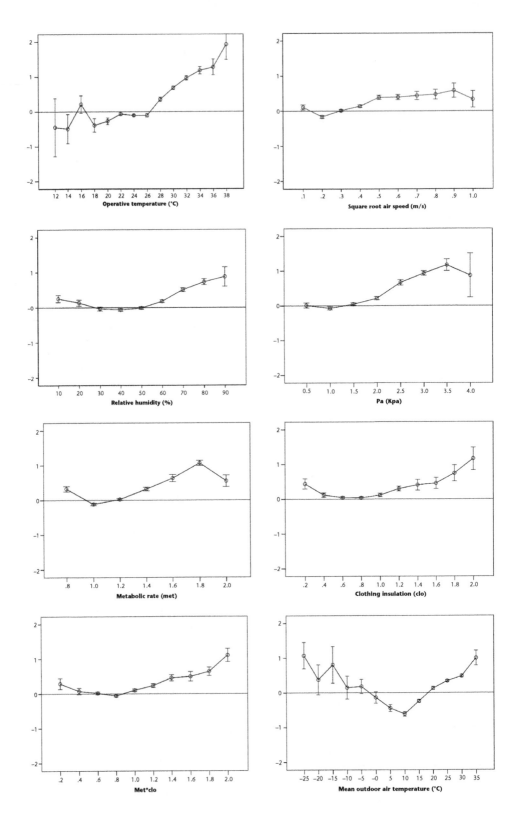

FIGURE 14.2 The bias in PMV against climate variables, showing an increasing bias in warm conditions. The error bars are the 95 per cent confidence limits.

Source: M. A. Humphreys.

The figure shows that PMV overestimated the subjective warmth if the operative temperature was high, the water vapour pressure high, or the clothing insulation high. The trends were less clear for the metabolic rate and for the air speed. There was a tendency to overestimate the subjective warmth when the outdoor temperatures were above 25°C and below minus 10°C.

In this analysis, all the data in the database were used rather than selecting only those with the highest standards of measurement – the 'class 1' surveys. The other classes of survey had less comprehensive measurements, lacking assessments of thermal asymmetry and vertical temperature gradients, but the measurements of the six principal variables were complete and sound. Their evidence could not be discounted.

A bias in PMV could also be shown from a database of summary statistics that Humphreys had been assembling over the years. Whenever a new field study of thermal comfort was encountered its summary statistics were added to the database. The database contains, among other data, the mean PMV and the mean thermal sensation for each block of survey data. By 2005, it had grown to include hundreds of blocks of data.[17] It was used to show how PMV differed from the actual mean vote over the range of mean indoor temperature encountered in the data (see Figure 14.3). Such group comparisons do not provide as rigorous a check as can be obtained from the individual observations, but they nevertheless show clear trends. When people are feeling warmer than neutral in a warm environment, PMV overestimates how warm they feel. When people are feeling cooler than neutral in a cool environment, PMV says – but here the evidence is weak – people should not feel as cold as they

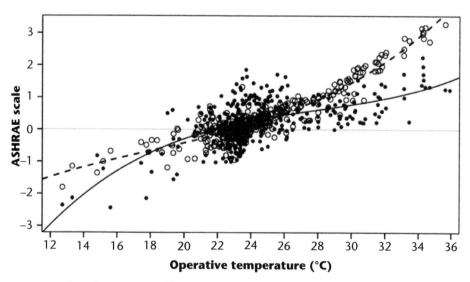

FIGURE 14.3 Actual mean vote (filled circles and solid line) and PMV (open circles and dashed line) against the mean operative temperature for the blocks of survey data (N> = 20) in the database of summary statistics. The lines are cubic fits.

Source: M. A. Humphreys.

do. These consequences are what would be expected from the use of the hypothetical heat load in the PMV equation, rather than the mean skin temperature.

Despite the introduction of adaptive PMV, a thorough revision of the equation is needed to address the biases that have been found from field surveys. Alternatively, PMV could be replaced in the Standards by one of the more recently developed thermal comfort heat exchange models. These would need to be subjected to the kinds of test that we have outlined in this chapter before being adopted.

It has become common in building thermal simulation to produce as an output a PMV contour map of a room. These maps have no meaning unless the assumed clothing insulation and metabolic rate are stated. These values must be held in the foreground of the awareness of the person studying the output. In daily life the clothing insulation is not fixed; it depends on the room temperature because people adapt to their thermal environment by adopting a different weight of clothing. It is also likely that the metabolic rate is to some extent temperature dependent. These adaptive processes make PMV contour maps unrealistic. The fundamental problem is that the contours lump together two quite different things: the building's indoor thermal environment (air temperature, mean radiant temperature, air speed, humidity) and the occupants' response to it (their clothing insulation and perhaps their metabolic rate). It would be wiser to keep these separate. Strictly speaking there can be no such thing as a PMV contour map of a room.

In summary what should be said about PMV in relation to field study results and its practical usefulness? PMV, in common with all multivariate indices of thermal comfort, correlates poorly with the actual thermal sensations reported in field studies. Using the best current measurement and assessment methods, PMV is found to be a poorer correlate that the simple air temperature. To the extent that this is attributable to the great difficulty in accurately assessing some of the input variables, particularly the clothing and the metabolic rate, the development of better measurement techniques will improve the correlation.

The predictions of the PMV are biased with respect to its input variables. The most important and best documented of these biases is its tendency to overestimate how warm people feel in warm indoor environments. This is a serious fault, because it tends to lead to the unnecessary cooling of buildings, with the economic and environmental costs that this entails. The biases are attributable to faults in the construction of the index, and need to be corrected by revising its formulation. Fanger and Toftum's PMV_e is a first step in this direction, but is based on professional judgement and lacks a secure foundation in the results of field research.

PMV is of dubious utility as a basis for standards of thermal comfort, and for the control of the thermal environment in buildings. This is because the clothing and the metabolic rate are needed to calculate PMV, and must be assumed in the standard and by the building's control system. Inspection of the standards shows that in fact the specified ranges are given in terms of the operative temperature rather than PMV. Room thermostats control for temperature, and know nothing of clothing or metabolic rate. This limits the usefulness of PMV as a control variable, and of any index that includes the clothing and the metabolic rate.

PMV is an internationally used method for assessing comfort and is a key tool of the HVAC industry. In light of the above concerns and in a warming world, it will be increasingly necessary to agree where it is, and is not, appropriate to use the PMV/PPD method for assessing and determining comfort conditions in buildings.

Notes

1 Fanger, P. O. (1970) *Thermal comfort*. Danish Technical Press, Copenhagen.
2 Nicol, J. F. (1974) An analysis of some observations of thermal comfort in Roorkee, India and Baghdad, Iraq, *Annals of Human Biology* 1(4), 411–26.
3 Doherty, T. J. and Arens, E. (1988) Evaluation of the physiological bases of thermal comfort models, *ASHRAE Transactions* 94(1), 1371–85.
4 Humphreys, M. A. (1994) Field studies and climate chamber experiments in thermal comfort research, in: *Thermal comfort: past, present and future*, Eds: Oseland, N. A. and Humphreys, M. A., Building Research Establishment Report, BRE, Watford, pp. 52–72.
5 Note that correlations cannot be compared *across* different surveys, where the ranges and distributions of the environmental variables differ.
6 Humphreys, M. A. (1994) Op. cit.
7 Busch, J. F. (1990) Thermal responses to the Thai office environment, *ASHRAE Transactions* 96(1), 853–8.
8 Schiller, G. E. (1990) A comparison of measured and predicted comfort in office buildings, *ASHRAE Transactions* 96(1), 609–22.
9 ET* includes air speed only as it affects the relative weights attributed to air and radiant temperatures. Thus if air and radiant temperatures are equal, ET* is independent of air speed. SET includes an additional air speed component – in effect the equation simulates a heated globe rather than an unheated one.
10 Humphreys, M. A. and Nicol, J. F. (1996) Conflicting criteria for thermal sensation within the Fanger predicted mean vote equation, *CIBSE/ASHRAE Joint National Conference Papers*, Harrogate.
11 Parsons, K. C. *et al.* (1997) A climatic chamber study into the validity of Fanger's PMV/PPD thermal comfort index for subjects wearing different levels of clothing insulation, *CIBSE National Conference Proceedings*, pp. 193–205.
12 de Dear, R. *et al.* (1997) *Developing an adaptive model of thermal comfort and preference – final report on RP-884*. Macquarie University, Sydney.
13 BSI (2007) BS EN 15251: 2007 *Indoor environmental input parameters for design and assessment of energy performance of buildings addressing indoor air quality, thermal environment, lighting and acoustics*. Comité Européen de Normalisation, Brussels.
14 Fanger, P. O. and Toftum, J. (2001) Thermal comfort in the future – excellence and expectation, *Proceedings: Moving Thermal Comfort into the 21st Century*, Cumberland Lodge, Windsor.
15 Fanger, P. O. and Toftum, J. (2002) Prediction of thermal sensation in non-air-conditioned buildings in warm climates, *Proceedings of the 9th International Conference on Indoor Air Quality and Climate*, Monterey, California.
16 Humphreys, M. A. and Nicol, J. F. (2002) The validity of ISO–PMV for predicting comfort votes in everyday life, *Energy and Buildings* 34, 667–84.
17 Humphreys, M. A. *et al.* (2007) Field studies of thermal comfort and the progress of the adaptive model, *Advances in Building Energy Research* 1, 55–88.

15

ADAPTATION AND THE ASHRAE RP-884 DATABASE

In the early 1990s, ASHRAE commissioned new field studies of thermal comfort across the world, studies in which the protocol for data collection was fixed and the accuracy of the physical measurements beyond reproach. The protocol required not only the measurement of the four main environmental variables (air temperature, mean radiant temperature, air speed and humidity), but also the measurement of radiant asymmetry. Measurements were to be taken at three heights so that the vertical temperature gradient could be quantified. Among the subjective responses would be the ASHRAE 7-point scale of warmth and the McIntyre 3-point scale of thermal preference. The recent activities of each respondent would be listed so that a sound estimate of their metabolic rate could be obtained. A standard clothing checklist would be used to enable an estimate of the clothing insulation to be made.

As described in Chapter 14, Richard de Dear and Gail Brager were awarded the ASHRAE contract for a meta-analysis of the data from these and other good-quality field studies. They began by assembling a database of all recent field studies for which the raw data were available, contacting researchers in the field to see whether they would be willing for their data to be included. These other studies did not necessarily completely conform to the ASHRAE protocol, but all had measurements comprehensive enough for PMV and SET to be calculated, using the newly available ASHRAE 'thermal comfort tool'. The collected data amounted to more than 20,000 sets of measurements. Each line in the database comprised the subjective data and the corresponding environmental measurements, for a single respondent on a single occasion. The data from the various surveys were first converted to a standard template. The various clothing checklists were converted into a common standard by an ingenious regression method. The mean radiant temperature was calculated from the air temperature, the globe temperature and air speed.

Richard de Dear sent us 40 Excel electronic files that comprised the database, when the work of assembling them was complete. The database was impressive. It is of course not possible for such a large database to be entirely error free, and Humphreys emailed Richard information on any errors found during our checking procedures.[1] Often it was obvious what the correct value should be. If no obvious correction could be made, the entry was deleted. There were, for instance, examples of people in offices in San Francisco having a zero clo-value. This turned out to be not a free-thinking adaptive response to a warm environment, but an error arising from different missing-value codes in different files of data. The coding of the McIntyre scale had also caused problems, because there is no fixed convention among researchers. Even after careful checking both by de Dear's team and by us at Oxford Brookes, various errors no doubt remain in the database, but the publicly available version certainly meets a high standard of reliability. The database has continued to provide the data from which to answer numerous research questions since it was made available in 1998. It would be hard to list all the research projects that have made use of it.

De Dear and Brager along with Donna Cooper produced a report that included both the description of the processes of compiling the database and a statistical analysis of the data.[2] Humphreys studied a copy in detail, because the analysis amounted to a more comprehensive version of his earlier meta-analyses of worldwide data in 1975–81.[3,4,5] The analysis by the ASHRAE team was wide ranging despite being produced under the pressure of a deadline. It takes a long time to become familiar with new data, to reflect on it, understand its import, and perform appropriate analyses. Deadlines can be counter-productive.

To what extent did the new analysis confirm the older findings, and what new findings might there be? Had the relation between the climate and the desired indoor temperatures remained stable over the decades?

The fundamental adaptive relation found in 1975 between the mean indoor temperature during a survey and the neutral or preferred temperature – a relation having a correlation coefficient of 0.96 – was weak in the new data. This cornerstone of the adaptive model seemed to have crumbled. The correlation (Pearson's r) in the new data was only 0.53. The R-squared value had fallen from 0.92 to 0.28. So instead of explaining more than 90 per cent of the variation in the comfort in 1975, the mean temperature explained only 28 per cent in the new data. The difference is seen by comparing Figure 15.1 for the new data with Figure 7.1 for the old data.

The other surprise was that the dependence of the neutral indoor temperature on the current mean outdoor temperature was much smaller than we had found in 1978 (see Chapter 8). The analysis given in the report yielded a gradient of only 0.255/K against the previous value of 0.534. So, while in 1978 it took a two-degree change in outdoor mean temperature to raise the preferred indoor temperature by one degree, it took four degrees in 1998.

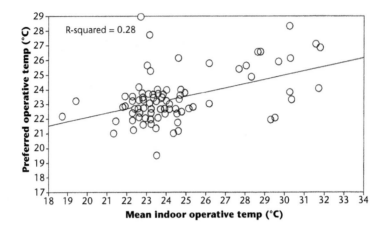

FIGURE 15.1 Relation between the preferred operative temperature and the mean indoor operative temperature from the 1997 database. Each point on the figure refers to a single building.

Source: M. A. Humphreys, after de Dear *et al.*, ASHRAE Report RP-884.

Assuming that the adaptive principle still held, what could explain the lower correlation between the preferred temperature and the mean temperature in the accommodation? And what could explain the reduced dependence of the preferred indoor temperatures on the outdoor temperature? De Dear and his team had used the same calculation procedure that we had used in 1975: obtaining the neutral temperatures from regression analysis, and the preferred temperatures from Probit analysis. They had also followed the procedure that we had published in 1978 to derive the relation between the temperatures for comfort indoors and the climate. So why were there substantially different conclusions?

It was possible that the differences could be attributed to the technological advances in data collection. The older data, from 1936 through to 1974, had almost all been collected laboriously by hand, using the instrumentation developed by Thomas Bedford and described in Chapter 2. Any substantial collection of field study data took a small research team some months to complete. The big copper globe thermometers then in use took about 30 minutes to reach equilibrium, so in an eight-hour day only about sixteen different locations could be visited and all the data compiled. A survey of modest size, having readings from some 300 workstation visits would take a single researcher about a month to complete. So it would follow that the mean operative temperature of the period of the survey would be representative of conditions that the occupants of the building experienced over the period.

With the advent of rapid data-logging and modern instruments this was no longer necessarily so. It had become possible to complete the survey of a large building in a day or two. The mean indoor temperature obtained during so brief a survey period might or might not be typical of the experience of the occupants – the survey-days might have been unusually hot, or unusually cold.

The data from each building had been separated by de Dear, because it was possible that the indoor environments within the different buildings in the same town and season differed systematically, and that therefore people would have adapted to different indoor conditions. So each point on Figure 15.1 represents the result from a different building. Separating the buildings meant that many of the batches of data had rather few observations, and therefore the estimates of the regression coefficients of subjective warmth on temperature would have a large margin of error. There was therefore a large margin of error on the estimates of the neutral temperatures that depended on the regression coefficients. The same uncertainty would be present in the estimates of the preferred temperature from Probit analysis. So the advantages of modern rapid methods of data acquisition had brought with them unforeseen analytical problems.

To overcome the difficulty of the often-brief data-collection periods and the small numbers of observations in many of the buildings, we pooled the data from the several buildings in each survey. This would help overcome the problem of the short duration of the data-collection period in many of the individual buildings. Then we used a fixed regression gradient (0.5 scale-units/K) that had been adjusted for error in the predictor variable to calculate each neutral temperature from the mean vote and the mean operative temperature, rather than using a separate regression gradient for each batch of data. This would overcome the problem of imprecision in the estimates of neutral temperature. Using this procedure, the strong adaptive relation between mean indoor temperature and the neutral temperature re-appeared with a satisfyingly high correlation.

For this chapter, we have recalculated the relation, keeping the buildings separate, as de Dear had done, but still using a standard regression coefficient (see Figure 15.2).

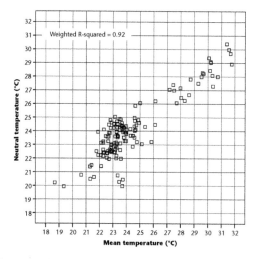

FIGURE 15.2 Relation between the neutral temperature and the mean temperature during the survey, recalculated using a standard regression coefficient.

Source: M. A. Humphreys.

The correlation is high (Pearson's r = 0.961, R-squared = 0.92). This suggests that much of the problem lay with the scatter of the regression coefficients. The matter is taken up in Chapter 28.

Also the relation between the indoor temperatures for thermal neutrality and the prevailing outdoor air temperature became very similar to the one obtained in 1978. Using the method just described to obtain the neutral temperatures, we found the gradient increased and the precision of the relation improved. The ASHRAE team had categorised the buildings slightly differently from the categories we had used in 1978. Their adaptive relation applied to all naturally ventilated buildings, whereas our old relation applied to buildings when no heating or cooling appliances were in use (the free-running mode). Data from naturally ventilated buildings that were heated at the time of the survey were removed. This raised the gradient to 0.546, satisfyingly close to its 1978 value of 0.534 (see Figure 15.3).

So, analysed in this manner, there was good agreement between what had been found in the 1970s and what could be deduced from the fresh data. Two things had become clear to us:

- The method of analysis could have important effects on the conclusions.
- The time-period of data collection needed to be considered when choosing the method of analysis.

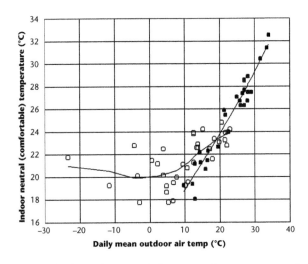

FIGURE 15.3 The recalculated relation between indoor neutral temperature and the mean outdoor temperature. Each point represents a separate survey in the ASHRAE RP-884 database. The gradient for the free-running mode (filled points) is 0.546 (trend lines Loess smoothed).

Source: M. A. Humphreys.

De Dear and Brager arranged a session at the 1998 ASHRAE summer meeting specifically to present the RP-884 adaptive model of thermal comfort, together with the findings of other recent thermal comfort field studies. De Dear presented a paper on the formation of the RP-884 database, and, with Brager, another on the results it produced, including the new adaptive relation with the climate. We presented a paper explaining the basic principles of the adaptive approach.[6] Other researchers presented new survey work that they had recently completed. It was a memorable session with lively discussion. The proceedings of the meeting were published as a separate book that is still cited.[7] The adaptive model was now 'on the map' for ASHRAE.

De Dear and Brager then began the process of incorporating the relation between the preferred indoor temperature and the mean outdoor temperature in ASHRAE Standard 55. ASHRAE had commissioned these surveys, so should not the resulting adaptive relation appear in the Standard? De Dear and Brager ably advocated the model and defended it against the technical criticism of some committee members and the reluctance of others. The relation was included in the 2004 revision of the Standard, but subject to strict limits on its use. It was a remarkable achievement considering how dismissive of the adaptive model many still were.

An adaptive model now appeared in one of the most widely used standards for thermal comfort. To be sure, it would need to be developed over the years as new experimental evidence and improved analytical methods became available. Humphreys and Nicol published in *ASHRAE Transactions* their suggestions for an alternative analysis of the data, showing that if analysed in this way the database yielded an adaptive relation close to that found in 1978.[8]

De Dear, Brager and Cooper did a great and lasting service to thermal comfort research by assembling the ASHRAE RP-884 database and making it available on their website. A relation between the climate and the preferred indoor temperatures may be evaluated from it. Some other findings that they drew from it are listed below:

- People were often comfortable beyond the limits of the 'comfort zone' given in the ASHRAE Standard – and they were frequently uncomfortable within it.
- The simple air temperature or the globe temperature correlated better with the thermal sensation than did either SET or PMV.
- Thermal neutrality on the ASHRAE scale does not necessarily equate to the desire for no change as indicated by the McIntyre scale.
- The systematic differences between thermal neutrality and thermal preference could be related to indoor and outdoor temperatures.

The beauty of having the database publicly available is that anyone with a reasonable grasp of statistical methods can check any analysis that is based upon it. Since the database was completed, a number of other surveys have been transformed into its protocol and are similarly available. At the time of writing, there are moves to assemble a single database to include the many field studies published since 1997, to form a 'Mark 2' database.

This is to be welcomed, despite the disadvantages of standard protocols for the collection of field research data, protocols that reflect the concerns of laboratory experiments. Field research should not be dominated by research protocols that rest on laboratory methods, where comprehensive measurements are needed to calculate the human heat exchanges. Heat exchange is but one facet of the adaptive approach to thermal comfort. Such protocols tend to restrict the researcher, inhibiting the conduct of creative new field research, where often quite simple and cheap instrumentation is sufficient to test and quantify new ideas.

Notes

1 Volume 1, Chapter 10 explains our methods of checking data.
2 de Dear, R. *et al.* (1997) *Developing an adaptive model of thermal comfort and preference.* ASHRAE RP-884 Final Report, Atlanta.
3 Humphreys, M. A. (1975) *Field studies of thermal comfort compared and applied,* Department of the Environment, Building Research Establishment, CP 76/75. (Reissued in: *J. Inst. Heat. & Vent. Eng.* 44, 5–27, 1976, and in *Physiological Requirements on the Microclimate,* Symposium, Prague, 1975.)
4 Humphreys, M. A. (1978) Outdoor temperatures and comfort indoors, *Building Research and Practice (J. CIB)* 6(2), 92–105.
5 Humphreys, M.A. (1981) The dependence of comfortable temperature upon indoor and outdoor climate, in: *Bioengineering, thermal physiology and comfort,* Eds: Cena K. and Clark J. A., Elsevier, Amsterdam, pp. 229–50.
6 Humphreys, M. A. and Nicol, J. F. (1998) Understanding the adaptive approach to thermal comfort, *ASHRAE Transactions* 104(1), 991–1004.
7 ASHRAE (1998) *Field studies of thermal comfort and adaptation,* Technical Bulletin 14(1).
8 Humphreys, M.A. and Nicol, J. F. (2000) Outdoor temperature and indoor thermal comfort – raising the precision of the relationship for the 1998 ASHRAE database of field studies, *ASHRAE Transactions* 106(2), 485–92.

16

GOING INTERNATIONAL

The SCATs project

The European Union (EU), as part of its strategy to foster key research fields and overcome differences and inequalities between member nations, allocates a considerable budget for joint scientific research that involves international research groupings. In line with the urgency with which the EU views the climate crisis, a good part of this budget is allocated to the conservation of energy and the development and encouragement of renewable energies. Given the importance of the built environment in energy use (variously estimated at between 40 and 50 per cent of the total) reducing energy use in buildings is an important factor in the overall conservation of energy.

The conservation of energy remains one of the motives for developing the adaptive approach to thermal comfort. It conserves energy in two ways:

1   Energy is saved because the adaptive approach allows the indoor temperature to drift in sympathy with the prevailing outdoor temperature, so reducing the difference between the indoor temperature and the outdoor temperature. This reduced difference reduces the energy needed for heating or cooling. Allowing the indoor temperature to drift during a day can also reduce the difference between the indoor and the outdoor temperature, again saving energy for heating and cooling. As early as 1986, Andris Auliciems demonstrated that cooling and heating loads could be reduced by up to 50 per cent if the indoor temperature followed an adaptive seasonal variation throughout the year.[1] This realisation inevitably attracted the interest of researchers keen to find ways to reduce the energy used in buildings.

2   The heat balance approach assumes that buildings with constant indoor temperatures are inherently superior to those with a variable indoor temperature and rates them more highly.[2] If, instead, buildings were rated according to their energy usage, they could be designed to operate in the free-running mode

for more of the year, and so rely less on mechanical heating and cooling.[3] A comfort standard that assumes an adaptive relationship between comfort and outdoor temperatures makes it easier for the designer to comply with the standard through climatic design, using freely available ambient energy. The adaptive approach was used by de Dear and Brager to develop ASHRAE standard 55 for naturally ventilated buildings[4] (see Chapter 15). It allows the indoor environment to change from month to month in response to the outdoor temperature, without implying that such buildings are inherently inferior.

The Smart Controls and Thermal Comfort project (generally referred to as the SCATs project) involved six academic and three non-academic organisations,[5] and was developed in response to an EU call for research into 'smart controls' to save energy. It built on the idea that, through the use of the adaptive relationship between indoor comfort and outdoor temperature, a smart control system could be developed for use in heated or cooled buildings to set indoor temperatures and reduce their energy use in much the way Auliciems had envisaged. A secondary aim was to create a database of indoor environment and comfort for the diverse climates of Europe, and from which a European Standard could be developed.

The basic plan was that five of the six universities would each conduct monthly surveys, in each of five or so local offices (both naturally ventilated and air-conditioned) over a period of a year. The results would be sent to Oxford Brookes University for analysis. The second UK university, the University of North London (now London Metropolitan University), was given the task of developing, in conjunction with an acoustic instrumentation consultancy, 01dB, a basic set of instruments which would be used by all the academic partners to carry out the surveys. Of the other non-academic partners, Colt Ventilation had been active in developing control systems for naturally ventilated buildings and Tour and Anderson were specialists in air conditioning controls.

The University of North London team developed a set of precision instruments that could automatically measure noise level, illuminance, air temperature, globe temperature, relative humidity and air speed, the instruments being connected to a laptop computer. The sensors were placed on a trolley to ensure that the instruments were at approximately desk height, and the illuminance sensor was placed on the desk of the subject at the time of the survey, along with the noise meter (Figure 16.1).

The main surveys were the monthly transverse surveys, where the subjects were approached by the researcher and asked a number of questions about their perception of the environment. Details were taken of the subjects' clothing, of their activity during the past hour, and notes made of the use of controls.[6] The questions in the survey addressed lighting and acoustics, productivity and overall comfort, as well as more conventional questions relating to the thermal environment, the humidity and the air movement. For noise, lighting, temperature, humidity and air movement there was a sensation scale and a preference scale. The survey questions were asked

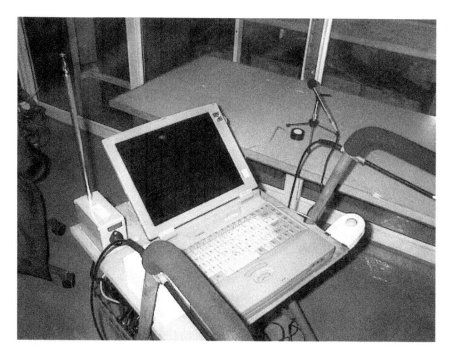

FIGURE 16.1 The SCATs trolley with instruments and laptop computer.

Source: photo by John Stoops.

while the measurements were being taken, except that the noise level was measured before the interview began. The answers were entered into the laptop computer along with the environmental measurements that were being recorded. In addition to the main monthly surveys, a sample of subjects in each building agreed to take part in a longitudinal survey. Each volunteer was given a paper survey form to fill in up to four times a day, asking for their comfort vote, their thermal preference vote and for information from which to assess their clothing insulation, metabolic rate, and their use of controls (windows, heaters, etc.).

At each building, a background survey was also undertaken to record the personal details of the subjects and the basic thermal characteristics of the building. Each potential subject was asked to fill in the survey form, but only about half were returned.

Because the surveys were undertaken in five countries with different languages (English, Greek, French, Portuguese and Swedish), care was needed to ensure that the scales were equivalent in the different languages. So in each country, one well-qualified person translated the questions from the English language version into the country's own language, and another such person independently translated them back into English, a procedure adopted to reduce the likelihood of unsuitable wording of the translated questions.

The project started in December 1997 and ended in December 2000. In all, some 850 subjects took part in the transverse surveys, yielding 4655 datasets between them over the twelve months of the survey. In the longitudinal survey, there were 109 subjects who together produced 27,284 datasets. The SCATs project has provided a unique set of data, collected over the same period in a wide range of European climates, using a standard set of equipment and the same basic questionnaire, in a number of languages.

The analysis of the data followed a number of lines of enquiry. First and foremost it was used to provide a potential algorithm for the adaptive control of heated and cooled buildings. To extract the relation between the climate and comfort indoors, comfort temperatures were estimated from the data, using the Griffiths method.[7] The values of the running mean of the outdoor temperature (T_{rm}) were calculated using the exponentially weighted running mean with a weighting-constant α of 0.8. It was found that, below an outdoor temperature of 10°C, the comfort temperature (T_c) was approximately constant, while above 10°C the comfort temperature increased with outdoor temperature. The suggested adaptive algorithm (see Figure 16.2) using data from the transverse surveys is:

$$\text{For } T_{rm} < 10°\text{C}, T_c = 22.9°\text{C}, \text{ otherwise } T_c = 0.302*T_{rm} + 19.4 \qquad (16.1)$$

Also shown in Figure 16.2 are the results calculated from the longitudinal surveys. The two types of survey gave similar results, but the equation derived from the transverse surveys is to be preferred, resting as it does on the responses of the larger number of people (850 rather than 109).

Equation 16.1 used data from all the monthly transverse surveys, whether from buildings operating in the free-running mode or from those in the conditioned mode (mechanically heated or cooled). Separating the two modes[8] gives two adaptive comfort equations – one for conditioned buildings:

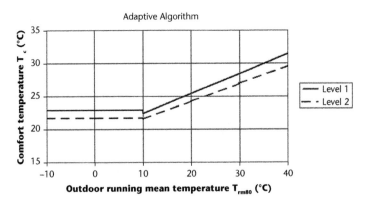

FIGURE 16.2 Values of the recommended indoor temperature as a function of the running mean outdoor temperature T_{rm80} (level 1). The results for the longitudinal survey (level 2) are shown as a dashed line.

Source: from Nicol and McCartney 2001.

$$T_c = 0.09*T_{rm} + 22.9 \tag{16.2}$$

and another for the free-running mode, with outdoor running mean temperature above 10°C:

$$T_c = 0.33*T_{rm} + 18.8 \tag{16.3}$$

These equations are illustrated in Figure 16.3. Equation 16.3 forms the basis of the adaptive standard for buildings operating in the free-running mode in European Standard 15251.[9] The low slope in equation 16.2 for the conditioned buildings can be attributed to the following causes: it includes all the buildings through the winter, when indoor temperature is often constant; and many of the air-conditioned buildings were set up to provide a constant indoor temperature during the summer. This result does not preclude the possibility of a substantial seasonal indoor temperature variation in air-conditioned buildings, but it does suggest that such a regime should be tested to ensure that the variation is not being achieved at the expense of comfort.

Part of task 7 of the SCATs project was to test whether the application of an adaptive control algorithm would compromise comfort, and whether it would save energy. The test took place in a large corporate building in south-west London. The building is divided into five different more-or-less comparable sub-buildings. The air conditioning of each sub-building could be separately controlled and monitored. Two sub-buildings were used for the test (A and O). A total of 64 subjects, 32 in each sub-building, took part in up to six transverse surveys and in all contributed 217 sets of subjective responses, while temperature and humidity were being measured. Eleven subjects took part in a concurrent longitudinal survey and contributed 1031 sets of subjective responses with their accompanying temperature measurements.

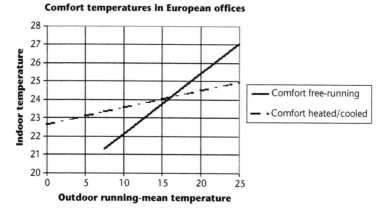

FIGURE 16.3 Comfort lines for the free-running and the conditioned modes.

Source: from CIBSE Guide A (2006).

The surveys took place over the period 9 August to 18 October 2000. On 5 September, sub-building A began to be controlled according to the adaptive algorithm shown in Figure 16.2 and equation 16.1. This allowed comparisons to be made with (a) the sub-building O which was normally controlled and (b) itself before the installation of the adaptive algorithm.

The proportion that was uncomfortable (voting 1, 2, 6 or 7 on the Bedford scale) was little different in the four scenarios (see Plate 16.1). The comfort votes were more symmetrically distributed about the neutral vote in the case of sub-building A after the implementation of the adaptive algorithm, and so the mean vote was nearer neutral. Little change was found in the preference vote, the overall comfort, the perception of indoor air quality, or the self-assessed productivity when the adaptive algorithm was being used. The report of this survey[10] demonstrated that the use of the adaptive algorithm did not result in an increase in discomfort or a loss of perceived productivity. This suggests that the savings to be expected from using a variable set-point regime for air conditioning systems need not be at the expense of the comfort of the building's occupants.

The conclusion of the report on energy use was that energy savings can be made, and were indeed made in the period of this survey in this building. The results from the experiment showed an appreciable energy saving over the period of the surveys, which multiplied up to an estimated 448 MWh annual saving for the whole building. This is equivalent to about 210 tonnes of CO_2 every year. The finding that a significant amount of energy was saved during the experiment by using an adaptive room temperature set point, without inconveniencing the occupants or reducing their productivity, confirms the overall aim of the SCATs project.

There is not space in this short chapter to describe all the uses to which the SCATs data have been put, but we can refer to papers that have reported on various investigations. The basic findings of the different parts of the project are covered in a set of reports available from the authors.[11] The basic description of the project can be found in McCartney and Nicol.[12]

Data from the SCATs project have been widely used to illustrate the ideas of adaptive comfort. In the first chapter of Guide A of the Chartered Institution of Building Services Engineers (CIBSE)[13] the discussion of the adaptive approach is based on the SCATs project. In formulating the thermal recommendations for buildings operating in the free-running mode in the European Standard CEN 15251,[14,15] the drafting committee used the results of the SCATs database to help define the acceptable indoor temperature range. The careful design of the comfort surveys, the use of standard equipment and the inclusion of surveys from the major climate zones throughout Europe were considered crucial for this European Standard. In the first volume of our book, we made use of a part of the SCATs database to introduce the statistical treatment of thermal comfort data. Further use is made of the SCATs database in Part II of this volume.

Other papers drawing on the SCATs database have considered the nature of standards,[16] and the formulation of maximum acceptable temperatures in offices.[17] The CIBSE Technical Memorandum on how to determine whether a new building is likely to overheat rests on the SCATs data.[18] The relationship between the results of comfort surveys and of post-occupancy evaluations[19] and the differences in interpretation of comfort scales in the different countries and languages represented among the SCATs subjects[20] are other subjects which have been explored using the database.

Because the database includes a full set of environmental variables, it has been possible to test the results against the predictions of the PMV/PPD index and its accuracy[21] and to investigate the dependence of subjective indoor air quality[22] and self-reported productivity[23] on thermal comfort and the thermal environment.

Particularly since the beginning of the twenty-first century, there has been an increasing concern to understand the adaptive mechanisms. Investigations to discover how people use adaptive opportunities have sought to answer the objection that adaptive thermal comfort is no more than a statistical 'black box'. Because data on the use of controls is a part of the SCATs database, it has been used in a number of projects to investigate their use.[24-30] The database has also been used in studies of acoustic and visual comfort.[31-33] Finally, the SCATs data are the basis of at least one doctoral thesis[34] and associated papers by a member of the research team. The database has also been used by other researchers,[35,36] is freely available to researchers, and will form a part of the new ASHRAE database of thermal comfort field studies that is now being assembled.

This brief story of the SCATs project illustrates the extent of the ongoing research into adaptive thermal comfort.

In Part I of this volume, we have told the story of the development of the adaptive approach to thermal comfort from its beginnings as an idea in the minds of two young researchers in the 1960s, and through the controversies it generated. Our story concludes with the adaptive approach being accepted as a model of how comfort is achieved, and as a topic of research in lands across the whole world. We hope that our story has helped to explain the principles of the adaptive approach, and that it will promote its further development. The adaptive movement has now become so vast and widespread that we lack the resources to tell its story. That must be a task for others.

The quantity and range of current research in many lands indicates that there is still much to be done, both in fundamental research and in its application. The physiological basis of thermal adaptation is not yet fully explained and is a topic of continuing research. The psychology of adaptation and how it affects the way people control their thermal environments needs to be better understood. This is why research on quantifying the use of controls continues to expand. Its results will need to be assimilated to the thermal simulation of buildings. The sociology of adaptation also needs careful research, and this has hardly begun. It is known that adaptive actions are constrained by social pressures, but perhaps social pressures also encourage adaptive responses? Much in this area needs to be understood and quantified if the full benefit of the adaptive approach is to be realised, with benefits both to people's comfort and to their care for the environment. There is also more to be done towards the formulation of adaptive standards. Current comfort standards have the adaptive approach as an 'add-on', the standards otherwise being based on data from laboratory experiments. There is surely a place for standards that start from an adaptive perspective, and are thus more readily applicable to the adaptive control of buildings and friendly to their climatic design.

Notes

1 Auliciems, A. (1986) Air conditioning in Australia III: thermobile controls, *Architectural Science Review* 33, 43–8.
2 Olesen, B. W. and Parsons, K. C. (2002) Introduction to thermal comfort standards and to the proposed new version of EN ISO 7730, *Energy and Buildings* 34(6), 537–48.
3 Nicol, J. F. and Humphreys, M. A. (2009) New standards for comfort and energy use in buildings, *Building Research and Information* 37(1), 68–73.
4 de Dear, R. J. and Brager, G. S. (2002) Thermal comfort in naturally ventilated buildings: revisions to ASHRAE Standard 55, *Energy and Buildings* 34(6), 549–61.
5 National and Kapodistrian University of Athens, Greece; Fundação Gomes Teixeira da Universidade do Porto (University of Porto), Portugal; Centre National de la Recherche Scientifique (ENTPE), Lyon, France; Oxford Brookes University, Oxford (Co-ordinator); University of North London, London; Chalmers University of Technology, Gothenburg, Sweden; Colt International, UK; Tour and Andersson Control AB, Malmo, Sweden; and 01dB, Lyon, France.
6 Details of the wording of the longitudinal and transverse questionnaires are available in McCartney, K. J. and Nicol, J. F. (2002) Developing an adaptive control algorithm for Europe: results of the SCATs project, *Energy and Buildings* 34(6), 623–35. For the background questionnaire, the questions are in the final report to task 2, available from the authors or from the NCEUB website.
7 See Volume 1, Chapters 3 and 7.
8 These results have been presented in CIBSE (2006) *Guide A environmental design.* Chartered Institution of Building Services Engineers, London.
9 BSI (2007) BS EN 15251: 2007 *Indoor environmental input parameters for design and assessment of energy performance of buildings addressing indoor air quality, thermal environment, lighting and acoustics.* Comité Européen de Normalisation, Brussels.
10 Brissman, J. et al. (2001) *Conducting monitoring and comfort surveys to verify the algorithm in practice: air conditioned buildings.* SCATs final report of task 7.
11 Nicol, F. and McCartney, K. (2001) *Final report (public) smart controls and thermal comfort (SCATs)* (also subsidiary reports to tasks 2 and 3 and contributions to reports to tasks 1, 6 and 7). Report to the European Commission of the Smart Controls and Thermal Comfort project (Contract JOE3-CT97-0066), Oxford Brookes University, Oxford (available from the authors or www.nceub.org.uk).
12 McCartney, K. J. and Nicol, J. F. (2002) Op. cit.
13 CIBSE (2006) Op. cit.
14 BSI (2007) BS EN 15251: 2007 Op. cit.
15 Nicol, J. F. and Humphreys, M. A. (2010) Derivation of the equations for comfort in free-running buildings in CEN Standard EN15251, *Buildings and Environment* 45(1), 11–7.
16 Nicol, J. F. and Humphreys, M. A. (2002) Adaptive thermal comfort and sustainable thermal standards for buildings, *Energy and Buildings* 34(6), 563–72.

17 Nicol, J. F. and Humphreys, M. A. (2007) Maximum temperatures in European office buildings to avoid heat discomfort, *Solar Energy* 81(3), 295–304.

18 CIBSE (2013) CIBSE Technical Memorandum TM52: *The limits of thermal comfort: predicting overheating in European Buildings.* Chartered Institution of Building Services Engineers, London.

19 Nicol, F. and Roaf, S. (2005) Post occupancy evaluation and field studies of thermal comfort, *Building Research and Information* 33(4), 338–46.

20 Humphreys, M. A. (2008) 'Why did the piggy bark?' Some effects of language and context on the interpretation of words used in scales of warmth and thermal preference, *Proceedings of International Conference on Air-Conditioning and the Low Carbon Cooling Challenge*, Windsor, July. Organised by the Network for Comfort and Energy Use in Buildings (NCEUB).

21 Arens, E. *et al.* (2010) Are 'class A' temperature requirements realistic or desirable?, *Building and Environment* 45(1), 4–10.

22 Humphreys, M. A. *et al.* (2002) An analysis of some subjective assessments of indoor air-quality in five European countries, *Indoor Air 2002*, Monterey.

23 Humphreys, M. A. and Nicol, J. F. (2007) Self-assessed productivity and the office environment: monthly surveys in five European countries, *ASHRAE Transactions* 113(1), 606–16.

24 Nicol, J. F. (2001) Characterising occupant behaviour in buildings: towards a stochastic model of occupant use of windows, lights, blinds heaters and fans, *Proceedings of the Seventh International IBPSA Conference*, Rio, Vol. 2, International Building Performance Simulation Association, pp. 1073–8.

25 Nicol, J. F. and Humphreys, M. A. (2004) A stochastic approach to thermal comfort, occupant behaviour and energy use in buildings, *ASHRAE Transactions* 110(2), 554–6.

26 Rijal, H. *et al.* (2007) Using results from field surveys to predict the effect of open windows on thermal comfort and energy use in buildings, *Energy and Buildings* 39(7), 823–36.

27 Rijal, H. B. *et al.* (2009) How do the occupants control the temperature in mixed-mode buildings? *Building Research and Information* 37(4), 381–96.

28 Tuohy, P. *et al.* (2009) Occupant behaviour in naturally ventilated and hybrid buildings, *ASHRAE Transactions* 115(1), 16–27.

29 Rijal, H. B. *et al.* (2012) Considering the impact of situation-specific motivations and constraints in the design of naturally ventilated and hybrid buildings, *Architectural Science Review* 55(1), 35–48.

30 Fisekis, K. *et al.* (2002) Prediction of discomfort glare from windows, *Lighting Research and Technology* 35(4), 360–71.

31 Wilson, M. and Nicol, J. F. (2001) Noise in offices and urban canyons, *Proceedings of the Institute of Acoustics* 23(5), 41–4.

32 Wilson, M. P. and Nicol, J. F. (2003) Some thoughts on acoustic comfort: a look at adaptive standards for noise, Paper to Institute of Acoustics conference 'Soundbite', Oxford 5/6 November, *Proceedings of the Institute of Acoustics* 25(7), 116–24.

33 Nicol, F. *et al.* (2006) Using field measurements of desktop illuminance in European offices to investigate its dependence on outdoor conditions and its effect on occupant satisfaction, productivity and the use of lights and blinds, *Energy and Buildings* 38(7), 802–13.

34 Stoops, J. L. (2000) *The thermal environment and occupant perception in European office buildings.* Licentiate thesis, Chalmers University, Sweden.

35 Haldi, F. and Robinson, D. (2010) On the unification of thermal perception and adaptive actions, *Building and Environment* 45(11), 2440–57.

36 Alexandre, J. *et al.* (2011) Impact of European Standard EN15251 in the certification of services buildings – a Portuguese case study, *Energy Policy* 39(10), 6390–9.

PART II

Analysis

17

INTRODUCING PART II

The reader will notice a radical change of style at this point in the book. In Part I, we used a narrative style to explore the foundations of the adaptive approach to thermal comfort, trusting that as the story unfolded the foundations of the model would become evident and its logic well understood. At this point in the book, we abandon the narrative mode and instead treat various topics of importance in field studies of thermal comfort and review their use in quantifying the adaptive thermal responses. We have not attempted a comprehensive survey of all the work on adaptive thermal comfort that has been undertaken in recent years, but rather have selected topics that we have worked on and that we consider to be important aspects of the adaptive approach. Part II of the book is perhaps much more demanding on the reader, and we recognise that many will find these chapters quite difficult. We have done our best to make them clear, and hope that the effort we have put into their writing has made them accessible.

Chapters 18 and 19 treat the properties of semantic scales, and discuss their construction and validation. We believe that the subjective scales we use in comfort research have not received sufficient critical attention in recent decades. The researcher tends to take them for granted rather than subject them to scrutiny. The problem has become acute with the international nature of the research effort, with research now being undertaken in so many different cultures and in so many different languages. It can happen that a scale that works well in one language and culture behaves strangely when translated into a different language and used in a different cultural setting. So we discuss the problems of translation and draw attention to the methods by which the behaviour of a semantic scale can be evaluated. Our treatment is necessarily incomplete, because it confines itself to the use of semantic scales in thermal comfort research, and the reader is referred to the literature of experimental psychology for a broader view of the topic.

It has also become evident in the last few years that scales of thermal preference, hitherto thought to give unambiguous information on the optimum conditions for thermal comfort, can behave strangely. An example is where a respondent – and especially a child – who sees the scale for the first time somehow misunderstands it and reverses its polarity. So we find people saying that they are 'much too warm' and at the same time would 'prefer warmer'. This seems to be because an assessment using such a scale entails two mental steps rather than one. Since many, and perhaps most, surveys of thermal comfort approach each respondent just once, the researcher and interviewer need to be aware of the potential problem. Another example is confusion between a desire for a different thermal sensation and a desire for a different lifestyle, a matter we explain in the chapters on scaling.

The adaptive approach to comfort, resting as it does on human thermal behaviour, does not depend on the use of any particular model of the heat exchange between the person and the thermal environment, and indeed much of the work can proceed with no reference at all to heat exchange. Nevertheless, we thought it wise to include a chapter that offers a much-simplified account of the heat exchange between a person and the environment so that the adaptive importance of clothing, air movement and activity become clearly evident. This we do in Chapter 20. The treatment also draws attention to what we believe to be an inconsistency in the PMV model. PMV is perhaps the most commonly used method of predicting thermal sensation, and is incorporated in international standards, so we believe that it is time its construction was re-examined.

We should perhaps explain our reasons for including in this volume four chapters on statistical methods. These are Chapters 21 to 24. There are many excellent books on this subject, so why add these small chapters to the literature? There are indeed many good books, but, in our experience, highly intelligent people without a mathematical education find them impenetrable. So we have taken a few topics that arise in adaptive thermal comfort research and have presented their statistical analysis in a largely visual manner. We have minimised the use of statistical notation, because it often presents a barrier to those with a background in architecture. Those with a mathematical background will find our exposition rather inelegant, but they will be able, very simply, to translate what we have written into the usual statistical notation. There are also some topics that are needed in analysing adaptive comfort data but are absent from the standard texts. So, in Chapter 22, we explain in some detail the effects of the presence of error in the predictor variable on regression statistics and how to correct for it, and, in Chapter 23, we consider geometrically the relation between the two regression lines that describe a scatter plot. Linear regression analysis is commonly used in adaptive comfort research, as also is the successive application of Probit or logistic regression to the several categories of the subjective warmth scale. What is not often realised is that there are formal equivalences between linear regression and the successive Probit regressions, and that a grasp of the equivalences can help in the extraction of information from published surveys of comfort where the original data are not available. This is the topic of Chapter 24.

The adaptive approach to thermal comfort has been greatly advanced by the existence of databases of thermal comfort surveys. The first of these was the database compiled for ASHRAE by Richard de Dear and colleagues in 1997. More recent is the database from the study of thermal comfort in European offices, which includes monthly surveys from five countries for an entire year. We have also compiled over the years a database of summary statistics extracted from some 700 surveys. We use these databases in Chapters 25–31 to quantify some of the main features of the adaptive approach. One of the surprises of recent research is that people are more sensitive to temperature changes during the working day than had been thought. They generally adapt very little during a day at work. Rather, their adaptation takes place from day to day and over longer periods of weeks and months, as we show in Chapter 25. This finding makes invalid any analysis of a survey that lasted several days that overlooks the distinction between within-day and day-on-day changes in room temperature. Quantifying the within-day sensitivity allows us to quantify the effects of within-day temperature variations on thermal comfort (Chapter 26).

We use the databases to explore in Chapter 27 the rapidity of thermal adaptation and the practical seasonal limits that are placed upon it. This leads to Chapter 28 on the derivation of neutral temperatures, and to Chapter 29 in which we explore and explain the relation between comfortable indoor temperatures and the climate, the relation that is commonly called the Adaptive Model. We show that current international standards considerably underestimate the dependence of the comfort-temperature on the climate, while overestimating the variation of temperature that is likely to be found comfortable within any single day at the office.

The databases also enable the exploration of the magnitude of any differences between the neutral temperature derived from a scale of warmth, and the preferred temperature derived from a scale of thermal preference. The large differences that have sometimes been reported seem to be attributable to idiosyncrasies in the wording of particular scales rather than to large systematic effects. We show that the differences are related both to the sensation of warmth and to the climate, but are cautious in quantifying what is a quite small effect.

We did not know where best to place the chapter on clothing behaviour. On the one hand clothing behaviour underlies so much of the adaptive approach that we wished to place it early in the volume. On the other hand its exploration makes much use of the databases that are introduced only in Chapters 25–30. So clothing behaviour has been placed in Chapter 31. In it we give prominence to the non-thermal functions of clothing, as these are often overlooked by researchers and can profoundly influence clothing behaviour. We then consider the trends visible in the databases and explore the logic that underlies the adaptive thermal change of clothing. This leads to the derivation of the exponentially weighted running mean temperature as a logical model to describe such changes – another reason why we would have liked the chapter to occur earlier in the book, for the running mean is also used to quantify the variation of the outdoor temperature in Chapter 29.

We had intended to write a chapter on the use of other controls such as fans and windows, and the mathematical models that were applied to their use, but we covered this topic in Volume 1, and a more detailed discussion would take us far beyond the exposition of the adaptive approach to thermal comfort and into the realms of the control of the thermal environment in buildings. This is a topic of current adaptive research, and we follow its progress with interest.

A study of adaptive comfort would be incomplete without some reference to the interactions that occur between the different aspects of the environment. They are numerous and some of them can be important. Much recent survey work gathers information not only on the thermal environment, but also on acoustics, lighting and air quality. The brevity of our treatment in Chapter 32 is not because the topic is unimportant, but rather because its full inclusion would require a whole further volume. Adaptive thermal comfort is but one aspect of environmental satisfaction, and the adaptive principle applies to these other aspects too.

In as far as it is possible to sum up these disparate strands, we do so in Chapter 33, where we also consider possible ways forward for research, and how the adaptive approach could best be formulated in standards.

Finally, a word about equations and symbols. We have numbered equations only on those rare occasions when we refer back to a previous equation. All symbols are locally defined instead of being collected in a comprehensive list of definitions. We believe this to be more convenient for the reader, and it avoids the need for total consistency throughout the volume, a consistency that would sometimes have required more cumbersome symbols than we have used.

18

USING THE METHOD OF SUCCESSIVE CATEGORIES TO EXPLORE THE PROPERTIES OF THERMAL COMFORT SCALES

Introduction

The first volume introduced the commonly used scales of warmth and comfort, explained the difference between nominal, ordinal, and interval data, and explained the implications of this distinction for the analysis and use of subjective scales. In this chapter, we trace the development of scales of warmth and comfort, starting with Thomas Bedford's method,[1] and then explain with various sets of data the application of the Method of Successive Categories to explore their properties.

The most direct way to discover how warm or cool or comfortable people are feeling is to ask them. In some circumstances it is possible to deduce whether people are comfortable by observing their actions, and this appears to have the advantage of objectivity. In the late 1960s, one of the authors (MAH) devised a way of inferring thermal comfort from observations of the clothing worn by school children, but he needed to check the result by using a supplementary questionnaire. This is because behavioural responses can be biased just as can subjective responses.[2] In that instance, some children continued wearing the winter clothing they preferred rather than wear a disliked summer alternative, despite the higher temperature. Their behaviour was to some extent constrained by considerations other than their comfort. They were adapting not just to their thermal conditions, but also to their social desires, sacrificing some thermal comfort for some gain in 'social comfort'.

An objective measure is superior only when quantifying objective data. Thermal comfort is essentially subjective, and therefore subjective responses have priority over objective measurements such as skin temperatures or sweat rates. Subjective scales must, of course, be constructed and presented in a way that is likely to yield unbiased answers, for it is possible for the way a question is put to bias the answer that is given.

The origin of scales of warmth and comfort: Bedford's scaling of subjective responses

Bedford, the pioneer of thermal comfort field studies, was investigating comfort in wintertime in light industry. He was aware of the danger of obtaining biased responses from the workers. It was the world economic depression of the early 1930s, and he thought his respondents might be unwilling to express dissatisfaction with their working conditions. So he put his opening question in a way that implied no criticism of the working environment. Rather than asking if the workplace was too hot or too cold, he asked them about themselves:

> Each worker was asked if she felt comfortably warm. If she did not, she was then asked to say whether she felt too warm or too cool, and then whether she felt just definitely too warm (or cool) or much too warm (or cool). If she replied that she felt comfortable, she was asked if she were really quite comfortable, or whether she would rather have the room slightly warmer or slightly cooler. In an enquiry dealing with the comfort of industrial workers, there is always the possibility that, from a desire not to appear disgruntled, operatives may hesitate to say that they are not perfectly comfortable. It is felt that by the method of questioning adopted in the present study errors from this cause were minimised.[3]

From the responses during these structured interviews, Bedford obtained seven categories. He gave them the following labels (the abbreviations are ours):

Much too warm	(mtw)	7
Too warm	(tw)	6
Comfortably warm	(cw)	5
Comfortable	(c)	4
Comfortably cool	(cc)	3
Too cool	(tc)	2
Much too cool	(mtc)	1

The categories form an ordinal scale. That is to say, the order of the categories is clear, but we cannot say that the gaps between successive categories represent equal psychological intervals. He numbered them from 1 to 7, with 'much too warm' being numbered 1. We reverse his numbering, since it seems more natural to associate warmer sensations with higher numbers: 'much too cool' becomes category 1 and 'much too warm' becomes category 7. The scale is a combined estimate of warmth and comfort, and has been criticised because the relation between the two is not necessarily constant.[4] The Bedford scale is often used in lands where British influence has been strong.

Other scales of warmth and comfort

As more researchers conducted field studies of thermal comfort in various countries, more and more subjective scales came into use. Some had as few as three categories,[5] and one study used a scale with twenty-five categories.[6] Scales with four, five, seven, nine and thirteen categories have also been used. Most scales are arranged symmetrically about a 'neutral' or 'comfortable' category, but asymmetrical scales have occasionally been used.[7,8] There may be an increased risk of bias when asymmetrical scales are used. Some comments on these various scales, together with references to the original papers, may be found in Humphreys' first meta-analysis of thermal comfort field studies.[9]

The use of seven categories of warmth had become normal practice by the mid 1970s. Two forms, with minor variations, are in common use today: the Bedford scale, which we have already discussed, and the ASHRAE scale. This scale is often used in lands where the influence of the USA has been strong. The categories of the early ASHRAE scale were originally:

Hot	7	
Warm	6	
Slightly warm	5	
Comfortable	4	(now: Neutral)
Slightly cool	3	
Cool	2	
Cold	1	

The centre category was later changed to 'neutral' to make the scale refer to thermal sensation alone. So the scale in its present form contains no reference to comfort, but it is usual to assume that the three central categories indicate thermal comfort. The assumption was used for construction of the PMV/PPD thermal index[10] and is therefore present in ISO 7730, which rests on the PMV/PPD equations.[11] The assumption is open to the criticism that it lacks experimental basis.

There is also a nine-category form of the ASHRAE scale, used mainly in physiological experiments where people may be well outside the limits of comfort. It adds the categories 'very cold' and 'very hot' to extend the range of the scale. It is sometimes used in thermal comfort fieldwork where extreme conditions are likely to be encountered.[12,13]

Thermal preference scales

Bedford's scale, as administered by interview, could be seen as a seven-category scale of thermal preference. Each category is not just a warmth sensation, but also an evaluation of it (for example, *too* warm, *comfortably* warm). The central three categories would today be called a 'thermal preference scale': 'Are you really comfortable, or would you rather have the room slightly warmer or slightly cooler?' So Bedford's central category (comfortable) indicated the preferred condition of his respondents. R. H. Fox used a preference scale during a study of body temperature among the elderly. It was a five-category scale: 'Would you prefer to feel: (1) very much warmer, (2) slightly warmer, (3) no change, (4) slightly cooler, (5) very much cooler?'[14]

During the 1970s, Griffiths and McIntyre used a simple three-category scale.[15] Seven-category preference scales have also been used, and there is no consensus on the best number of categories to have. Preference scales are usually used to supplement the ASHRAE scale or some other scale of warmth, but they may also be used independently. The use of the scales over many years has shown that people often prefer a sensation that differs from thermal neutrality, and that there is a tendency for people in warm climates to prefer states that are cooler than neutral, and for people in cold climates to prefer a sensation warmer than neutral. The topic is complex and is the subject of Chapter 30.

The three-category McIntyre scale may be presented in the form:

I would prefer to be:
1   Warmer
2   No change
3   Cooler

A five-category version known as the Nicol scale may be presented in the form:

I would prefer to be:
1   Much warmer
2   A bit warmer
3   No change
4   A bit cooler
5   Much cooler

Numbering systems vary. For survey work and for some kinds of analysis, it is convenient to avoid negative numbers, and it is always wise to number the scale so that the warm end (prefer to be cooler or much cooler) has the higher numbers. This avoids confusion. The attraction of ascribing 'zero' to the 'prefer no change' category is also obvious, and many researchers do so. In this volume, we adopt either convention as seems convenient.

Some problems with preference scales

Absence of votes in an end category

Problems have been encountered when using the three-category version. Sometimes in a hot climate it happens that nobody chooses the 'prefer warmer' option, making it impossible for the researcher to locate a preferred state.[16] The opposite problem has been found in a survey in a cold climate in winter – no-one wanted to be cooler.[17,18] The use of a five- or seven-category version can avoid these problems, because it divides the psychological range more finely.

Mistaken polarity

The scales are counter-intuitive. 'Warmer' implies that it is cooler than desired, and the casual respondent on occasion misinterprets the scale, using the scale in the reverse way to that intended by the researcher. Thus a person may sometimes give the response 'much too warm' to the Bedford scale, followed by the response 'warmer' on a preference scale – hardly a plausible response.[19,20] A mistake as extreme as this is possible to notice and correct, but less extreme instances will pass unnoticed and are difficult to correct; for example, it is logically possible, if unusual, to give the response 'warm' on the ASHRAE scale and 'prefer warmer' on the preference scale. (I'm feeling quite warm but I would like to be even warmer.)

Comfortable rooms or comfortable people?

This question applies both to thermal sensation scales and to thermal preference scales. Thermal comfort and thermal sensation are, strictly speaking, properties of the respondent and not of a building or of a room within it. So the questionnaire should make clear that the enquiry is about the respondent's own thermal condition and not about the room. The response can be misleading if the respondent thinks the question refers to the room. For example, a person who is perfectly comfortable in a cool room while wearing several layers of clothes may say they would 'prefer warmer' – meaning that they would prefer the room to be warmer so that they could wear fewer layers of clothing and still be comfortable. The reverse may also occur. People may be comfortable in very light clothing in a warm room, yet say that they would 'prefer cooler' and so be able to wear a business suit in comfort. Such responses express a social or cultural preference rather than a preferred thermal sensation. Social and cultural thermal preferences are worthy of investigation, but they are not to be confused with preferred thermal states of the body.

Exploring the properties of a warmth scale: the method of successive categories

The method of successive categories enables the comparison of the 'psychological widths' of the categories in a scale. It is a well-established research method and was devised by the pioneers of psychometric scaling. In Chapter 10 of *Psychometric Methods*, Joy Guilford traced its history back to the work of Louis Thurstone and Rensis Likert.[21] Likert later abandoned the method because it was simpler and in his view sufficiently accurate to treat the ordinal values as if they were real numbers.[22] This indeed is often true, but it cannot be relied upon. We will show that the method of successive categories is useful for exploring the properties of thermal comfort scales. This is because it can be used to explore the psychological properties of a subjective scale without reference to the particular thermal environments encountered during the survey.

The method is perhaps unfamiliar to all but a few thermal comfort researchers. We first used it in 2007 to explore changes in a scale's properties when it is administered in translation,[23] but we have not explained the method fully, and we have only recently seen just how useful it can be in thermal comfort research. The method enables a direct comparison of the psychological widths of the scale categories within any survey of sufficient size. Further, if more than one scale has been used in the same survey it can yield a direct quantitative comparison of these scales, both of the psychological widths of their categories and their relative positions. This allows, for example, the direct comparison of thermal neutrality on the ASHRAE scale with 'prefer no change' on a thermal preference scale. The method also allows real numbers to be assigned to the categories, which changes the scale from an ordinal to an interval scale. It shows the change a scale undergoes when it is translated from one language to another. We also suggest methods for adjusting the numbering, so that the translations of scales from different surveys become comparable. This step however requires further assumptions whose validity must be considered before they are accepted.

The first step is to assume that there exists some kind of psychological continuum of thermal sensation. That is to say, our sensations of warmth and coolness have an inner representation of some kind that is continuous. We may draw a line representing this psychological continuum and mark on it the place that corresponds to our current thermal sensation. This is what is assumed when the ASHRAE scale is used in continuous form, that is to say, when it is presented as a line with the category labels attached. The respondent can mark his thermal sensation anywhere along the line, and is not restricted to the labelled points. This is the form of the scale often used by de Dear, and we have occasionally used it too.[24] The scales are often presented this way in climate laboratory work too.

The assumption of a psychological continuum of this kind can break down. If some parts of the body feel unpleasantly cold while other parts feel unpleasantly warm, the integration of the sensations onto a common continuum becomes impossible. For example, one of us (MAH) recalls feeling very uncomfortable on a long car journey in the South African winter of 1956:

> I was one of three people on the back seat of the car, and was sitting squashed up against a door. The seal round this door was defective, leaving a gap. This produced a continuous cold draught on my left thigh and over time I became severely uncomfortable, although other parts of my body were comfortably warm. I could not have expressed an overall sensation of warmth on a single continuum.

Even if no integrated thermal sensation can be expressed because of differences of sensation among the parts of the body, the person is able to rate, say, the warmth of the feet on a continuum.

If we have a considerable quantity of data, from one person or from many people, the distribution of thermal sensation on the psychological continuum tends towards the Normal form. The reason for this can be stated fairly simply, though the argument needs careful reflection for it to be understood.

The differences in thermal sensation among individuals (and the differences within the same person on different occasions) result from numerous influences, both internal to the body and external. They include the levels of activity and of clothing insulation, the posture, body build, recent thermal history, and the thermal environment in which people are placed: air temperature, air movement, thermal radiation, humidity and any environmental asymmetries. These various influences result in diverse inputs to the nervous system from the numerous thermal sensors for cold and heat in the various body parts. These inputs produce, when integrated by the mental processes, an overall thermal evaluation.

The presence of multiple influences is the classic circumstance that produces a Normal distribution, and is the basis for example of the classical theory of measurement errors. Put more formally, when many variables, *almost whatever their individual distributions*, are combined to form another variable, the distribution of this other variable tends to the Normal form. This is a well-established deduction from Central Limit Theorem: under a very wide range of conditions, when several random variables are added, the distribution of their sum tends to Normal. We would therefore expect the distribution of responses on the psychological continuum of thermal sensation to be Normal under the conditions pertaining to Bedford's data, and that this would be true of the data from all field surveys of thermal comfort where there had been no deliberate selection of the thermal environment, no manipulation of it by the researcher, and where the data were sufficiently numerous.

Comparing the psychological widths of the categories of Bedford's scale using his original winter data

We illustrate the method by applying it first to Bedford's original data. He may have been aware of the method, for there is a footnote on p. 19 of his 1936 report that suggests he considered using some such procedure, but decided against it:

> The use of an arbitrary scale cannot be avoided, but it may be thought that a more reasonable scale could be constructed by assuming a normal distribution of the personal feeling of warmth. This point has been examined, but it is found that the use of a scale based on this assumption does not significantly affect any of the conclusions in this Report. It has, therefore, been thought desirable to use the simple scale set out above.

Bedford's data are shown in Table 18.1. We draw a cumulative Normal distribution of unit standard deviation on the psychological continuum (see Figure 18.1). The proportion of responses in category 1 (mtc) is entered on the vertical axis (see the lowest horizontal dotted line on the figure), and this will correspond, via the curve, to a certain value on the horizontal axis (see the vertical line furthest left on the figure). The cumulative proportion in categories 1 and 2 (mtc, tc) taken together are next entered on the vertical axis, and will correspond to the next value on the horizontal axis, and so on for the successive categories. We have now identified the upper boundaries of the successive scale categories on the psychological continuum.

TABLE 18.1 The numbers of votes in each of Bedford's seven categories.

Label	Number	Cumulative number	Cumulative proportion	Probit
mtc	24	24	.009	−2.35
tc	157	181	.070	−1.47
cc	206	387	.151	−1.03
c	1803	2190	.852	1.04
cw	246	2436	.948	1.62
tw	123	2559	.995	2.60
mtw	12	2571	1.000	−

Source: from Table XL on p. 97 of Bedford's report.

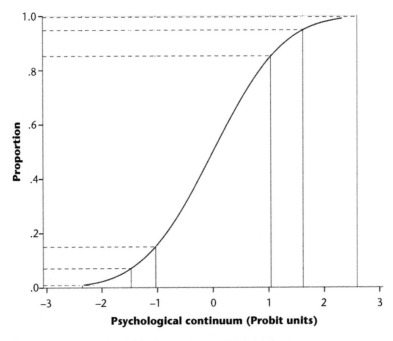

FIGURE 18.1 Illustration of the procedure to transform cumulative proportions into Probits, using Bedford's data.

The positions of these boundaries on the continuum are true numbers, apart from a margin of error arising from the limited quantity of data in the survey. The process has transformed the scale from ordinal status to interval status. The effect of the cumulative Normal curve in the figure is to transform the cumulative proportions into Probits, and so equal intervals on the horizontal (Probit) axis represent equal intervals on the psychological continuum. (This is not to assert that the transformed scale is necessarily linear with respect to any particular environmental variable or index. For example, thermal sensation bears a curvilinear relation to the air speed.)

A practical note for researchers: Most of the calculations in this volume have been done using the statistical package SPSS (versions 15 through 19). The values for the category upper boundaries can conveniently be calculated directly from line–by–line records of thermal sensations by using the SPSS Ordinal Regression procedure, without a covariate, and setting the link function to Probit, which means that the Normal distribution is used. (The default link function is the Logistic distribution. It differs little from the Normal, and its use will in practice lead to almost identical conclusions. Figure 18.2 shows the almost linear relation between the Probit and the Logit between probability 0.05 and 0.95.) Alternatively, if cumulative proportions have already been calculated, they may be transformed into Probits, either using tables, or using the Probit transformation facility in the statistical package. Tables of Probits usually add 5 to the Probit to avoid the nuisance of handling negative numbers. If the Probit function does not appear to be available in the statistical package, the inverse normal distribution function (IDF (normal) in SPSS) is used, setting the mean to zero and the standard deviation to unity. It is the same. Finney (1971) discusses alternative distributions that can be used.

Figure 18.3 displays at a glance the relative widths of the categories on the psychological continuum. The end categories have no definite width because the Probits of the probabilities $p = 0$ and $p = 1$ are indefinitely large. We see that the category 'comfortable' is very wide in Bedford's data, while categories 'comfortably cool' and 'comfortably warm' are narrow. It follows that the psychological differences between the steps of the scale are in this example unequal.

It does not follow that the relative widths will be the same whenever the scale is used. The very narrow categories for 'comfortably cool' and for 'comfortably warm' could have arisen from Bedford's method of questioning his respondents. To be classified as 'comfortably cool' or 'comfortably warm' the respondent needed to change her first reply that she was comfortable. She had already replied 'yes' to the question 'are you comfortably warm?' Then Bedford asked if really she would 'prefer the room slightly warmer, or slightly cooler', so inviting her to change her first reply. She may have been reluctant to do so.

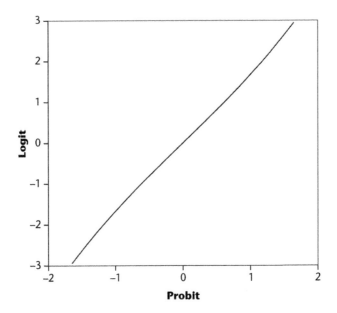

FIGURE 18.2 A comparison of the cumulative Normal and Logistic distributions.

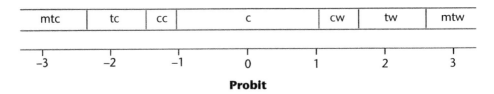

FIGURE 18.3 The scale categories along the psychological continuum (Probit scale). Categories
mtc and mtw have no defined outer limits.

Source: Bedford's data.

Hickish's summertime survey

Hickish, who had worked with Bedford, some twenty years later conducted a summertime survey using the same procedures.[25] He says in his report that he used the
same method of questioning his respondents, but we wonder if his initial question
was exactly the same. Bedford in his wintertime survey had assumed without discussion that his respondents wished to feel comfortably warm: 'Each worker was asked
if she felt comfortably warm.' To ask someone in summertime 'Are you feeling
comfortably warm?' does not seem right. So it may be doubted that Hickish used
precisely the same words in his interviews. His data are shown in Table 18.2.

He too obtained narrow categories for 'comfortably cool' and 'comfortably warm'
compared with 'comfortable', but the difference is not as great as for Bedford's data.
The category widths are shown in Figure 18.4. (We also notice that zero on the

TABLE 18.2 The numbers of votes in each of Hickish's seven categories.

Label	Number	Cumulative number	Cumulative proportion	Probit
mtc	1	1	.001	−3.22
tc	16	17	.011	−2.29
cc	69	86	.056	−1.59
c	714	800	.521	0.05
cw	421	1221	.794	0.82
tw	181	1402	.912	1.35
mtw	135	1537	1.000	–

Source: from Table 6 of Hickish's 1955 paper.

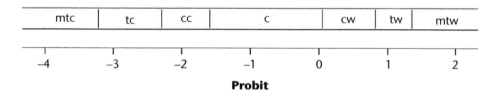

FIGURE 18.4 Category widths derived from Hickish's data for factory workers in summertime.

Probit scale corresponds not to the centre of the 'comfortable' category, but is close to the division between 'comfortable' and 'comfortably warm', so the median postal worker (Probit = 0) felt a little bit warm).

It might have been wiser of Bedford and Hickish to put the follow-up question differently, perhaps: 'How would you describe your comfort? Are you comfortable and warm, or comfortable and cool, or between the two?' One of the authors (MAH) used Bedford-style interviews rather than a paper questionnaire when collecting data in the North West Frontier Province in Pakistan. A good command of English was not widespread in the region, but the term 'OK' was widely used and understood both in the regional language and in Urdu. If the respondent indicated that he was thermally 'OK', the follow-up was: 'Are you "OK warm", "OK cool", or "OK-OK"?' Unfortunately the data were too few to enable a reliable estimate of the relative category widths in either season.[26]

The uneven nature of the categories of the Bedford scale, as administered by Bedford and later by Hickish, has implications for the analysis of the data. The use of the rank-values of the scale as the dependent variable in regression analysis, as Bedford did, and as has been normal practice ever since, is to some extent incorrect, because regression analysis requires that the dependent variable be an interval variable – a true number. The unequal steps on the scale will to some extent affect the correlation coefficients and regression equations derived from the data, although the effect is small unless the irregularities are extreme. More obvious would be the effect on the outcome of a Probit analysis, where it could distort the estimates of the proportions found to be in the comfort zone.

Comparing the behaviour of different scales within the same body of survey data

It is common nowadays for more than one scale to be used in a single survey. For example, a survey questionnaire may include either the ASHRAE scale or the Bedford scale, and either the McIntyre or the Nicol preference scale. The method of successive categories allows us to compare directly the psychological category widths of scales that are used in the same body of data, since they refer to the same thermal environments, come from the same respondents, and relate to the same psychological continuum. The psychological widths of the categories of the scales may therefore be directly compared.

An interesting example comes from a web-based survey of winter comfort in the living-room of more than a thousand Japanese homes, a project in which two of us (MAH, JFN) collaborated with the project leader, Dr Hom Rijal, a former colleague of ours.[27] It has become well known that the ASHRAE scale, when translated word-for-word into Japanese, is not entirely satisfactory because some of the words in the translation are not naturally used in Japanese to refer to thermal sensation. We understand that the problem is with the translations of 'warm' and 'cool'. So SHASE (Society of Heating, Air-conditioning and Sanitary Engineers of Japan) developed a form of the scale with labels more suitable for use in Japan. When translated word-for-word back into English the SHASE scale becomes (the abbreviations are ours):

very cold	(vc)
cold	(c)
slightly cold	(sc)
neutral	(n)
slightly hot	(sh)
hot	(h)
very hot	(vh)

It is interesting to compare the categories of the word-for-word Japanese translation of the ASHRAE scale with those of the SHASE scale.

Rijal asked his respondents to rate their thermal sensation using the SHASE scale, and shortly afterwards asked them to rate their thermal sensation again, this time using the ASHRAE scale (the word-for-word translation into Japanese). There were 1030 respondents, ample for a comparison to be made. Next he invited them to give their thermal preference on a five-point scale, a translation of the Nicol scale. Applying the method of successive categories, and assuming that all three scales refer to the same psychological continuum of warmth sensation, we can compare their properties. Table 18.3 shows the cumulative percentages of votes on each of the scales, together with the Probit transforms of these values. The Probit transforms are compared in Figure 18.5.

The ASHRAE and the SHASE scales behave differently. 'Cool' on the ASHRAE scale represented a very narrow width on the psychological continuum, for the respondents in this survey. 'Neutral' on the SHASE scale covers almost the same range as 'neutral' and 'slightly warm' taken together on the ASHRAE scale. Also, if we use the usual convention that the three central categories of the scale define the comfort zone, we find that the comfort zones are different, largely because 'slightly hot' on the SHASE scale is about equivalent to 'warm' on the ASHRAE scale. 'Warm' on the ASHRAE scale is conventionally considered to be outside the comfort zone, while 'slightly hot' on the SHASE scale is within it. So the two scales yield seriously different conclusions about

TABLE 18.3 Comparison of the categories of the SHASE scale, ASHRAE scale (in Japanese translation), and the Nicol preference scale (N = 1030) in Rijal's data.

SHASE scale			ASHRAE scale			Nicol preference scale		
Label	Cumulative proportion	Probit	Label	Cumulative proportion	Probit	Label	Cumulative proportion	Probit
vc	.004	−2.66	cold	.030	−1.88	pmw	.043	−1.72
c	.023	−1.99	cool	.045	−1.7	pw	.296	−0.54
sc	.232	−0.73	sl.cl	.190	−0.88	pnc	.981	2.09
n	.920	1.40	n	.677	0.46	pc	.998	3.10
sh	.997	2.59	sl. wm	.941	1.56	pmc	1.000	−
h	1.000	−	warm	.997	2.76			
vh	1.000	−	hot	1.000	−			

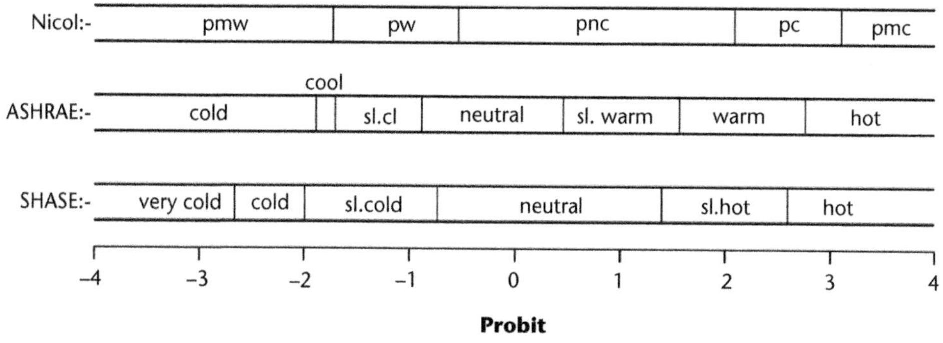

FIGURE 18.5 Comparing the SHASE, ASHRAE and Nicol scales in Japanese.

Source: Rijal's data.

Note: Key: Lower scale: SHASE: very cold, cold, slightly cold, neutral, slightly hot (no 'very hot' votes). Middle scale: ASHRAE: cold, cool, slightly cool, neutral, slightly warm, warm, hot. Upper scale: Nicol: prefer much warmer, prefer warmer, prefer no change, prefer cooler, prefer much cooler.

comfort. In our opinion it is preferable to use the SHASE scale for these data, because it was developed for Japanese language and culture. The analysis shows beyond doubt the importance of the particular words used for the scale categories.

Turning our attention to the Nicol preference scale, we see that the centre of the 'prefer no change' category falls within the 'neutral' category on the SHASE scale, slightly towards the warmer end, while it falls within the 'slightly warm' category on the ASHRAE scale.

The median respondent in the survey (the zero-point on the Probit scale) was slightly on the cool side of the 'neutral' category of the SHASE scale, which is not unusual in a winter survey. Had we used the ASHRAE scale we would have concluded that the median respondent was slightly on the warm side of the 'neutral' category – a different result.

The method of successive categories thus provides a way to obtain the 'semantic offset' – the difference between the preferred condition and the neutral condition – without first calculating 'neutral' and 'preferred' temperatures – often sources of considerable uncertainty. The semantic offset is discussed in Chapter 30.

Another example of the different behaviour of two seemingly similar scales is provided by the data of Feriadi and Wong in their study of thermal comfort in Indonesian dwellings.[28] Unusually, they used both the Bedford scale and the ASHRAE scale, and also the McIntyre scale of thermal preference. We presume that the scales were translated from English into the local language for the survey, although translation is not mentioned in their paper. They collected 525 sets of observations, each from a different person, from more than 300 households. We have available only the summary data from the publication. Table 18.4 shows the data, with the calculated Probits. The scale categories are shown in Figure 18.6.

The categories of the three scales may be compared directly, because they all apply to the same body of data. Notice the very different scale categories of the ASHRAE and Bedford scales. The usual assumption that the two scales are in practice

TABLE 18.4 Cumulative proportions in the categories together with their Probits (525 observations).

Category	ASHRAE scale		Bedford scale		McIntyre scale	
	Cumulative proportion	Probit	Cumulative proportion	Probit	Cumulative proportion	Probit
−3	.053	−1.61	.002	−2.89	–	–
−2	.211	−0.80	.038	−1.77	–	–
−1	.360	−0.36	.137	−1.09	.042	−1.73
0	.516	0.04	.524	0.06	.259	−0.65
1	.701	0.53	.707	0.54	1.000	–
2	.751	0.68	.970	1.87	–	–
3	1.000	–	1.000	–	–	–

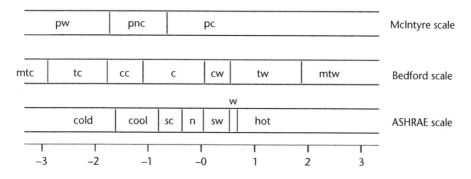

FIGURE 18.6 Comparison of the ASHRAE scale, the Bedford scale and the McIntyre scale from Feriadi and Wong's data, showing the strange behaviour of the ASHRAE scale.

equivalent has broken down in these data, showing that the assumption can be incorrect. Feriadi and Wong did their regression analysis using the ASHRAE scale, but applying the method of successive categories casts much doubt on its behaviour – the widths of the categories are very irregular, and it seems very compressed compared with the Bedford scale. A rather different set of regression statistics would have been obtained had they based their analysis on the Bedford scale.

The centre of the 'prefer no change' category of the McIntyre scale corresponds to the margin between 'comfortable' and 'comfortably cool' on the Bedford scale, an offset between the scales of about half a scale category. The offset between the McIntyre scale and the ASHRAE scale is much greater, being some two scale-points, and the centre of 'prefer no change' corresponds to the centre of the 'cool' category on the ASHRAE scale – an unusually large offset.

The stability of the ASHRAE scale when translated

The difficulty of translating the ASHRAE scale into Japanese, and the strangely different behaviour of the Bedford and ASHRAE scales in the Indonesian data, alert us to the possibility that when the scale is translated into other languages the relative width of the categories on the psychological continuum may change.

In the winter of 2005–06, Pitts explored the result of translating the ASHRAE scale from English into Chinese, Greek and Arabic.[29] He asked 140 people, mainly his university students, to translate the labels on the scale (cold, cool, slightly cool, neutral, slightly warm, warm, hot) into their own language and then mark on a line representing the continuum of thermal sensation the positions corresponding to the scale-points they had translated. They also marked the position that represented maximum comfort. He found that there were some significant differences from the English version in the locations of the scale points on the psychological continuum and noticed, as had other researchers before him, that thermal neutrality and maximum comfort did not necessarily coincide. He concluded:

Researchers might infer from these results that not only should translation be carried out with great care when using English originated phrases in other languages but also that there may be systematic variations to be expected between different language groups and populations, and perhaps between different language writing systems. The analysis of such factors requires further research.

A second and extremely important finding is that conditions of thermal neutrality and comfort are significantly different for those respondents in the English language grouping; indeed the sensation of slightly warm is more indicative of comfort according to the [winter] data collected. [Our insertion.]

Such differences are to be expected, because concepts easily expressed in one language can be difficult to express in another. Seemingly equivalent words commonly have different semantic ranges in different languages, and their meaning can be influenced by their cultural context. Wierzbicka and Harrison are among those who have explored these properties of language.[30,31] The expression of concepts in different languages, and the theory of translation, are topics on which linguists hold a range of opinions. Bellos gives a scholarly but accessible account of the philosophical debates about translation, and how translators approach their task.[32] We confine ourselves to that which is generally agreed among linguists. There is a further comment on translation and language at the end of this chapter.

Granted that differences of meaning can and do occur on translation, the practical question for us is how to quantify the behaviour of a translated scale when it is used in thermal comfort research. The problem applies equally to laboratory and field investigation, but we take our examples only from field research. It is evident from the Japanese scales analysed above that different translations of a scale into the same language can have different properties. It follows that when we quantify the properties of a translated scale, the results apply only to that particular translation. A different translation would probably have different properties.

We use the database of thermal comfort surveys from the SCATs project (Smart Controls and Thermal Comfort) to explore some of these differences. As explained in Chapter 16, the SCATs project was a year-round study of the workplace environment in European offices. Studies were conducted in France, Greece, Portugal, Sweden and the UK. Researchers visited the respondents at their workstation once a month for a year or more. On each occasion, an environmental questionnaire was used that included among other items the ASHRAE scale and the Nicol scale. There existed no standard translations, so the scales were translated into the national languages especially for the surveys. The translations are shown in Tables 18.5 and 18.6. The thermal environment was measured (air temperature, operative temperature, relative humidity and air speed), as were other aspects of the environment (horizontal illuminance, sound pressure level, concentration of carbon dioxide). Further information about the project can be found in the papers by McCartney and Nicol, and Nicol and McCartney.[33,34]

TABLE 18.5 The translations of the question introducing the ASHRAE scale.

Language	Question introducing the ASHRAE scale
English	How do you feel at this time?
French	Quelle est votre sensation thermique en ce moment?
Greek	Πως αισθάνεστε αυτη τη στιγμή?
Portuguese	Como se sente neste momento?
Swedish	Hur känner du dig just nu?

TABLE 18.6 The translations of the ASHRAE scale-categories.

English	French	Greek	Portuguese	Swedish
Cold	Froid	Ελαφρύ κρύο	Frio	Kall
Cool	Frais	Πολύ δροσερά	Fresco	Sval
Slightly cool	Légèrement frais	Αναπαυτικά λίγο δροσερά	Ligeiramente fresco	Något sval
Neutral	Neutre	Ανετα ιδανικά	Bem	Neutral
Slightly warm	Légèrement chaud	Αναπαυτικά λίγο ζεστά	Ligeiramente quente	Något varm
Warm	Chaud	Πολύ ζεστά	Quente	Varm
Hot	Très chaud	Πάρα πολύ ζεστά	Muito quente	Het

The experimental protocol was the same in all five countries. There are data from over 4500 workstation visits, each visit providing a set of environmental measurements and a corresponding set of questionnaire responses. From these records, we quantify the category-widths of the ASHRAE scale in the English, French, Greek, Portuguese and Swedish versions.

We use the same method as before to compare the psychological widths of the scale categories. The data are given in Table 18.7, and Figure 18.7 shows the resulting category widths.

TABLE 18.7 Cumulative proportions of the usage of the categories of the ASHRAE scale in the five languages, together with the total numbers of responses.

Language	English		French		Greek		Portuguese		Swedish	
Category label	Cum prop	Probit	Cum prop	Probit	Cum prop	Probit	Cum prop	Probit	Cum prop	Probit
1 cold	.022	-2.02	.012	-2.27	.028	-1.92	.015	-2.18	.008	-2.40
2 cool	.086	-1.37	.033	-1.84	.053	-1.62	.042	-1.73	.027	-1.93
3 sl. cool	.211	-0.80	.198	-0.85	.293	-0.55	.129	-1.14	.145	-1.06
4 neutral	.492	-0.20	.539	0.10	.619	0.30	.756	0.69	.829	0.95
5 sl. warm	.821	0.92	.843	1.01	.948	1.62	.923	1.42	.971	1.88
6 warm	.959	1.75	.984	2.16	1.000	–	.992	2.37	.998	2.87
7 hot	1.000	–	1.000	–	1.000	–	1.000	–	1.000	–
N:	1285		516		325		1559		970	

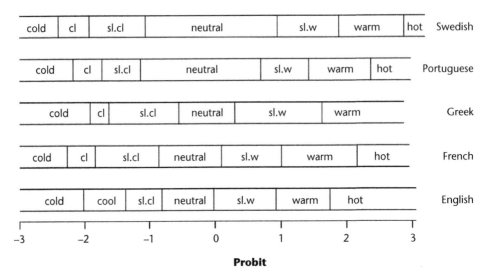

FIGURE 18.7 Comparing the category widths on the psychological continuum when the ASHRAE scale is administered translated into different languages.

The scale administered in English had reasonably equal category widths. In French, the categories were reasonably equal except for the rather narrow 'cool' category. The upper bound of the 'warm' category cannot be located in the Greek surveys because no-one voted 'hot'. The 'cool' category is very narrow, and the 'slightly warm' is quite wide, but the Greek data were not numerous, and so the result is not as precise as for the other languages. In both Portuguese and Swedish, the 'neutral' category is very wide compared with the other categories.

The findings show that it is unwise to assume that a translated scale means the same as the original scale. So it is advisable to check the relative widths of the categories of a scale when it has been used in translation, rather than assume that the translated scale has the same properties as it had in the source language.

We cannot say from these data whether the different behaviour of the scales is caused only by the translation used, or whether there was also some cultural difference that caused the scales to exhibit uneven category widths. European office culture has become more uniform in recent decades, so perhaps translation is the more important reason for the differences. That the effect is not wholly due to cultural differences is clear from the Japanese data. The two different translations of the ASHRAE scale gave, in the same cultural setting, different widths for the scale categories. A partition between the effect of language and culture, although of theoretical interest, is not important for thermal comfort survey work; what we need to know is how the scale behaves in that particular translation and in that particular social and cultural setting. So although it would be interesting to partition the differences between culture and translation, it is not necessary in practice.

Applying the method to the behaviour of the Nicol scale of thermal preference shows diversity of the category widths between languages for this scale too. Both the Nicol and the ASHRAE scales were used in all five languages, so we may directly compare the properties of the two scales in each of the countries, in their particular translations.

Accuracy of the locations of the divisions between the scale points

The output from an ordinal regression program normally provides the standard error of the estimates of the locations of the boundaries between the successive categories on the Probit scale. The accuracy of the estimate of the location is higher towards the centre of the data, and accuracy can be low towards the margins. We again use the SCATs data to illustrate this. We choose the data from Portugal (Table 18.6). It contains 1559 observations. The category boundaries and their standard deviations are shown in Table 18.8, in Probit units. The data are from the output from an SPSS Ordinal Regression program. The boundary between 'cold' and 'cool' is not very precisely defined (-2.18 ± 0.082), nor is that between 'warm' and 'hot' (2.37 ± 0.098). But the boundary between 'slightly cool' and 'neutral' is quite precisely located (-1.14 ± 0.040); so is that between 'neutral' and 'slightly warm' (0.69 ± 0.035).

The program assumes that the data are independent. But it will be recalled that in the SCATs project each respondent was, as far as was practical, visited monthly, and so their reported thermal sensations may not be independent. It has been found experimentally that the variability of a person's comfort vote from occasion to occasion is little less than the variability of the votes among a group of people. A discussion of this finding can be seen on p.135ff in McIntyre's book.[35] The finding is true of these data too – an analysis of variance showed that variation within-subject from visit to visit has a standard deviation of 0.9 scale units, while that between-subjects within visit is 1.1 units. So it is reasonable to treat the responses as if they were independent despite the monthly repeat visits.

TABLE 18.8 The accuracy of the estimates of the category boundaries for the Portuguese data.

ASHRAE category	Portuguese translation	Cumulative proportion	Category upper margin (Probit units)	Standard error (Probit units)
Cold	Frio	.015	−2.18	0.082
Cool	Fresco	.042	−1.73	0.057
Slightly cool	Ligeiramente fresco	.129	−1.14	0.040
Neutral	Bem	.756	0.69	0.035
Slightly warm	Ligeiramente quente	.923	1.42	0.047
Warm	Quente	.992	2.37	0.098
Hot	Muito quente	1.000	–	–

TABLE 18.9 The estimates of the category upper boundaries together with their standard errors, for Rijal's Japanese winter data.

SHASE scale category (English translation)	Category upper margin (Probit units)	Standard error (Probit units)
Very cold	−2.66	0.168
Slightly cold	−1.99	0.085
Cold	−0.73	0.043
Neutral	1.40	0.057
Slightly hot	2.59	0.154
Hot	–	–
Very hot	–	–

A second example is taken from Rijal's Japanese winter data (Table 18.9). Here each observation is truly independent, since each respondent contributed data on just one occasion. We consider the accuracy of the category margins of the SHASE scale. The upper margins of the 'cold' and 'neutral' categories are well defined, while those of 'very cold' and 'slightly hot' are only approximate. The reason for the relatively low accuracy towards the extremes of the distribution can be easily understood from Figure 18.1. A small change in the proportions in the more extreme categories causes a large change in the value of the Probit. The accuracy of the placing of the margin also depends on the number of observations contributing to the distribution. The Portuguese data had about 50 per cent more observations than did the Japanese data, and its accuracy is correspondingly greater, as is evident from the central regions of the scales. The margins of the more outlying categories can never be accurately located with the quantity of data that is usual in thermal comfort surveys.

The standard error of a category margin can also be estimated directly from the properties of the binomial distribution. This is useful if the line-by-line datasets are not available, and we have only the numbers of responses in each of the categories, as is common when obtaining the data from published research reports.

Example: The cumulative proportion (p) of the upper margin of the 'neutral' category in the Portuguese data (Table 18.8) is 0.756, and the total number of responses (n) in the dataset is 1559. The standard error of a binomial proportion is

$$\sqrt{(p(1-p)/n)}$$

Entering the values of (p) and (n) we obtain

$$\sqrt{(0.756{\times}0.244/1559)} = 0.011.$$

Adding one standard error we have $0.756+0.011 = 0.767$.

The Probit of 0.756 is 0.694, while that of 0.767 is 0.729.

The difference between them is 0.035, as in Table 18.8.

The stability of a scale within a survey

We may desire to test whether a scale has a stable meaning. For example, we may wish to know whether, in a survey of long duration, the scale has the same properties at the beginning of the survey as it has at the end. Or we may wish to test whether a scale has the same meaning for different batches of data within a shorter survey. We give two illustrations:

Example: *The Portuguese data from the SCATs project.* The SCATs project was, as will be recalled, a year-round survey. We may wonder whether the ASHRAE scale (in Portuguese) had the same characteristics in the first six months of the survey as in the second six months. We split the data into two batches – the first half and the second half of the survey period. We calculate the cumulative proportions and their Probits and the standard deviations of the Probits for each half of the data as previously. Table 18.10 compares the Probits of the category boundaries from the two sections.

The agreement between the two halves is illustrated in Figure 18.8. The gradient of the line on the figure is the geometric mean of the gradients of the two regression lines ('y on x' and 'x on y') because both axes have equal status. The use of such lines is explored in Chapter 23. We do not expect the gradient of the line to be exactly unity because the environmental conditions were not exactly the same in the two batches. But if the scale categories have the same meaning in either batch there will be a high correlation between the two sets of category boundaries. The correlation coefficient is high ($r = 0.999$), indicating an extremely high degree of consistency between the two batches.

Example: *The SHASE scale for Japanese dwellings with different heating systems.* Rijal's data were collected from people in dwellings with either warm floor heating or warm air heating and were obtained over the same few winter days. We may wish to compare the behaviour of the SHASE scale in the two batches. We can use the same method as for the Portuguese data, but split the data according to the type of heating. Table 18.11 gives the Probit values, together with their standard deviations, and Figure 18.9 compares the placing of the category boundaries in the two batches of data. Recall that no one voted 'very hot' in Rijal's data, so the upper boundary of the 'hot' category is missing.

Again, the gradient of the line need not be exactly unity because the thermal conditions in the two batches were not exactly the same. The only measurement made was of the air temperature, one metre above the floor and in the centre of the room. Its mean was the same for the two types of heating ($21.0°C$), but because of the different radiant fields and thermal gradients characteristic of the two kinds of system, the equality of the air temperatures does not imply that the environments were thermally equivalent.

The point representing the upper margins of the 'neutral' category on Figure 18.9, if taken by itself, has a statistically significant departure from the line ($t = 2.22$, $p < 0.05$), but, when all five points are considered together, the departures fail to reach significance. So overall there is no significant inconsistency in the behaviour of the categories between the two batches. The correlation between the two sets of category margins is very high (r = 0.997) and indicates a high degree of consistency.

Both examples show a high degree of consistency in the relative category widths of the tested scale within the same survey. If this finding proves to be generally true, it suggests that a scale with particular wording is likely to have stable characteristics for the duration of a long survey and across various types of heating system.

TABLE 18.10 Checking the stability of the ASHRAE scale (Portuguese translation) by means of a split-half test.

	First batch		Second batch	
Scale-point	Probit	s.e. of Probit	Probit	s.e. of Probit
1	−2.04	.103	−2.37	.139
2	−1.67	.077	−1.80	.084
3	−1.09	.056	−1.18	.058
4	.76	.050	.63	.048
5	1.45	.067	1.39	.065
6	2.42	.148	2.32	.132

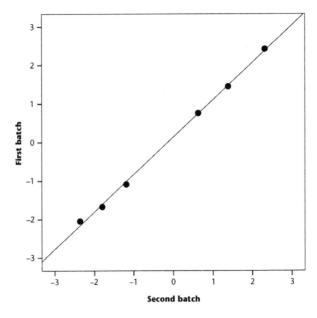

FIGURE 18.8 Comparing the psychological category widths of the Portuguese version of the ASHRAE scale in the first and second halves of the year-round survey. 781 and 778 votes respectively. The numbers (Probit units) are drawn from Table 18.10.

TABLE 18.11 Checking the stability of the SHASE scale with warm floor heating and with warm air heating.

SHASE category	Warm air heating		Floor heating	
	Probit	s.e. of Probit	Probit	s.e. of Probit
Very cold	−2.52	.202	−2.89	.314
Cold	−1.86	.109	−2.16	.140
Slightly cold	−0.64	.060	−0.83	.063
Neutral	1.28	.075	1.54	.087
Slightly hot	2.66	.238	2.52	.202
Hot				
Very hot	No votes of 'very hot'			

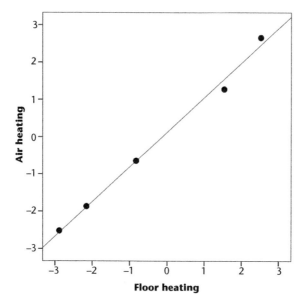

FIGURE 18.9 Comparing the psychological category widths of the SHASE scale for the two types of heating (515 votes in each group). The numbers (Probit units) are drawn from Table 18.11.

Attributing numerical values to the categories

For certain kinds of analysis, it is more convenient to have true numbers attached to the centres of the categories than to the margins between them. To do this, the mid-points of the categories are located on the Probit scale (not on the cumulative proportion scale) and their numerical values obtained. Each value is the mean of the Probits of the upper and lower margins of the category. For an example, we use the SHASE data from Table 18.9. The upper margin of the 'neutral' category has a Probit of +1.40 while the lower margin has a value of −0.73. The centre of the category therefore has a Probit value of 0.34. This procedure may be applied to all except the extreme categories. Thus we have transformed the SHASE ordinal scale into a scale that has true numbers attached to the categories – numbers that may be legitimately added or subtracted. It is now an interval scale. Attributing true numbers to the centres of the categories permits the valid use of statistical procedures such as regression analysis that assume interval scales rather than ordinal scales.

We may go a step further. Some environmental scales have a meaningful zero point. For example, it is reasonable to attach a value of zero to the centre of the 'neutral' category, because it is generally taken to represent zero thermal sensation, or at least a balance of warm and cool sensations from the various parts of the body. So for the SHASE data we would subtract 0.34 from the numbers attributed to the categories. The subtraction changes the interval scale into a ratio scale in terms of thermal sensation. (That is to say, for example, that 2 on the scale feels twice as warm as 1.)

There is a complication. If we compare the SHASE and ASHRAE scales (Figure 18.5) we see the centres of the 'neutral' categories are different. The anomaly (0.7 units on the Probit scale) is much too big to be attributed to chance, and it is impossible that both can indicate the body condition of zero thermal sensation. A judgement must be made about which (if either) scale should be used. Our judgement is to trust the SHASE scale rather than the ASHRAE scale, because it was developed with due regard for Japanese concepts and the usage of Japanese words.

The fact that 'neutral' is differently placed on the psychological continuum alerts us to problems that may arise from very asymmetrical category widths on either side of the 'neutral' category. In the word-for-word translation of the ASHRAE scale into Japanese, there is a strong asymmetry. The very narrow width of 'cool' compared with the width of 'warm' is likely to 'nudge' the neutral category away from its true position. Scales having a natural zero should be symmetrical in the number of categories either side of the 'neutral' category, in their category labels, and also as far as practical the categories should be symmetrical in their widths on the psychological continuum.

But what are we to do about the end categories, as they have semi-infinite widths on the psychological continuum, and therefore cannot have a centre? Two methods of attributing notional numbers to them are suggested.

First method

Consider the category 'very cold' on the SHASE scale (vote 1 in Table 18.3). It represents a proportion of 0.004 of the votes cast. Take the centre of the category as half this number, 0.002, (the category median) and use the Probit of this as the value for the centre of the 'very cold' category. This would attribute the numerical value −2.9 to the category. (No great accuracy can be expected because of the few votes in the 'very cold' category – just four out of 1030.) The objection to this procedure is that the median does not truly represent the centre of the category because the relationship between the percentage and its Probit is non-linear, and so the result is biased.

> *Note*: Sometimes this method has wrongly been used to obtain the centre points of all the scale categories. It suffers from a bias that increases the further the category is from the overall survey median, as the reader may verify by considering Figure 18.1.

Second method

Again consider the 'very cold' category of the SHASE scale. Attribute to the end category the average width (on the Probit scale) of all those categories for which widths can be obtained. In this example, we can obtain the widths of four categories. The mean width is 1.30 Probit units, from Table 18.11: ($\{2.523 + 2.662\}/4$). Take the Probit value representing the centre of this notional category as the number to allocate to 'very cold'. This results in an attributed number of −3.2 on the Probit scale. The objection to this procedure is that it is just an informed guess. In effect, it assumes that the end category would be the width that is the average of all the other categories, had an extra category been added at that end of the scale.

The two methods give different answers. The first method is perhaps to be preferred. It has the merit of simplicity, and takes more regard of the numbers of votes cast in the category.

The number of votes in the extreme categories in all ordinary surveys is very small, and will not much affect the results of a statistical analysis that uses the entire dataset, so it does not matter much which method is used. This would not be true of surveys in extreme conditions, where a large proportion of the population were hot or cold. In these circumstances, the scales ordinarily used are unsuitable, and a more appropriate scale would need to be found or developed. The matter is discussed in Chapter 19.

Adjusting the range of a scale when it has been re-numbered

Further adjustments may be desirable after the scale has been re-numbered, to make it comparable in range with the original nominal scale. It is not obvious what we mean by this, so we explain by giving an example. Consider again the Portuguese data from the SCATs project (see Table 18.12). From the numerical values of the category upper margins (column 4 in the table) we calculate the numerical values of the centres of the categories −2 through to +2 (column 5 in the table), as explained in the previous section. The values for the end categories were estimated using method 1 above. The new numbers for the category centres differ considerably from those conventionally attributed to the ASHRAE scale (−3 through +3).

We wish to make the ranges of the scales before and after re-numbering to be in some useful sense comparable. How might this be achieved? A procedure is shown

TABLE 18.12 Scaling the re-numbered categories in the Portuguese data.

ASHRAE category	Conventional value	Portuguese translation	Category upper margin (Probit units)	Category centre (Probit scale)	Putting 'Bem' = zero	Adjusting for the s.d. of the votes
Cold	−3	Frio	−2.18	−2.43	−2.20	−2.11
Cool	−2	Fresco	−1.73	−1.96	−1.73	−1.66
Slightly cool	−1	Ligeiramente fresco	−1.14	−1.44	−1.21	1.16
Neutral	0	Bem	0.69	−0.23	0	0
Slightly warm	1	Ligeiramente quente	1.42	1.06	1.29	1.24
Warm	2	Quente	2.37	1.90	2.13	2.04
Hot	3	Muito quente	–	2.65	2.88	2.76

in columns 6 and 7 in Table 18.12. We attribute the number zero to the centre of the neutral (*Bem*) category, on the ground that this best represents the state of zero thermal sensation. This is done by adding 0.23 to all the values. The new values are shown in column 6 of the table. If the two scales are to be comparable in their extent, the spread of votes should be similar on each. This spread of the votes can be represented by the standard deviation of the responses. This should be the same before and after the re-numbering. With the original numbering (−3 through +3) the standard deviation of the responses is 0.910. Using the new numbers instead of the original ones, the standard deviation becomes 0.949. So we need to multiply the values in column 6 by (0.910/0.949), or 0.959. This gives the values in column 7. The re-numbered scale has the same standard deviation as the original scale, and has 'neutral' as zero. In effect, the total process has 'jiggled' the values attributed to the categories, while keeping their overall spread of responses unchanged. Table 18.13 compares the category numbers before and after re-numbering and re-scaling.

Making re-numbered scales comparable across different sets of data: a suggestion

The method of successive categories enables us to compare the category widths only within a particular survey. It cannot tell us whether a particular category had the same width in, say, Portuguese and Swedish, because that would depend upon the

TABLE 18.13 The result of re-numbering the scale and adjusting for equal standard deviations of the comfort vote.

Original number	−3	−2	−1	0	1	2	3
New number	−2.11	−1.66	−1.16	0	1.24	2.04	2.76

thermal environments represented in the data. Such a comparison would be very useful, and we consider a possible procedure.

It could be argued that the true inter-respondent variation of thermal sensation among human populations is likely to be much the same in all groups of respondents. If this is so, the different scatter of thermal sensations among the particular datasets is attributable to the effects of the translations of the scales and to the different ranges of environmental conditions encountered. If we accept this argument, we would proceed to estimate the residual standard deviation of the thermal sensation in each dataset. (This may be done by multiple regression analysis of the sensation of warmth on all relevant predictor variables, using dummy variables for the various distinct groups, or by an equivalent procedure.) We would then apply an appropriate multiplying factor to each set of category numbers so as to obtain the same residual in each.

We might choose as the residual standard deviation, not the value from the particular survey, but the average value found from analysis of field study databases. This value is about 0.92 scale units for the SCATs data and the ASHRAE database of field experiments combined, and it can be considered to be typical of field data generally, as we explain in Chapter 24.[36] The value applies to people whose activity is sedentary or nearly so, such as everyday office work.[37]

The attribution of equal residual variation of thermal sensation would standardise the translated scales across differing surveys. So if the assumption of equal residual variation of thermal sensation is considered to be safe in the circumstances of the study, it becomes possible to compare the numerical values of the categories of the ASHRAE scale in different translations and across the various surveys. The residual is among the more stable statistics in thermal comfort surveys, but may prove not to be consistent enough for this application; applying a correction that itself contains uncertainty can be self-defeating.

Much remains to be done in the area of standardisation across languages. Our suggestion of using the residual standard deviation is tentative. A more direct way would be to find and test bilingual respondents in randomised trials using the same thermal environments.

Chapter summary

In this chapter, we have shown the usefulness of the method of successive categories for exploring scales of thermal sensation and thermal preference, and have applied it to data from surveys and from various sources.

The virtue of the method is its ability to provide immediate insight into the properties of a scale, and to alert the researcher if a scale is behaving strangely:

- It sheds light on the changes that can occur when a scale is translated into another language.
- It enables a direct comparison of different scales in the same body of data.

- It enables an estimate to be made of the 'semantic offset' between neutrality and preference without first calculating neutral and preferred temperatures.
- The stability of a scale within a body of survey data can be checked, and the accuracy of the boundaries of the categories calculated.

The method of successive categories is a valuable exploratory tool for thermal comfort research. In the next chapter, we discuss the formation of new scales, where the researcher suspects that the special circumstances of the project make it inadvisable to use any of the standard scales.

A note on the art of translation and the cultural context of words

It has long been recognised that translation is a difficult art and that a translated text may make an impression on the reader that differs from that of the original. Around 132 BCE, Jesus ben Sirach's grandson translated some of his grandfather's writings into the common Greek of the Roman Empire, so that they could have a wider audience. In the prologue to his translation he writes (when translated into twentieth-century English):

> You are invited . . . to be indulgent in cases where, despite our diligent labour in translating, we may seem to have rendered some phrases imperfectly. For what was originally expressed in Hebrew does not have exactly the same sense when translated into another language. Not only this book, but even the Law itself, the Prophecies, and the rest of the books differ not a little when read in the original. (Source: New Revised Standard Version.)

These different effects would of course have been perceived only by those bilingual in Hebrew and Greek. Two difficulties faced the translator: it was hard to find the right word or phrase in the new language, and even the best translation might not convey 'exactly the same sense' as the original. So at best something is lost in translation and at worst quite the wrong impression is given.

Here is a seventeenth-century English translation of the same passage:

> Wherefore let me intreat you . . . to pardon us, wherein we may seem to come short of som [sic] words we have laboured to interpret. For the same things uttered in Hebrew, and translated into another tongue, have not the same force in them: and not only these things, but the law itself, and the prophets, and the rest of the books, have no small difference, when they are spoken in their own language. (Source: King James Version.)

We notice that even these two English translations convey different impressions. For example, 'have not the same force' seems a stronger phrase than 'not . . . exactly the same sense', and the modern translation seems to have in mind a private reader, while the older translation seems to be thinking of someone reading aloud to an audience. The source of the difficulty runs deep. Anna Wierzbicka, an Australian linguist whose native language is Polish, gives examples of concepts that are taken for granted in English but do not exist in other languages.[38] And, because English is so widely spoken, English speakers tend to assume that their concepts are the human norm rather than being specific to 'Anglo' culture. Wierzbicka cites the English concepts of 'reasonableness' and 'fairness' as concepts absent from other languages. Indeed, linguistic research has shown that surprisingly few concepts are universal and therefore easily translatable into all major languages.

The point is reinforced by K. David Harrison in his study of languages that are dying out.[39] These languages often have concepts that are unique, and so, when a language dies, the concept is also lost. Ways of looking at things (e.g. in mathematics, biology, geography, philosophy, agriculture and linguistics) depend on culture and are reflected in the language of that culture. So the difficulty in translation arises not only from the differing semantic range covered by the apparently equivalent word in the other language, but also from concepts being construed differently in different languages.

Since language and culture are closely intertwined, we should expect that the particular language used would influence people's feelings. For example, people who are bilingual sometimes notice that their personality seems to shift when they switch languages. The effect has been confirmed experimentally. Ramirez-Esparza found that people fluent in both Spanish and English had slightly higher average scores for extraversion, conscientiousness and agreeableness when completing the widely used Big Five Inventory in English, but a significantly higher score for neuroticism when completing it in Spanish.[40]

Daniel Gross has argued persuasively that emotional responses to stimuli are deeply affected by their social and cultural context, rather than being physiologically determined, as current brain scan researchers sometimes seem to presume.[41] A young grandson of one of the authors (MAH) was scared of dogs and went with his mum to visit a friend who owned a miniature French Bulldog. Matthew showed no fear of the dog. Eventually the dog barked. Matthew looked puzzled: 'Mummy, why did the piggy bark?'

The visual, tactile and auditory stimuli alone did not determine the child's emotional and physical responses – his classification of the French Bulldog as a 'piggy' rather than a 'doggy' was decisive.

The meaning associated with sensory information changes as a culture changes over the centuries. Chris Woolgar has provided overwhelming evidence that information from the senses was interpreted differently, in the culture of late medieval England, from the way it is interpreted today, having different mental connotation and emotional force.[42] One might therefore expect differences in the interpretation of sensory information across different societies today.

So there is reason to believe that the verbal scales commonly used in environmental research may change their meaning when translated into a different language. The question has become more pressing now that studies of environmental comfort are conducted worldwide in diverse languages and cultures, both in the laboratory and in the field.

Notes

1 Bedford, T. (1936) *The warmth factor in comfort at work,* Medical Research Council Report no. 76, HMSO, London.
2 Humphreys, M. A. (1973) Classroom temperature, clothing and thermal comfort – a study of secondary school children in summertime, *J. Inst. Heat. & Vent. Eng.* 41, 191–202.
3 Quoted from p. 18 of Bedford's report.
4 Davies, A. D. M. (1972) *Subjective ratings of the classroom environment: a sixty-two week study of St. George's School Wallasey.* University of Liverpool, Liverpool.
5 Lane, W. R. (1965) *Education, children and thermal comfort.* University of Iowa, Iowa.
6 Peccolo, C. (1962) *The effect of thermal comfort on learning.* Digest of PhD Thesis, University of Iowa.
7 Ambler, H. R. (1955) Notes on the climate of Nigeria with reference to personnel, *Journal of Tropical Medicine and Hygiene,* 58, 99–112.
8 Goromosov, M. S. (1965) *The microclimate in dwellings.* State Publishing House for Medical Literature, Moscow.
9 Humphreys, M. A. (1975–76) *Field studies of thermal comfort compared and applied.* Department of the Environment, Building Research Establishment, CP 76/75. (A paper prepared for the symposium: *Physiological Requirements on the Microclimate,* Prague, 8–10 September 1975, and published in the proceedings. Also published in: *J. Inst. Heat. & Vent. Eng.* 44, 5–27, 1976.)
10 Fanger, P. O. (1970) *Thermal comfort.* Danish Technical Press, Copenhagen.
11 ISO (2005) ISO 7730: *Ergonomics of the thermal environment – analytical determination and interpretation of thermal comfort using calculation of the PMV and PPD indices and local thermal comfort criteria.* ISO, Geneva.
12 Rijal, H. B. and Yoshida, H. (2006) Winter thermal comfort of residents in the Himalaya region of Nepal, *Proceedings of International Conference on Comfort and Energy Use in Buildings – Getting Them Right,* Windsor. Organised by the Network for Comfort and Energy Use in Buildings.
13 Rijal, H. B., Yoshida, H. and Umemiya, N. (2002) Investigation of the thermal comfort in Nepal, *Proceedings of International Symposium on Building Research and the Sustainability of the Built Environment in the Tropics,* Indonesia, pp. 243–62.
14 Fox, R. H. *et al.* (1973) Body temperatures in the elderly: a national study of physiological, social and environmental conditions, *British Medical Journal* 1, 200–6.
15 McIntyre, D. A. (1980) *Indoor climate.* Applied Science Publishers, London.
16 Heidari, S. (2000) *Thermal comfort in Iranian courtyard housing.* PhD Thesis, University of Sheffield, UK.
17 Yan, H. (2011) Personal communication: Yan, H., Yang, L. and Zhu, X. *Indoor thermal conditions and thermal comfort in residential buildings during the winter in Lhasa, China* (unpublished research report).
18 Yang, L. *et al.* (2013) Residential thermal environment in cold climates at high altitudes and building energy use implications, *Energy and Buildings* 62, 139–45.
19 Humphreys, M. A. and Hancock, M. (2007) Do people like to feel 'neutral'? Exploring the variation of the desired sensation on the ASHRAE scale, *Energy and Buildings* 39(7), 867–74.
20 Teli, D. *et al.* (2012) Field study on thermal comfort in a UK primary school, *Proceedings of 7th Windsor Conference: The changing context of comfort in an unpredictable world,* Cumberland Lodge, Windsor, April. Network for Comfort and Energy Use in Buildings, http://nceub.org.uk.
21 Guilford, J. P. (1954) *Psychometric methods,* Second Edition. McGraw-Hill, New York.
22 Likert, R. (1932) A technique for the measurement of attitudes, *Archives of Psychology* 140, 1–55.

23 Humphreys, M. A. (2008) 'Why did the piggy bark?' Some effects of language and context on the interpretation of words used in scales of warmth and thermal preference, *Proceedings of International Conference on Air-Conditioning and the Low Carbon Cooling Challenge*, Windsor, July. Organised by the Network for Comfort and Energy Use in Buildings.

24 Humphreys, M. A. and Hancock, M. (2007) Op. cit.

25 Hickish, D. E. (1955) Thermal sensations of workers in light industry in summer. A field study in southern England, *Journal of Hygiene* 53(1), 112–23.

26 Humphreys, M. A. (1994) An adaptive approach to the thermal comfort of office workers in North West Pakistan, *Renewable Energy* 5(5–8), 985–92.

27 Rijal, H. B. *et al.* (2012) A comparison of the winter thermal comfort of floor heating systems and air conditioning systems in Japanese homes, *5th International Building Physics Conference, Kyoto, Japan,* May 2012 (IBPC).

28 Feriadi, H. and Wong, N. H. (2004) Thermal comfort for naturally ventilated houses in Indonesia, *Energy and Buildings* 36(7), 614–26.

29 Pitts, A. (2006) The languages and semantics of thermal comfort. *Proceedings of the NCEUB Conference: Comfort and Energy Use in Buildings – Getting Them Right*, Cumberland Lodge, Windsor.

30 Wierzbicka, A. (2006) *English: meaning and culture*. Oxford University Press, Oxford.

31 Harrison, K. D. (2007) *When languages die. The extinction of the world's languages and the erosion of human knowledge.* Oxford University Press, Oxford.

32 Bellos, D. (2011) *Is that a fish in your ear? Translation and the meaning of everything.* Particular Books (Penguin), London.

33 Nicol, J. F. and McCartney, K. (2001) *Final report (public) smart controls and thermal comfort (SCATs)* (also subsidiary reports to tasks 1–7). Report to the European Commission of the Smart Controls and Thermal Comfort project (Contract JOE3-CT97-0066) Oxford Brookes University, Oxford.

34 McCartney, K. J. and Nicol, J. F. (2002) Developing an adaptive control algorithm for Europe, *Energy and Buildings* 34(6), 623–35.

35 McIntyre, D. A. (1980) Op. cit.

36 Humphreys, M. A. *et al.* (2013) Updating the adaptive relation between climate and comfort indoors; new insights and an extended database, *Building and Environment*, 63, 40–55, http://dx.doi.org/10.1016/j.buildenv.2013.01.024.

37 A value of 0.76 scale units underlies the PMV/PPD equation. This lower value applies to steady-state conditions in which everyone has the same level of activity and wears the same clothing. In a group of people in everyday life there is some diversity of activity, and this tends to increase the residual standard deviation.

38 Wierzbicka, A. (2006) Op. cit.

39 Harrison, K. D. (2007) Op. cit.

40 Ramirez-Esparza, N. *et al.* (2006) Do bilinguals have two personalities? A special case of cultural frame switching, *Journal of Research in Personality* 40, 99–120.

41 Gross, D. M., 2006. *The secret history of emotion. From Aristotle's rhetoric to modern brain science.* University of Chicago Press, Chicago.

42 Woolgar, C. M. (2006) *The senses in late medieval England.* Yale University Press, London.

19

DEVELOPING, ADAPTING AND TESTING THERMAL SUBJECTIVE SCALES

Sometimes none of the established thermal subjective scales is suitable for the particular group of respondents in a research project. When this is so, a new scale must be developed or an existing one adapted. This is not a matter of jotting down a few words to form the new scale, and we emphasise the importance of consulting appropriate books or seeking expert advice before compiling and administering even a simple questionnaire. We have from time to time used Oppenheim's book: *Questionnaire design, interviewing and attitude measurement*. It is a good practical handbook for those needing to construct scales and use them. It first appeared in 1966 and a revised and expanded edition was published in 1992.[1]

The procedures for developing and adapting scales are probably best explained by giving examples. For this we draw first on our own work in schools, explaining and criticising the procedures we used. We then draw on the work of other researchers to supplement the discussion.

In the 1950s and 1960s, many new schools were built throughout the UK. Unlike the older school buildings, many had large areas of glass in their facades, and some overheated on sunny days. But there were no criteria by which to judge overheating in schools. How hot was too hot? And what, apart from discomfort, were the effects of overheating on the children and their teachers? The Building Research Establishment was asked to investigate these matters, both in primary schools (children aged 5–11) and in secondary schools (children aged 12 and over).[2]

These projects needed new scales and new methods.

Example 1. Constructing warmth scales to be used with very young children

There were two reasons why a new scale was needed. First, as far as the researchers could discover, no-one had used thermal comfort rating scales with children as young as six and seven years old, and some psychologists doubted whether children so young were capable of giving meaningful data by using rating scales. Using a rating scale requires a maturity of thinking that some believed to come only later in a child's mental development. Second, the grown-up language of the Bedford scale categories might not be easy for young children to understand. So it was decided to develop and test a scale especially for this project. The section head at the time was Dr John Langdon, an eminent environmental psychologist, whose advice we (the research team) were fortunate to have.

Developing the new scale

Members of the team spent time with children in the age range, talking with them one by one about their classroom and whether they were comfortable in it, and about heat, cold and comfort more generally. The class teacher was present with each child to reassure them and to help create a familiar atmosphere. From our interview notes, we made a list of words and phrases that the children used to describe their thermal conditions, and from the list we picked out those that concerned the children's thermal sensation and comfort. Figure 19.2 shows a typical classroom scene.

We wrote these descriptions on cards about 100×40 mm in size, shuffled them well, and gave the pack of cards to a child. We asked him or her to arrange them in order, either on the floor or on a table, as they chose. The child was allowed to put cards side by side if he or she could not decide on a correct order. This was repeated with many children. The procedure enabled us to identify virtual duplicates, and retain only the card with the simpler words. For some of the cards, there was little agreement among the children about where they fitted on the sequence. These cards were rejected, because the procedure showed that their words or phrases did not have a well-defined meaning for the children, in relation to the other cards. This left some 20 cards, and we went back to the children with this reduced pack. We found that children could rank them in order almost as consistently as could adults given the same task. From these 20 cards, we chose seven that appeared to us to be about equally spaced along the psychological continuum, and used simple words or brief phrases. We also picked out three cards that expressed a thermal preference on a three-category scale.

We again checked with the children whether they could arrange the cards in sequence, and also whether, when asked to do so, they could pick the card that best described how they were feeling at that time. They did so with little or no hesitation. The chosen descriptions are shown in Figure 19.1.

Notice that the preference scale wrongly asks about the room rather than the child. The work took place before we were aware of the ambiguity that arises from this form of question. It might have been preferable to use: I wish I was warmer/I feel just nice/I wish I was cooler. Also it would have been better to again enlist the help of the children in selecting the final words and phrases from the pack rather than relying solely on the judgement of the research team.

How do you feel now ?

much too hot ☐
too warm ☐
nice and warm ☐
just nice ☐
nice and cool ☐
too cool ☐
much too cold ☐

day ....
time ....

What do you wish ?

I wish it was warmer in here ☐
I like it just as it is ☐
I wish it was cooler in here ☐

day ....
time ....

FIGURE 19.1 The scales under development as formatted for the pilot study.

Source: BRS experimental records.

FIGURE 19.2 A group of young children working in their classroom during the main survey.

Source: BRS project folders. Photographer: J. F. Russell.

Testing the scales

The pilot study

We tested both scales on a class of children in a pilot experiment that continued over a period of 12 school days in January 1971. Each child had a pad of 48 cards, either of the three-point scale or of the seven-point scale shown in Figure 19.1. The two groups were balanced for ability, as ranked by the class teacher, and for the numbers of girls and boys. The children gave their thermal response four times a day by ticking the chosen response box. The four times were: just before the mid-morning playtime; just before lunchtime; just before the mid-afternoon playtime; and just before it was time to go home. At these times the children had usually been in the classroom for an hour or more. They were encouraged to keep their responses secret, and the teacher was careful not to influence them. To ensure this she used the form of words that we had suggested to her. Each time a child made a response, the child detached it from the pad and placed it in his or her personal envelope. At the end of the day the children sealed their envelopes and 'posted' them in a box in the school hall. This procedure reduced the likelihood of a child looking back to see their previous response, and so helped make the responses independent.

We recorded the classroom temperature during the experimental period at three locations in the room at a height of about 0.7 m above the floor – the height of the children's tables. We used bimetallic thermographs, which were then the usual temperature-recording instruments, and placed them in white-painted steel boxes (bread bins). The boxes served to protect the thermographs from inquisitive fingers and also ensured that the recorded temperature would be somewhere between the air temperature and the mean radiant temperature. During the pilot survey, the weather was unseasonably mild, and the heating system provided a fairly steady temperature with a mean of 17°C. The standard deviation of the room temperature was a mere 0.7 K for the whole experimental period. We had been expecting colder weather and more varied temperatures. The small variation meant that the test of the scales was more stringent than we had intended.

The very first session was used to train the children in the use of the scales and its result was discarded. Statistical analysis (linear regression and correlation) of the results of the remaining 47 sessions showed that children's responses on the seven-category scale were positively correlated with the room temperature ($r = 0.33$, $p = 0.024$), but were more scattered than had normally been found among adults. The increased scatter perhaps reflected the diversity of activity that was usual among a class of young children, or perhaps indicated a less restrained use of language.

In order to check whether the significant result was attributable only to more able children, the data from the least able third of the class were removed and the correlation re-calculated. There was a slight drop in the correlation, contrary to our expectation. So the ability to use the scale was not confined to the more able children.

The significant overall correlation justified the examination of the individual records of each child. Many of these records showed no obvious temperature trend. Those that showed a clear trend were not confined to the most able in the class. This suggested that it was the narrowness of the temperature range, rather than the ability of the children to use the scale, that had limited the number of statistically significant records. Indeed, when working with adults, and with a similar number of observations per person, we had found that significant individual records were not always obtained if the standard deviation of temperature was less than 1 K.

Although positive, the correlation of the responses with room temperature on the three-category preference scale was not statistically significant ($p > 0.10$). A preference scale is conceptually more complex than a direct warmth scale, since its logic requires two mental steps rather than one before an answer can be given (How am I feeling? How does this differ from what I would like to feel?). This may have confused some of the children. We had tested both the scales because we were not sure that children could handle seven categories. Perhaps it was the conceptual structure implicit in the scale rather than the number of categories that mattered.

So we decided to use only the seven-category version (see Figure 19.3) for the full experiments that followed. The scale was written in the scripts the children had been taught to use. There were two types of script because some schools used a special

How are you feeling?

much too hot ☐
too warm ☐
nice and warm ☐
just nice ☐
nice and cool ☐
too cool ☐
much too cold ☐

FIGURE 19.3 The seven-category scale as presented to the children in the main surveys.

Source: BRS experimental records. Drawing: Barbara Buller.

phonetic system to teach children to read and write – the Initial Teaching Alphabet which has long ago ceased to be used.[3]

The experiments took place in several schools during the summers of 1971 and 1972, in Hertfordshire, in the south of the UK.[4] The project was described in Chapter 4. The weather was cool both years during the experimental period, and the results told us little about overheating – but that is an unavoidable risk with fieldwork in the UK's variable climate.

The account we have given illustrates the work that is entailed in developing a new scale. We have gone into some detail because the whole process is usually condensed to a line or two of a journal article, yet it is essential to the validation of any new scale. It is a process of continual refinement, always with repeated reference to the target population.

So the new scale could be understood and used by the children, and their responses could be correlated with the room temperature. But how do the categories in this scale behave? We can use the method of successive categories, as explained in Chapter 18, to explore its properties. We had not done so at the time of the project, being unaware of the usefulness of the method.

We consider the data both from the pilot study and from the main project. Table 19.1 gives the cumulative proportions in the scale categories, together with their Probits, first for the pilot study and then for the main study.

TABLE 19.1 Responses of young children using the new seven-category scale.

A: The pilot study.

Scale category	Cumulative numbers	Proportions	Probits
Much too cold	50	.0706	−1.47
Too cool	125	.1766	−0.93
Nice and cool	233	.3291	−0.44
Just nice	381	.5381	0.10
Nice and warm	529	.7472	0.67
Too warm	631	.8912	1.23
Much too hot	708	1	–

B: The experimental periods.

Scale category	Cumulative numbers	Proportions	Probits
Much too cold	311	.0332	−1.84
Too cool	733	.0782	−1.42
Nice and cool	1807	.1927	−0.87
Just nice	5183	.5529	0.13
Nice and warm	7231	.7713	0.74
Too warm	8202	.8749	1.15
Much too hot	9375	1	–

Figure 19.4 displays the data from the pilot survey and shows the relative widths of the scale categories in Probit units. The categories were remarkably even, indicating that the scale was almost linear on the children's psychological continuum, a highly satisfactory outcome from the scale development procedure.

When the scale was used in the main project (Figure 19.5), we see that the 'just nice' category has become wider than the other categories. The difference in relative widths between the pilot survey and the main surveys is too great to be attributed to chance; its source is unknown. There is a tendency with rating scales for people who are uncertain of their response to tick the central category, but we would expect that to have applied equally to the pilot study and the main surveys if it were occurring. It would be interesting to see whether the scale had a wide 'just nice' category in both summers, to check the stability of the scale, but separate data for the two summers are no longer available.

The result warns us that a scale may not function exactly the same in general use as it did during its development tests. The pilot data were obtained under the careful instruction and close supervision of a teacher who felt personally involved in the research. This would have been an unreasonable burden to place on the class teachers during the main project, where after initial instruction by the teacher, the children completed their questionnaires unaided. It is, of course, the properties of the scale in normal use that are definitive, rather than its properties under the more ideal conditions of the pilot study.

The frequency of occurrence of words changes over the decades, as new words and phrases enter our language and become commonplace and other words and

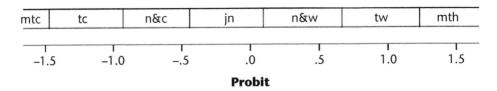

FIGURE 19.4 Category widths for the new scale for young children (pilot study). Key: From left: much too cold; too cool; nice and cool; just nice; nice and warm; too warm; much too hot.

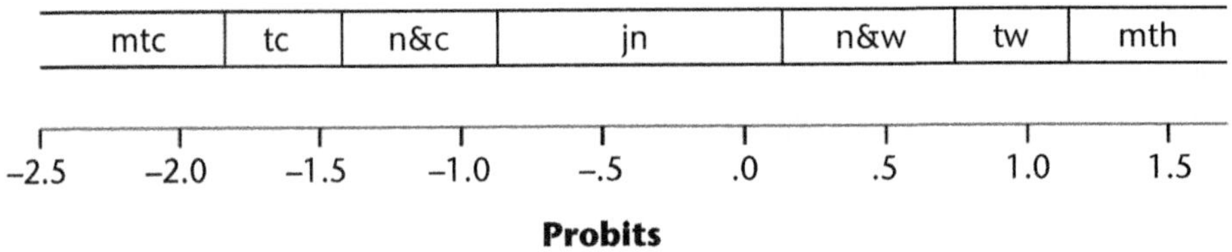

FIGURE 19.5 Category widths for the new scale for young children (main survey). Key: From left: much too cold; too cool; nice and cool; just nice; nice and warm; too warm; much too hot.

expressions fall into disuse. So there is no reason to expect a scale developed in 1968–69 to be optimal for use today. Children may use different phrases or attach a different shade of meaning to the same phrase. For example, in current UK English usage the adjective 'cool' has developed a new area of meaning, describing things that are approved of, and perhaps stylish, or to be desired because they differ from the ordinary. One wonders how this might affect its use in scales of subjective warmth.

Example 2. Constructing a scale for children aged 7–11 years

We use for our second example a well-conducted summertime field survey from a primary school (children aged 7–11) in Southampton, which is also in the south of the UK. Teli and her team gathered 1314 sets of responses from the children.[5] The survey included questions on the children's thermal sensation and preference, over-all comfort and tiredness. They adapted the ASHRAE scale, re-wording it on the advice of the teachers so that it would be more in the everyday language of the chil-dren. The revised categories were: cold; cool; a bit cool; OK; a bit warm; warm; hot.

Applying the method of successive categories to the published distribution of the children's 'comfort votes' shows that the categories of the scale have quite regular widths, so the scale is likely to behave well.

It was good to obtain advice from teachers who were in daily contact with the children and who would be familiar with their use of English. Reference to the children themselves would also have been good, had it been possible. The chosen wording is interesting. 'OK' is unambiguous as an expression of approval, but it has different degrees of approval according to the intonation. They range from reluctant concession to enthusiastic approval. This comment applies to current UK English usage; we do not know how wide the semantic range of 'OK' is when it is used in other versions of English or in other languages.

'A bit warm' or 'a bit cool' could be taken to indicate either comfort or discomfort – but this question can be answered by comparison with the scale of thermal preference in the questionnaire, and by the direct questions about comfort that were also asked.

The questionnaire was attractively presented in colour (see Plate 19.1). Coloured questionnaires have become very common, both on paper and on the internet. Children's and student textbooks abound in colour illustrations, as do many official forms, and children probably nowadays expect questionnaires to be in colour.

There has been research into the effects of colour on the speed of comprehen-sion of text and on the effectiveness of advertising, but we know of none on the effect of colour on the use of rating scales of warmth and comfort. There are reasons to be cautious about the effects of colour on a questionnaire, for there are cultural differences in the subdivision of the colours in the visible spectrum and of the meanings associated with the colours in different languages and societies. Guy Deutscher has recently set out the evidence for these differences in an accessible yet scholarly account.[6]

We cannot assume that questionnaires function in the same way with and without colouring, or with and without coloured icons.

Blue is associated with coldness and red with warmth in contemporary western culture, and this convention is used on Teli's scale of subjective warmth (see Plate 19.1, question 1). The colouring should be seen as an integral part of the scale, and when we noted above that the scale had regular category widths, the finding applied to the scale as coloured. A black and white version could have different properties.

We wonder to what extent young English children associate red with warmth and blue with coldness, and whether children who come from other cultures share the convention. Colours have other associations: red is associated with danger and green with safety; red means 'stop', orange means 'be careful' and green means 'go'. The matter is complex and invites further research.

We wondered to what extent, in question 2, the blue snowflakes and yellow suns might affect the responses to the thermal preference scale. The snowflakes should be regarded as integral to the scale, in that the scale could have different properties without them, or if they were coloured differently. The response 'I wish it was a lot colder' has three blue snowflakes beside it, yet the question implies that the child is warm. Thus the blue corresponds to the *desired* thermal state rather than to the *actual* thermal state. It is not at all clear how a thermal preference scale should be coloured – another topic that invites exploration.

These examples raise questions about the use and interpretation of colour and of icons on thermal comfort questionnaires – a question that needs to be answered through further research.

Example 3. A visual scale of subjective warmth

A further interesting example of scale development is provided by the work of D. Stafford Woolard.[7–9] He was conducting field research into the relation between the thermal environment within dwellings and the thermal sensation of indigenous people in the Solomon Islands. The climate is hot and humid (mean indoor air temperature $27.7°C$; globe temperature $28.0°C$; wet–bulb $24.4°C$; air-speed $0.29\,m/s$). The buildings were of varied construction and levels of sophistication, but all had copious natural ventilation through shaded or louvred windows. Only a small proportion of his respondents could speak and read English, so he could not use a written version of the ASHRAE scale in his surveys. As well as the several local languages, most of his respondents could speak and understand Pidgin, a trade language with a vocabulary derived chiefly from English, but with a simpler grammatical structure.

With the help of respondents who were fluent in both English and Pidgin, he translated the ASHRAE scale into Pidgin. The Pidgin version of the scale is shown in Table 19.2. It was to be administered orally because many of his respondents were unable to read. We discussed the translation of thermal scales in the previous chapter

TABLE 19.2 Woolard's Pidgin translation of the ASHRAE scale and his graphic-scale icons.

	1	2	3	4	5	6	7
ASHRAE scale:	Cold	Cool	Slightly cool	Neutral	Slightly warm	Warm	Hot
Pidgin scale:	Cold tumas (cold too much)	Cold lelebit (cold little bit)	Cold lelebit but hemi no cold tumas (but not too much)	Hemi gud (good)	Hot lelebit but hemi no hot tumas (but not too much)	Hot lelebit (hot little bit)	Hot tumas (hot too much)
Icons:							

Source: icons drawn by Hugh Stewart-Kilick.

and, examining Woolard's translation of the scale into Pidgin, we notice that it has become similar to the Bedford scale: every category description now includes an evaluation of its desirability. For example, 'Neutral' on the ASHRAE scale has become 'Hemi gut' (good).

Woolard assumed that the categories of the Pidgin version of the scale had ordinal properties, believing that their order is guaranteed by the wording (there is no distinction in Pidgin corresponding to warm/hot in English). This overlooked the possibility of 'Hemi gut' being identified with feeling 'Cold lelebit but hemi no cold tumas' (a little bit cold but I'm not too cold) or 'Hot lelebit but hemi no hot tumas' (a little bit hot but I'm not too hot).

Noting the power of cartoons (icons) to convey information across the boundaries of language and culture, Woolard wondered if it would be possible and advantageous to make a thermal sensation scale comprising a series of icons, each portraying a particular thermal state, ranging from severe shivering to copious sweating. So he constructed a graphic (visual) scale of seven icons designed to portray thermal sensations from 'cold' to 'hot'. The scale is shown in the bottom row of Table 19.2.

The icons convey their meaning through posture, facial expression, and the portrayal of shivering or of sweating. The cold icons have arms folded and legs together, while the warm icons have the arms hanging and the legs apart. The central icon is smiling, while the others look progressively less content. The colder icons indicate progressively more severe shivering, while the hotter icons indicate progressively more copious sweating.

A new scale such as this needed validation. Would the respondents recognise the gradations of hot and of cold portrayed by the icons? Was each icon distinct from its neighbour?

Woolard's testing was thorough, using the method of paired comparisons. He made up cards showing pairs of icons side by side, the icons being quite large (110 mm tall). Twenty-one cards are needed to cover all the combinations of the seven icons, disregarding the order in which the pair appears on the card. He presented all the cards one by one, in randomised order, to 200 of his respondents, asking the respondent to point to the icon that indicated the hotter condition. Each time he used exactly the same question, spoken in Pidgin. The results of the paired comparisons are shown in Table 19.3.

The entries in the table are the number of 'correct' responses from the 200 respondents, 'correct' meaning that the icon pairs conformed to the order intended by the researchers, as shown in Table 19.2. Thus icon 2 was judged 'hotter' than icon 1 by 173 out of the 200 respondents (87 per cent); icon 7 was judged 'hotter' than icon 1 by 194 of the 200 respondents (97 per cent). (A few respondents took the shaking to mean shaking with fever instead of shaking with cold.) The results showed a high degree of agreement with the intended order of the icons, so validating the ordinal nature of the icon scale. From such a table of paired comparisons it is possible, if a few assumptions are made, to derive an interval-scale for the icons, so giving a numerical value to the place of each icon on the psychological continuum. Woolard did this using the 'analysis techniques of Edwards',[10] presumably Thurstone's method.[11] He found that the distances between the successive icons were 'approximately regular' (0.26, 0.36, 0.33, 0.49, 0.50, 0.61).[12]

Woolard also tested how well the same 200 respondents could arrange the icons in rank order. He did this by giving the respondent a pack of seven cards, one for each icon, the cards being presented in random order. He asked the respondent to arrange the cards in order of ascending hotness: 166 did so correctly, 26 transposed one pair

TABLE 19.3 Results of the paired comparisons (from Woolard's data).

Icon	1	2	3	4	5	6	7
1	–	173	172	173	173	182	194
2		–	183	173	176	183	194
3			–	172	186	189	192
4				–	183	185	189
5					–	185	183
6						–	186
7							–

TABLE 19.4 Regression statistics for the comparison of the three scales (Woolard's data).

Scale	Constant	Coefficient	R^2	N
Pidgin/ASHRAE	0.26	0.95	0.86	200
Pidgin/icons	0.00	1.00	1.00	1764
ASHRAE/icons	0.23	0.98	1.00	120

of icons, while 8 had further displacements. This result indicated a good degree of agreement between the intended and the perceived rank-order of the icons.

During the subsequent survey people gave their responses to the thermal environment on two scales: the visual scale and either the ASHRAE scale (120 occasions) or the Pidgin translation (1764 occasions, presented orally). A further 120 gave responses on the ASHRAE scale and on the Pidgin translation. Woolard correlated the responses. The correlations were very high, and the regression statistics are shown in Table 19.4.

Two perfectly equivalent scales would, sampling error apart, have a zero constant, a unit regression coefficient, and an R-squared of unity. The relation between the Pidgin scale and the series of icons completely fulfilled these three criteria, showing that these two scales, as presented in the survey, were entirely equivalent. It follows that Woolard did not need to use both the Pidgin scale and the graphic scale, because they convey the same information.

Only if the two scales have identical properties is there an expectation of zero for the constant and unity or the gradient. If a pair of scales is less than perfectly related, the regression of one upon the other no longer indicates the structural relation between them, but instead gives an equation that predicts the value on one scale from a value on the other. Thus the first equation in Table 19.3 predicts the value on the Pidgin scale from the value on the ASHRAE scale. The structural relation between the two scales is better approximated by the geometric mean of the two regression coefficients, as described in Chapter 23.

The three examples we have given well illustrate the problems that can be encountered when developing a new scale of warmth or modifying an existing one. They show the care needed when developing scales of thermal comfort for use in field surveys. We have not given much technical detail about the mathematical and statistical procedures that can be used. For these the reader is referred to the standard psychometric textbooks.

Note on humidity scales

We conclude with a brief note about constructing scales for assessing the humidity of the atmosphere. The almost uniform finding from thermal comfort field research is that the humidity of the environment has little if any effect on the sensation of warmth, provided that the humidity is expressed as the pressure of the water vapour in the atmosphere or some equivalent metric. An apparent effect may be found if the humidity is expressed as the relative humidity, but this arises from the dependence of relative humidity on the air temperature. It is the temperature part of the relative humidity that affects the thermal sensation, and not the water vapour. Further, if people are asked to assess directly how dry or humid the air is, their answers fail to correlate or correlate only very weakly with the humidity as physically measured.

Yet people remain convinced that they can sense the humidity of the atmosphere, and it seems unlikely that what they say is devoid of content. We wonder if the problem is a mismatch between the scientific measurement of humidity and the meaning of the words related to humidity in common speech in English. Perhaps the question should be recast: what do people mean when they speak of the humidity of the environment? Investigating this question would entail listing and grouping the adjectives commonly used to describe 'humidity' and exploring the correlations among them. It would be necessary to test whether there was substantial agreement among people about how the adjectives mapped onto a variety of thermal environments, and only then seeking the physical correlates. We feel that much remains to be done to clarify the psychological concept of humidity.

Notes

1 Oppenheim, A. N. (1992) *Questionnaire design, interviewing and attitude measurement.* Second edition, Continuum, London.
2 See Chapter 3.
3 Children learned to read earlier using this logical phonetic alphabet, but as nothing they encountered outside the classroom used the script, it proved to be no better in the longer term.
4 Humphreys, M. A. (1977) A study of the thermal comfort of primary school children in summer, *Building & Environment* 12, 231–40.
5 Teli, D. *et al.* (2012) Field study on thermal comfort in a UK primary school, *Proceedings of 7th Windsor Conference: The changing Context of Comfort in an Unpredictable World,* Cumberland Lodge, Windsor, April. Network for Comfort and Energy Use in Buildings, http://nceub.org.uk.
6 Deutscher, G. (2010) *Through the language glass: how words colour your world.* Heinemann, London.
7 Woolard, D. S. (1979) Thermal habitability of shelters in the Solomon Islands. PhD Thesis, University of Queensland, Australia.
8 Woolard, D. S. (1981) The graphic scale of thermal sensation, *Architectural Science Review* 24(4), 90–3.
9 Woolard, D. S. (1981) Thermal sensations of Solomon Islanders at home, *Architectural Science Review* 24(4), 94–7.
10 Edwards, A. L. (1957) *Techniques of attitude scale construction.* Appleton Century Crofts, New York. (A standard textbook reprinted many times: most recently, 1983, Irvington, New York.)
11 Thurstone, L. L. (1927) A law of comparative judgement, *Psychological Review* 34, 273–86.
12 For a recent overview of paired comparison methods see, Tsukida, A. and Gupta, M. R. (2011) How to analyse paired comparison data. *UWEE Technical Report,* http://www.ee.washington.edu.

20

A SIMPLE HEAT EXCHANGE MODEL FOR THERMAL COMFORT CONDITIONS

The previous two chapters considered how to assess the properties of subjective scales of thermal comfort and how to construct new scales should they be needed. In this chapter, we consider the objective physical relation between a person and the room's thermal environment. This continues a theme from Volume 1, where we outlined the principles of heat exchange between the human body and its thermal environment and drew attention to some of the heat-exchange models that are in common use. Here we construct a simplified model that helps explain how a person relates to the thermal environment.

A good modern computer thermal simulation of the human body calculates, with varying degrees of completeness, the distribution of temperatures within the body tissues and on the skin surfaces. It can do this for the various activities undertaken and the clothing ensembles worn, and for the multidimensional thermal environment in which the person is situated. All this can be done, not only for the steady state, but also for ever-changing levels of activity, clothing and environment. The development of such models is limited only by our knowledge of thermal physiology and the processes of environmental heat exchange.

Some models go further, and from the physiological body-states seek to estimate the psychological result. They predict the sensation of warmth and perhaps also make an assessment of comfort. The psychological aspects are much less developed in thermal modelling than are the physiological and physical aspects, because our understanding of the complex 'brain software' processes that lie between the thermal state of the body and its perception and affective outcome is still incomplete. Perhaps the endpoint of such models is to produce a virtual thermal human being that can make known the sensation of the warmth of its various body parts and its degree of contentment with the thermal environment.

It would be good to have such a virtual thermal human being – and no doubt we shall have one soon – but its very complexity would limit its usefulness as an aid to understanding. It is difficult to check the correctness of a complex virtual model. We have to take on trust the physiological and psychological research on which it rests and its translation into computational algorithms. Ultimately the only way to test such a virtual human is by comparison with the responses of people in the thermal environments to which they are exposed in daily life.

In this chapter we take a different approach. We build a thermal model of the human being that is simple enough to understand. It cannot make accurate predictions of the result of a particular thermal environment on the person, and has a limited range of application, but it shows clearly how the basic processes are related to each other. We develop it using only basic algebra and a simple heat flow equation. An earlier version of the model was developed at the BRS as a teaching aid on courses for teachers of architectural and building science at colleges and universities.[1] It rests on still earlier models, especially that of Burton and Edholm.[2] It is not comprehensive, its usefulness being restricted to the condition that the dominant paths of heat flow from the clothed body are by convection and radiation rather than by evaporation. That is to say it ceases to apply in very hot environments where the evaporation of sweat becomes the chief means of cooling the body. We know of no equivalent simplified model for hot environments, where thermal regulation is dominated by sweating. For such environments it is usual to use a complex model such as Gagge's Standard Effective Temperature.

Figure 20.1 illustrates the basic model. It traces the flow of heat from its source deep within the body.[3] From there it flows through the body tissues to the skin, then through the clothing layer, and thence to the room air and the room surfaces (furnishings, walls, ceiling and floor). Each layer has a thermal resistance (thermal insulation) which is adjustable, so giving a wide range of room temperatures that are compatible with thermal comfort.

We now consider the flow through each of these layers and set up their heat flow equations.

Metabolic heat is generated deep within the body by the digestion of foodstuffs and transported around it by the circulation of the blood. A small proportion of this heat, perhaps some 10 per cent in moderate thermal environments, is dissipated directly to the room air by the breathing process. We breathe in dry air and breathe out moist air. The evaporation of moisture from the internal surfaces of the lungs releases the latent heat of evaporation to the environment. We breathe in air at room temperature and breathe it out at a temperature a little below body temperature, and so there is a small convective heat loss too. But the larger proportion of the metabolic heat, some 90 per cent, flows out through the peripheral body tissues to the skin surface. We call this proportion k_t.

It is usual to express the metabolic heat generation (M) in terms of the surface area of the skin, so M has the units W/m^2. The metabolic rate increases with increased

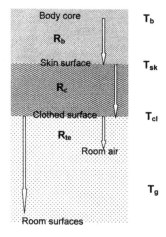

Key: The arrows represent components of
heat flow from the body core to the room.
Other symbols as in the text.

FIGURE 20.1 Schematic representation of the heat flows from the body core to the room.

physical activity. A person who is sitting down has a metabolic rate of about $60\,W/m^2$ and when standing still about $70\,W/m^2$, while someone strolling at $0.7\,m/s$ has a rate of about $110\,W/m^2$.

We will take the body core temperature (T_b) to be constant at $37°C$. This is a good enough approximation for our purpose, because the variation of the deep body temperature during the diurnal cycle and the differences caused by everyday variations in activity amount to no more than a fraction of a degree. These small changes in temperature are crucial in activating the body's thermoregulation system, but are unimportant in the heat flow equations we will be setting out. We represent the skin surface by its mean temperature (T_{sk}) – a good enough approximation for our present purpose.

We divide the body into just two zones: a uniform temperature core and the peripheral tissues – the layers beneath the skin. Within the comfort zone, the body defends its core temperature from change by increasing or restricting the circulation of blood in the peripheral tissues. It does so by means of dilation or constriction of the blood vessels in the peripheral tissues. This has the effect of decreasing or increasing the thermal insulation afforded by these peripheral tissues. It is as though we have a built-in automatically adjustable layer of clothing. We give this thermal insulation (thermal resistance) the symbol R_b, the thermal insulation of the body's peripheral tissues. R_b does not have a fixed value – it is a control variable. Its units are m^2K/W.

Heat flow from the core to the skin

We can now write the equation for the flow of the metabolic heat (M) from the body core to its surface:

$$k_tM = (T_b - T_{sk})/R_b$$

Rearranging the equation we have:

$$(T_b - T_{sk}) = k_tMR_b$$

That is to say, the temperature drop from the core to the skin is obtained by multiplying the heat flow (k_tM) by the thermal insulation (R_b) afforded by the peripheral tissues. Notice that if the metabolic rate increases and the core temperature remains (very nearly) the same, the skin temperature falls. So an active person has a lower mean skin temperature. This is counter-intuitive but readily confirmed by measurement. The person will feel warmer because of a scarcely measurable increase in the core temperature, but the skin becomes cooler. The cooler skin is obtained by wearing less clothing, or by evaporating sweat from the skin, or by lowering the room temperature, or by increasing air movement in the room, or by some combination of these.

The insulation provided by the peripheral tissues increases from perhaps 0.04 m²K/W when vasodilated to about 0.09 m²K/W when vasoconstricted (about 0.3 clo[4] to 0.6 clo). The values are approximate, and different researchers have found various values.[5] The increase of 0.05 m²K/W (about 0.3 clo) would cause a drop in the mean skin temperature. For a seated person ($M = 60$ W/m²), multiplying the increase in the tissue insulation (0.05 m²K/W) by the heat flow though it (k_tM) we have:

0.9*60*0.05 = 2.7 K

This is about equivalent to putting on a jacket.

Heat flow from the skin and through the clothing

We now consider the second stage of heat flow. At the skin surface, some heat is released to the environment by the evaporation of moisture from the skin. This evaporative heat loss has two parts. One part comes from the continuous slow diffusion of moisture through the skin. The other part comes from the evaporation of any sweat that may be produced. For a thermally comfortable person at rest and unacclimatised to heat some 20 per cent of the metabolic heat is dissipated directly to the environment by these processes, rising to some 40 per cent if the person is more active. An active person is comfortable when sweating lightly, but the sweat is not perceived because, if the clothes are vapour permeable, the skin stays dry. We will assume that 30 per cent of the heat flowing from the skin surface is dissipated by evaporation. This is a rather rough approximation, but sufficient for our simple model.[6] The larger part of the heat, some 70 per cent, passes by conduction through the clothing layer. We will call this proportion k_c.[7]

Clothing can be regarded as an insulating layer. The body heat flows from the skin surface through this insulating layer to the surface of the clothed body. The heat flow equation for this is:

$$k_t k_c M = (T_{sk} - T_{cl})/R_c$$

where R_c is the thermal insulation of the clothing layer (m²K/W). Rearranging the equation we have:

$$(T_{sk} - T_{cl}) = k_t k_c M R_c$$

That is to say, the temperature drop across the clothing layer is obtained by multiplying the heat flowing through it ($k_t k_c M$) by its thermal insulation (R_c). For a seated person who is wearing a suit and normal underwear the thermal insulation of the clothing will be about 1 clo unit. The temperature drop will be:

$$0.9*0.7*60*0.155 = 5.9$$

So for someone sitting down a suit affords a thermal 'buffer' of about 6K. A seated person would therefore be comfortable in a room 6K cooler than would a naked person.

For someone more active at 120 W/m², for example sitting and packing books, the same clothing insulation would provide a thermal buffer of some 12K. However, the insulation afforded by a clothing ensemble is usually reduced by the activity, because the air between and within the layers of the fabric is disturbed. Also the activity disturbs the layer of air around the person, reducing its effective insulation.

Heat flow from the surface of the clothed body to the room

We now consider the third stage of the heat flow. Heat flows from the clothing surface to the room. It does so by two separate processes. There is heat flow by *convection* from the surface, and there is heat flow by thermal *radiation* to the surrounding room surfaces. The heat flow equation thus has two distinct parts:

$$k_t k_c M = (T_{cl} - T_a)h_c + (T_{cl} - T_{rs})h_r$$

T_a is the temperature of the air in the room and T_{rs} is the mean temperature of the room surfaces (strictly the mean radiant temperature). The surface heat transfer coefficient by convection is h_c and h_r is the radiant heat transfer coefficient (W/(m²K). Surface heat transfer coefficients are convenient to use when considering heat transferred across a boundary, such as between the surface of the clothed body and the surrounding air, rather than the flow through a layer of insulating material. With some algebraic manipulation the equation can be re-written as:

$$k_t k_c M \{1/(h_c + h_r)\} = T_{cl} - \{(h_c T_a + h_r T_{rs})/(h_c + h_r)\}$$

This equation looks complicated but it is informative and can be simplified. First we examine the expression:

$$(h_c T_a + h_r T_{rs})/(h_c + h_r)$$

It is a *temperature* – the weighted average of the air temperature and the mean radiant temperature. The weights are the heat transfer coefficients by convection and by radiation from the clothed body surface. Notice that it is the ratio of the two coefficients that matters, and not their absolute values. This is useful because it can be shown, rather surprisingly, that a ping-pong ball has at low air speeds nearly the same relative weighting for convection and radiation heat transfer as does the clothed body.[8] So we can replace:

$$(h_c T_a + h_r T_{rs})/(h_c + h_r)$$

by T_g, the temperature of a ping-pong ball globe thermometer (°C). The temperature of such a globe is therefore a valid measurement of the room temperature when we are considering heat flow from a person to the room. It combines the air temperature and the mean radiant temperature into a convenient single value.

As explained in Volume 1, the smaller an object, the closer its temperature is to the air temperature, compared with the radiant temperature. A ping-pong ball seems too small to represent the clothed human body. The representation works because parts of the human body radiate to each other rather than directly to the room. For example, the arms exchange radiation with the trunk, and the insides of the legs exchange radiation with each other. So the effective area available for radiation to the room is less than that available for convection. The ping-pong ball has its entire surface available both for radiation and for convection and, despite its smaller size, gives the correct balance of convection and radiation.

No one sphere diameter is correct for the entire range of indoor air speeds, and at air speeds above 0.2 m/s theory favours the use of a larger sphere. At 1 m/s, a speed that might be found under a ceiling fan, a sphere of about 80 mm diameter would be suitable. In practice the difference in the temperature recorded by a 40 mm sphere and an 80 mm sphere would be very small, and the smaller sphere is more convenient. Its widespread use is therefore justifiable.

Next we examine the expression:

$$1/(h_c + h_r)$$

It is a *thermal insulation*, that afforded by the thermal environment of the room. It is the combined insulation of the layer of air in contact with the clothed body, and the 'insulation' afforded by radiation to the room surfaces. We shall call it R_{te}, the insulation (thermal resistance) of the thermal environment.

The convection coefficient h_c depends on the air speed. The greater the air speed, the greater it becomes. So if the air speed increases, R_{te} decreases. If the air speed is very great, R_{te} tends to zero and the environment offers no thermal insulation to the person. That would be thermally equivalent to taking off a layer of clothing. What insulation does this 'coat' have? We can provide values for h_c and h_r and hence for R_{te}.

The convection coefficient is proportional (to a good approximation) to the square root of the air speed. A reasonable expression for this coefficient is:

$$h_c = 12\sqrt{v} \quad (W/(m^2K))$$

where v is the air speed (velocity) in m/s. At very low air speeds, natural convection dominates the heat exchange, so the convection coefficient has a minimum value, normally around $4\,W/m^2$. At one metre per second, the convection coefficient would be about $12\,W/m^2$.

The radiant transfer coefficient is calculated directly from the physical laws of radiation heat exchange. Taking 0.9 as the emissivity of the surfaces, the radiation coefficient is $5.0\,W/(m^2K)$ at $17°C$ and 5.5 at $27°C$, so we take a value of $5.2\,W/(m^2\,K)$ as suitable for usual room temperatures. But we need to remember that the surface area available for radiation exchange is only about 0.8 of that available for convection, so we reduce the value of the coefficient accordingly to $4.2\,W/(m^2K)$.

We can now insert these numbers and arrive at approximate values for the thermal insulation afforded by the room. When the room air is still, R_{te} becomes $\{1/(4 + 4.2)\}$ which equals $0.12\,m^2K/W$. This is some 0.8 clo, perhaps the insulation afforded by a winter overcoat. If the air speed is 1 m/s, R_{te} becomes $0.06\,m^2K/W$, or 0.4 clo. So increasing the air speed to 1.0 m/s reduces the insulation by about 0.4 clo. It is equivalent to taking off a jacket, a worthwhile adjustment to be able to make. And if the clothing is very air-permeable, as it often is in warm climates, the effect is still greater.

We return to the rather complicated equation we derived for the heat flow from the surface of the clothed body to the room:

$$k_t k_c M\{1/(h_c + h_r)\} = T_{cl} - \{(h_c T_a + h_r T_{rs})/(h_c + h_r)\}$$

We can replace:

$$\{(h_c T_a + h_r T_{rs})/(h_c + h_r)\}$$

by T_g, the temperature of a ping-pong ball globe thermometer, and we can also replace

$$\{1/(h_c + h_r)\}$$

by R_{te}, the thermal insulation of the room environment. The equation for the heat flow from the clothed surface to the room becomes much simpler:

$$k_t k_c M = (T_{cl} - T_g)/R_{te}$$

and the temperature drop from the clothed surface to the room is:

$$(T_{cl} - T_g) = k_t k_c M R_{te}$$

Combining the three parts

Now is the time to put it all together. We have obtained simple equations governing the temperature fall across each of the three layers of 'insulation' between the body core and the room:

1 From the core to the skin: $\qquad\qquad (T_b - T_{sk}) = k_t M R_b$
2 From the skin to the clothing surface: $\qquad (T_{sk} - T_{cl}) = k_t k_c M R_c$
3 From the clothing surface to the room: $\quad (T_{cl} - T_g) = k_t k_c M R_{te}$

If we add these three together we get the equation describing the temperature fall from the body core and the room:

$$(T_b - T_g) = k_t M(R_b + k_c R_c + k_c R_{te})$$

Hence:

$$T_g = T_b - k_t M(R_b + k_c R_c + k_c R_{te})$$

We have put this equation into the gradient-intercept form. The intercept is the body's core temperature. If we insert a value for each of the three insulations $(R_b, k_c R_c, R_{te})$ and a value for each of the two constants (k_t, k_c) we obtain a straight line relating the globe temperature to the metabolic rate. We can draw lines for the vasoconstricted state and for the vasodilated state, to represent the limits of comfort, and we can do so for various levels of clothing insulation. These are shown in Figure 20.2. We can then alter the air movement to see what effect that has.

The figure rewards careful study. The comfort zones are seen to radiate from a single point defined by the body core temperature of 37°C and a metabolic rate of zero (not of course a real-life situation!). It turns out that this rather approximate treatment reveals much about the effects of metabolic rate, clothing and air movement on the temperatures that are likely to be comfortable; that is to say, that lie within the zones indicated by the processes of vasoconstriction and dilation, the body's first defences against cold and heat.

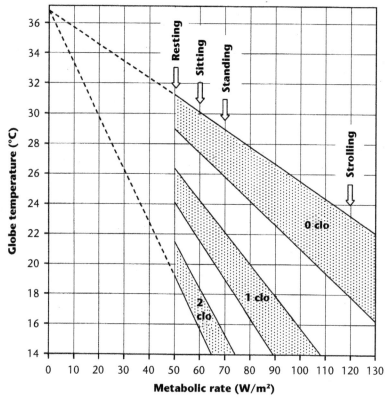

Note: The upper margin of each represents the vasodilated state
while the lower magin represents the vasoconstricted state.

FIGURE 20.2 Schematic diagram of comfort zones for still air and for various levels of clothing
insulation ($k_c = 0.7$).

Consider a vertical line at metabolic rate $50\,\mathrm{W/m^2}$, representing a person at rest.
The line cuts off equal portions of temperature at each of the three levels of clothing
insulation. The figure has a powerful optical illusion caused by the different slopes
of the lines, so it is necessary to look carefully at the temperature scale to verify this
equality. Each zone, whether for 0, 1 or 2 clo, has, at a metabolic rate of $50\,\mathrm{W/m^2}$,
a temperature-width of just over 2 K. This means that people are equally sensitive to
temperature changes whether they are naked or quite heavily clad. This may come
as a surprise. If the clothing is heavier, it takes longer for a change in room temperature to affect the person's sensation of warmth, so we seem to be less sensitive to the
change. But if a person is wearing heavy clothing and the room temperature drops
a few degrees below the comfort temperature, after an hour or two the person feels
the cold. The figure shows (and the equations confirm) that if thermal regulation is
by vasoconstriction and dilation, rather than by increasing the sweating or shivering,
then the sensitivity of people to room temperature change is independent of their
clothing insulation. In a later chapter we will show from field data, where people are

clothed to suit the room temperature and have fans available for use when required, that this constant sensitivity to changes of room temperature holds quite well over the entire range for which field survey data have been collected, and in mean room temperatures as high as 34°C.[9]

Second, we see that the comfort zones 'spread out' as the metabolic rate increases. So when the metabolic rate is 100 W/m² (a gentle stroll) the globe-temperature width is just over 4 K, no matter what the clothing insulation. The width has doubled. So people who are more active, and who therefore have a higher metabolic rate, are less sensitive to room temperature changes than are people who are resting. For the same clothing, the more active people do of course need a cooler room temperature for comfort.

Third, we notice that the range of metabolic rate that is contained within the comfort zone depends on the clothing insulation. A person seated with no clothes would be comfortable at about 29°C, and the comfort zone at this room temperature

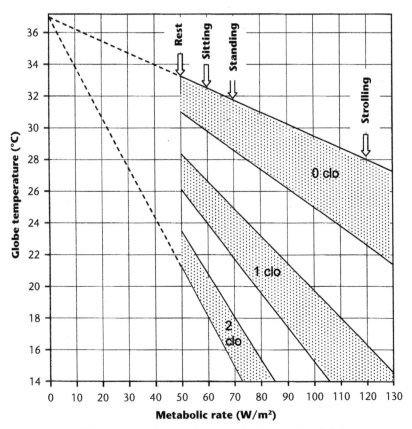

Note: The upper margin of each represents the vasodilated state while the lower margin represents the vasoconstricted state.

FIGURE 20.3 Schematic diagram of comfort zones for air movement 1 m/s and for various levels of clothing insulation.

stretches from about 50 to 70 W/m², a range of some 20 W/m². But a seated person wearing 2 clo and comfortable at about 17°C would have a range of less than 10 W/m². This person would be much less tolerant of a change in metabolic rate. This explains why, when we are outdoors in cold weather and wearing a warm coat, we get hot when walking to the bus stop and then get cold while waiting for the bus.

We now re-draw the figure for an air speed of 1 m/s to see the effect of increasing the air movement (see Figure 20.3). The general appearance of the figure is unchanged. All three zones have been 'rotated' slightly in the anticlockwise direction. Everyone's comfort temperature has been raised by the increased air movement, the amount depending only on the metabolic rate. For each clothing-level, the extent of the zone in terms of globe temperature remains the same as in the previous figure, so people's sensitivity to a change in temperature is unaffected by the change in air movement. The effect of increasing the air speed is like removing a garment.

Figure 20.3 suggests that a seated person (metabolic rate 60 W/m²) will be thermally neutral at about 29°C, if the clothing insulation is 0.5 clo and fans are providing air movement of about a metre per second. Allowing for a reduction of the clothing insulation caused by effect of the air movement penetrating the clothing, the person would be thermally neutral at perhaps 31°C, and 'slightly warm' at 33°C. Above this room temperature, the simple model will become gradually less applicable as sweating becomes increasingly important for heat balance.

Necessary complexity of thermal indices

The equation and the figures have also shown something useful about thermal indices. Provided the skin is not wet, that is to say that any sweat that may be produced immediately evaporates, the globe temperature and the air speed together provide an adequate description of the thermal environment of the room. There is no need for anything more elaborate. Further, if there are no air currents, the globe temperature alone is sufficient. And very often the air temperature and the globe temperature are practically the same, so just the air temperature will suffice. So the rather despised room thermostat that responds chiefly to the air temperature in the room is usually adequate to control the room for comfort.

We can formalise this and make a table to show how complex an index needs to be if it is to perform as an adequate measure of the thermal environment of a room, from the point of view of overall thermal effects. (We are not considering discomfort caused by strongly asymmetrical conditions such as large temperature gradients from floor to ceiling, or very strong thermal radiation from a particular direction.) Table 20.1 shows how complex an index of the thermal environment needs to be if it is to offer a sufficient description of the thermal environment.

Our analysis has shown that there is no need for separate measurement of either the air temperature or the radiant temperature. The globe temperature is sufficient, unless it is desired to describe the effects of asymmetric thermal radiation.

TABLE 20.1 Variables needed to adequately describe the thermal environment, and the conditions for their valid use.

Set of measurements	Set is sufficient provided that:
Air temperature	The difference between the air temperature and the mean radiant temperature is small (<2 K)
	Air speed is very slight (<0.2 m/s)
	Sweat is freely evaporated from the skin
Globe temperature*	Air speed is very slight (<0.2 m/s)
	Sweat is freely evaporated from the skin
Globe temperature*, air speed	Sweat is freely evaporated from the skin
Globe temperature*, air speed, humidity	

*If the difference between air temperature and mean radiant temperature is small, the air temperature can be used instead of the globe temperature.

Globe temperature and air speed

It is sometimes thought that the globe thermometer allows for air speed. It does not do so fully. All it does is to give the correct relative weight to the air temperature and to the mean radiant temperature, over a range of air speeds. If the air temperature and the mean radiant temperature are equal, the globe temperature is unaffected by the air speed, as is evident from its formula:

$$T_g = \{ (h_c T_a + h_r T_{rs})/(h_c + h_r) \}$$

Putting the mean radiant temperature equal to the air temperature, T_g reduces to:

$$T_g = T_a = T_{rs}$$

The globe temperature therefore needs to be supplemented by an effect of air speed, as is seen in the equation for the heat flow from the clothed surface to the room.

Consider a room where the mean radiant temperature is higher than the air temperature. Increasing the air speed will have two effects: it will reduce the globe temperature, because the ratio of the convection and radiation coefficients has changed; and it will increase the heat flow from the clothed surface of the body because, unlike the globe, the human body has an internal source of heat.

Although this effect is implicit in the equation, it is useful to make it more evident. We can deduce from the overall heat flow equation the effect of the air movement, and so make explicit the adjustment to apply when the room air is motion. It is a temperature difference, that between the globe temperature required for comfort in the presence and absence of the air movement:

$$k_t k_c M(R_{te(still\ air)} - R_{te(moving\ air)})$$

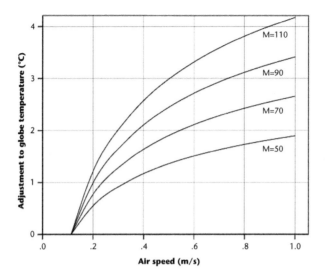

FIGURE 20.4 Adjustment to the globe temperature for air movement (assumptions: $k_t = 0.9$, $k_c = 0.7$, $h_c = 12\sqrt{v}$ W/(m²K), $h_r = 4.2$ W/(m²K)).

Figure 20.4 shows this quantity for the air speed range 0 to 1.0 m/s and for four metabolic rates. The room will feel cooler by this amount in the presence of air movement. The amount is proportional to the metabolic rate, as we saw from Figure 20.3.

Summary

In closing this chapter, we remind the reader that our purpose has been to show the principles of human heat exchange with the environment rather than to make accurate estimations of the temperatures for comfort. The equations should not be used for this purpose, because more refined assumptions are necessary if accurate answers are to be obtained.

The simple model has shown that the clothing has no effect on the temperature-width of the comfort zone – just on its temperature location. It has also shown that:

- the width of the comfort zone is proportional to the metabolic rate;
- lighter clothing gives a greater width of the comfort zone in terms of metabolic rate;
- increasing the air movement is analogous to removing a layer of clothing.

It has further shown that, in a wide variety of normal conditions, the globe temperature is a sufficient index of warmth.

Notes

1 Humphreys, M. A. (1970) A simple theoretical derivation of thermal comfort conditions, *J. Inst. Heat. & Vent. Eng.* 38, 95–8.
2 Burton, A. C. and Edholm, O. G. (1955) *Man in a cold environment.* Arnold, London.

3 It is usual to think of the room temperature affecting the body's thermal state and the sensation of warmth, and so also the clothing insulation and perhaps the level of activity. From the point of view of the model it is more convenient to look at it the other way round, and ask, for a given body thermal state and level of clothing and activity, what room temperature would be needed for comfort?

4 A 'clo' is defined as $0.155\,\mathrm{m^2K/W}$, and is approximately the insulation afforded by a business suit with the usual underwear. It is used because it is easier to imagine than is the value in $\mathrm{m^2K/W}$.

5 We have drawn our values from Burton and Edholm (1955), but see also the discussion in Parsons, K. (2003) *Human thermal environments*. CRC Press (Taylor and Francis), London.

6 A more refined portrayal of this proportion can be inserted, and leads to predictions of thermally neutral conditions that agree well with those predicted by Fanger's heat balance equation.

7 Heat acclimatisation results in the onset of sweating at lower body core temperatures, so making hot environments less stressful. It is therefore likely to reduce k_c and raise the upper limit of the comfort zone.

8 Humphreys, M. A. (1977) The optimum diameter for a globe thermometer for indoor use, *Annals of Occupational Hygiene* 20(2), 135–40.

9 In terms of heat exchange the humidity of the thermal environment has little effect on the room temperatures that are found to be comfortable. However, if the air temperature rises above the temperature of the skin, the humidity can become a matter of life or death. A person with a supply of drinking water can survive indefinitely in a hot dry environment because of the heat lost through the evaporation of sweat. But if the humidity is too high the evaporation of the sweat becomes insufficient for thermal balance. People in such a condition experience serious discomfort that can become dangerous. For example, no evaporation would be possible in an atmosphere at or above skin temperature and saturated with water vapour. Such an environment is lethal because the body core temperature becomes dangerously elevated.

21

REGRESSION ANALYSIS

General features and effects of data-selection and binning

Note on the statistical chapters

We have written the following chapters (21–24) on statistical methods for those who find a visual presentation an aid to understanding, and so we have made liberal use of scatter plots and histograms. It can be difficult to relate the statistical treatment of data in the textbooks to the particular context of thermal comfort field research, particularly if the researcher lacks education in mathematics and statistics, as many do who come from an architectural background. To help overcome this difficulty, we choose topics that arise during the analysis of thermal comfort field study data. The chapters are intended to supplement the standard textbooks and not to replace them. Those who have a first-rate mathematical and statistical background can omit these chapters without losing the argument of the book.

In Volume 1, we took a set of data from the SCATs project and used it to illustrate some of the methods of statistical analysis that can be applied to field study thermal comfort data. Here we use a different approach. We have built a computer-generated dataset and from it bring out some features of correlation and regression analysis that have a bearing on the analysis and understanding of our field data.

We have constructed a set of data of 5000 'observations'. This is large enough that the effects of random variation on its statistics are tiny. The dataset could be imagined as the data collected during a survey in a building with 5000 respondents, each interviewed once during a field study that continued over a period of days. The room temperature during any day, and the temperature in the different rooms, are subject to variation. However, there is no day-on-day drift of temperature over the period of the fieldwork. There would therefore be no need for the occupants to alter their clothes in response to day-on-day drifts in indoor temperature, and so there is no behavioural adaptation, beyond that which normally occurs during a single working day.

We formed an idealised dataset by building columns of data to represent the room temperature and the subjective thermal responses of the occupants of the building. We use the columns of data to form an overall picture of how regression operates for such data. We built the dataset using the SPSS statistical package version 19, but any good statistical package would do.

> On a procedural note, although there is SPSS syntax for creating data, it is simpler to start by creating a column of 5000 zeros in Excel and pasting it into a new SPSS data-file. We then used the menu procedure: 'Transform – Compute Variable – Random Numbers – Rv.Normal' to produce the required variable. In SPSS, random numbers are generated from an algorithm which will always produce the same sequence of numbers if using the same 'seed value' to start the sequence. The seed value is itself chosen at random, so repeating the command will normally produce a different sequence, uncorrelated with the first. Very occasionally the same seed is selected so replicating the same sequence of random numbers. Syntax is available to set the seed value, but it does not appear in the menu.

The first column represents the room temperature. Each of its 5000 values is drawn from a Normal distribution having a standard deviation of 2 K. (A quantity that is the sum of many independent processes often has a distribution that is close to Normal. Because many factors affect room temperature, its distribution is often found to be approximately Normal.) A standard deviation of 2 K is about what would be found in a medium to lightweight building of not especially good design and without especially close temperature control. We set the mean temperature to zero for the sake of simplicity, and this is equivalent to expressing the room temperature in terms of its departure from the mean condition. The distribution and its statistics are shown in Figure 21.1. The computer has generated for us a sample with a mean of –0.01 and a standard deviation of 1.981, well within the sampling error of the specified values of zero and 2.0.

A second column of data expresses the variation of subjective warmth from person to person, had there been no variation in room temperature. It therefore represents the individual differences in response to the room temperature, for not everyone responds to the room temperature to the same extent, nor is anyone's response consistent over a period of days. Again we require a Normal distribution, and we set its standard deviation to 0.9 scale units, typical of field thermal comfort data when the ASHRAE scale or the Bedford scale is used, and again we choose a zero mean. The distribution with its statistics is shown in Figure 21.2. The computer has generated a sample with mean 0.02 and a standard deviation of 0.899, within the expected sampling error.

The third column represents the mean subjective warmth in relation to changes in the room temperature. It is 0.4 times the temperature. It gives a straight line of gradient 0.4 scale units per degree, when plotted against the temperature. This gradient

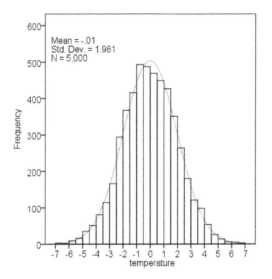

FIGURE 21.1 The computer-generated distribution of room temperature.

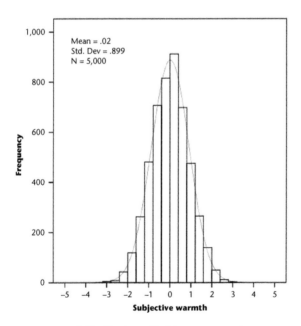

FIGURE 21.2 Computer-generated distribution of subjective warmth.

is chosen as typical of the sensitivity of people to temperature changes within a working day at the office, as derived from the various field studies of thermal comfort included in the principal databases.[1] Each of the 5000 values of the temperature results in a point on this line. The line of points is shown as Figure 21.3.

The points shown on Figure 21.3 have a distribution on the temperature axis shown in Figure 21.1. The gradient of the line causes the points *also* to

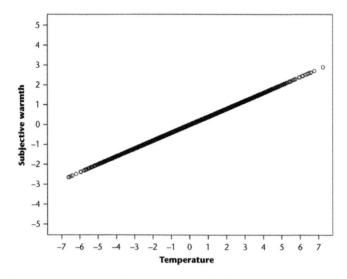

FIGURE 21.3 Computer-generated points representing the dependence of the mean subjective warmth on the temperature (gradient 0.4 scale units/K).

have a distribution on the subjective warmth axis. This distribution is shown in Figure 21.4. It has a standard deviation of 0.792. This value is the standard deviation of the temperature multiplied by the gradient of the line in Figure 21.3 ($1.981*0.4 = 0.792$). The value in our specified parent population would have been 0.8 ($2.0*0.4 = 0.8$).

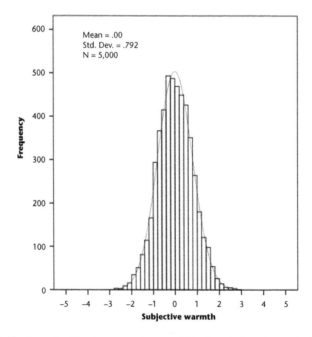

FIGURE 21.4 Distribution of subjective warmth caused by the relation shown in Figure 21.3.

The next column in our database is the sum of the previous two. To each of points on the line in Figure 21.3 we add the corresponding value from the distribution of the individual differences in the subjective warmth (Figure 21.2). The result of this procedure is shown in Figure 21.5. This is a scatter diagram relating the subjective warmth and the temperature – a very common figure in field comfort research. It looks unfamiliar because the values of the subjective warmth have not yet been grouped as values on the ASHRAE or the Bedford scale.

Through the points is drawn the calculated regression line that predicts the mean thermal sensation at any particular temperature. Its gradient is found to be 0.400 scale units/K as in Figure 21.3, and equal to its expected value. Also drawn on it are the lines indicating the standard deviation of the subjective warmth (0.899 scale units). This is the residual standard deviation of the subjective warmth, and is the same as in Figure 21.2. The square of the correlation coefficient (R–squared) between subjective warmth and the room temperature is 0.437.

It should be noticed that the spread of the points on the subjective warmth axis is made up of two parts: the variation due to the individual differences among the respondents, shown in Figure 21.2, and the variation caused by the regression relation (Figure 21.3) shown in Figure 21.4. In Figure 21.5, the overall standard deviation of the subjective warmth is now 1.199.

It is helpful now to make a table of the various values we have from our computer-generated sample of 5000 'observations'. They are shown in Table 21.1. The first column shows the sources of the variation of the warmth sensation. The second column contains their associated standard deviations. Standard deviations cannot be

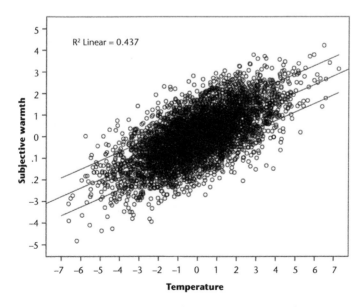

FIGURE 21.5 Computer-generated scatter plot of subjective warmth and temperature.

TABLE 21.1 Combining the standard deviations (computer-generated sample values).

Source	s.d	Variance	Comments
Individual differences	0.899	0.808	See Figure 21.2
Regression	0.792	0.627	See Figure 21.4
	1.199	1.435	

Note: R-squared = 0.627/1.435 = 0.437.

TABLE 21.2 Combining the standard deviations using the specified population values.

Source	s.d.	Variance
Individual differences	0.9	0.81
Regression	0.8	0.64
	1.2042	1.45

Note: R-squared = 0.64/1.45 = 0.441.

added; we must square them first to obtain the variances. These are in column 3 of the table. The variances may then be added to obtain the value (1.435) in the bottom row. The square root of this value is the overall standard deviation of the warmth sensation (1.199) shown in the bottom row of column 2.

The R-squared value of 0.437 shown in Figure 21.5 is the variance due to regression (0.627) divided by the total variance of the subjective warmth (1.435). It is the proportion of the total variance 'explained' by the regression relation – by the fact that subjective warmth depends partly on the temperature. R-squared is therefore a very informative statistic.

It is interesting to compare the values in Table 21.1 with those that would have arisen had the computer-generated data been precise rather than subject to a sampling variance. The values are shown in Table 21.2. The computer-generated data are very close to the specified 'population' values. Generating a dataset of 5000 'observations' has ensured that the sampling errors in the statistics are small. From Table 21.2, the R-squared value is 0.441, compared with the value of 0.437 from the generated sample. The sample value is within 1 per cent of the population value.

Exploring the regression model

Having constructed our computer-generated thermal comfort database we can explore some of its features.

Selection on the temperature axis

We begin by considering some quite extreme and unrealistic selections of the data, and thereby show clearly their result.

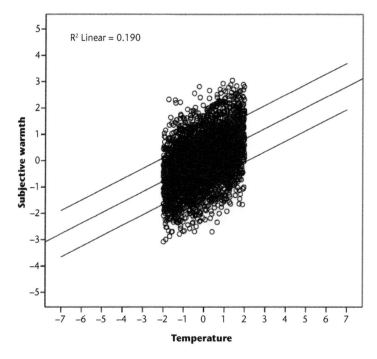

FIGURE 21.6 Scatter plot restricting the value of room temperature (the *x*-variate).

Suppose we restrict the temperatures to ±2 as in Figure 21.6. That is to say, we use only those values that lie within this range. What then happens to the regression statistics?

The regression line is unchanged, as is also the residual standard deviation of the subjective warmth. A little thought will show that this must be so. However, the R-squared value has gone down from 0.437 to 0.190. The regression line now explains a smaller proportion of the total variance.

What happens if we make the opposite selection, and only choose temperatures *outside* the range ±2, as in Figure 21.7?

The regression line and the residual standard deviation remain the same, but the R-squared value has increased to 0.666.

So (as the textbooks tell us) selection of values on the *x*-axis (in our case the temperature) makes no systematic difference to the regression line of *y* on *x* (in our case the subjective warmth on the temperature). Nor does such selection affect the residual standard deviation of *y*. However, the *correlation* in general changes when a selection is made.

That the regression line is unaffected by selection on the *x*-axis is important for the analysis of comfort survey work. It means that the distribution of the room temperature does not affect the regression line, so it does not matter whether its distribution is Normal.

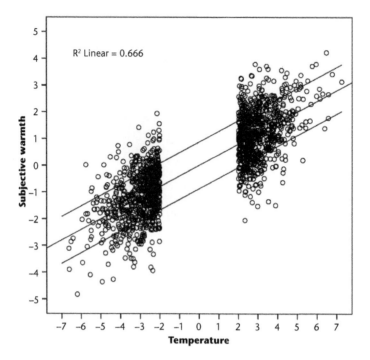

FIGURE 21.7 Scatter plot with the central values of the *x*-variate removed.

We consider an example. The room temperatures drawn from the many different surveys in a large database may have a distribution very different from Normal. This is true of the ASHRAE RP-884 database, because the focus of research interest has been on buildings with controlled indoor temperatures, usually in the low 20°Cs. There have also been field studies in buildings without air conditioning in hot climates, where the indoor temperatures are typically in the low 30°Cs. There is in the database a relative shortage of data from buildings lying between these two groups. This leads to a strange bimodal distribution of room temperature (Figure 21.8). The Normal curve is shown for comparison. The strange distribution of the room temperatures would not invalidate a regression of the comfort votes on the room temperature. It would, however, make us consider whether an overall regression of subjective warmth on room temperature had meaning for these data, because people would be adapted to different temperatures in the different buildings. This would lead us to analyse the data building by building, before looking for overall patterns, as in fact was done by de Dear and his co-workers in their analysis.[2]

Grouping (binning) on the temperature axis

Our exploration so far suggests that binning the temperatures would not affect the regression analysis. There are, however, some small effects that we should be aware

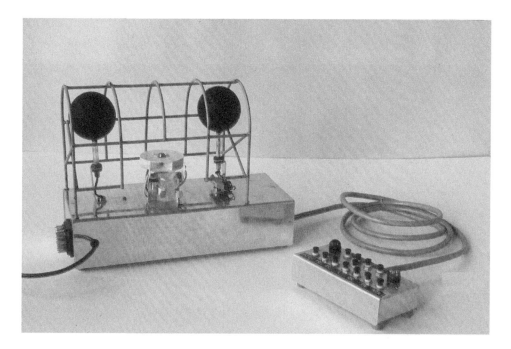

PLATE 3.1 The revised desk instruments, 1968 version. One of the black 50 mm spheres is a warm-body anemometer, the other a simple globe thermometer. The small centrifugal fan blows air over the wet and dry thermo-junctions. The response box has two rows of buttons. The front one has seven buttons representing the Bedford scale. The back row has four for the skin moisture scale. Between the rows is a 'cancel' button to enable a correction to be made. The length of the instrument box was about 240 mm.

Source: BRS photograph.

PLATE 4.1 Primary school children in a Hertfordshire school. The classroom teaching methods were informal and child-centred. At that time (circa 1972) young children had a free bottle of milk (about 150 ml) to drink at school every day.

Source: BRS photograph.

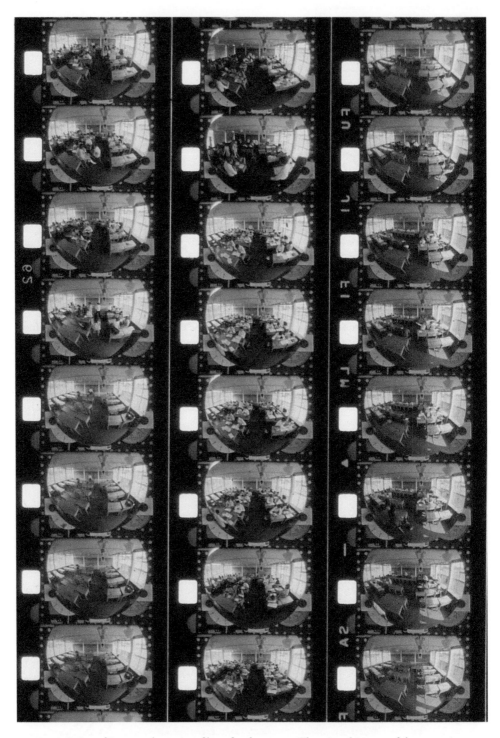

PLATE 4.2 Strips of 8 mm colour cine-film of a classroom. The virtual image of the room is seen in the circular convex mirror. At the bottom of each shot can be seen the time and temperature dials. Shots were at four minute intervals.

Source: M. A. Humphreys.

PLATE 4.3 Enlargement of a single frame of the cine-film. The dial on the left shows the temperature (27°C), on the right the time of day (3 pm). The film has deteriorated over the 40 years.

Source: M. A. Humphreys.

PLATE 4.4 Children completing the response cards.

Source: BRS photograph.

PLATE 11.1 Assembling the handsets, Islamabad, July 1993.

Source: M. A. Humphreys.

PLATE 11.2 The guard outside the Mingora bank, July 1993. He is wearing a shalwar-kamis suit and a Pathan hat.

Source: M. A. Humphreys.

PLATE 11.3 Mingora, July 1993. The Himalayan foothills are visible in the distance.

Source: M. A. Humphreys.

PLATE 11.4 Transporting the handsets by taxi, Islamabad, July 1993.

Source: M. A. Humphreys.

PLATE 11.5 The Brookes research team outside the Raffles Hotel, Murree, July 1993. Clockwise from the left: Sue Roaf, Ollie Sykes, Mike Humphreys, Fergus Nicol, Mary Hancock.

Source: M. A. Humphreys.

PLATE 11.6 Summer clothing in the Mingora bank: a light shalwar-kamis.

Source: M. A. Humphreys.

PLATE 11.7 Winter clothing in the Mingora bank: sleeved under-vest, winter-weight shalwar-kamis, pullover, lined tweed waistcoat and Pathan hat. It was not uncommon to add a scarf (muffler).

Source: M. A. Humphreys.

PLATE 12.1 Group photo: 2004 Windsor Conference.

PLATE 12.2 Group photo: 2014 Windsor Conference.

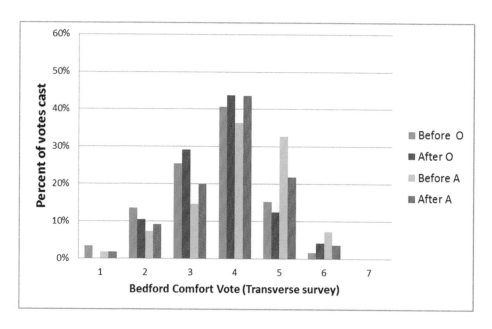

PLATE 16.1 The votes cast in the survey, showing the percentage of votes on the Bedford scale for each sub-building and time period, before and after the implementation of the adaptive algorithm in sub-building A.

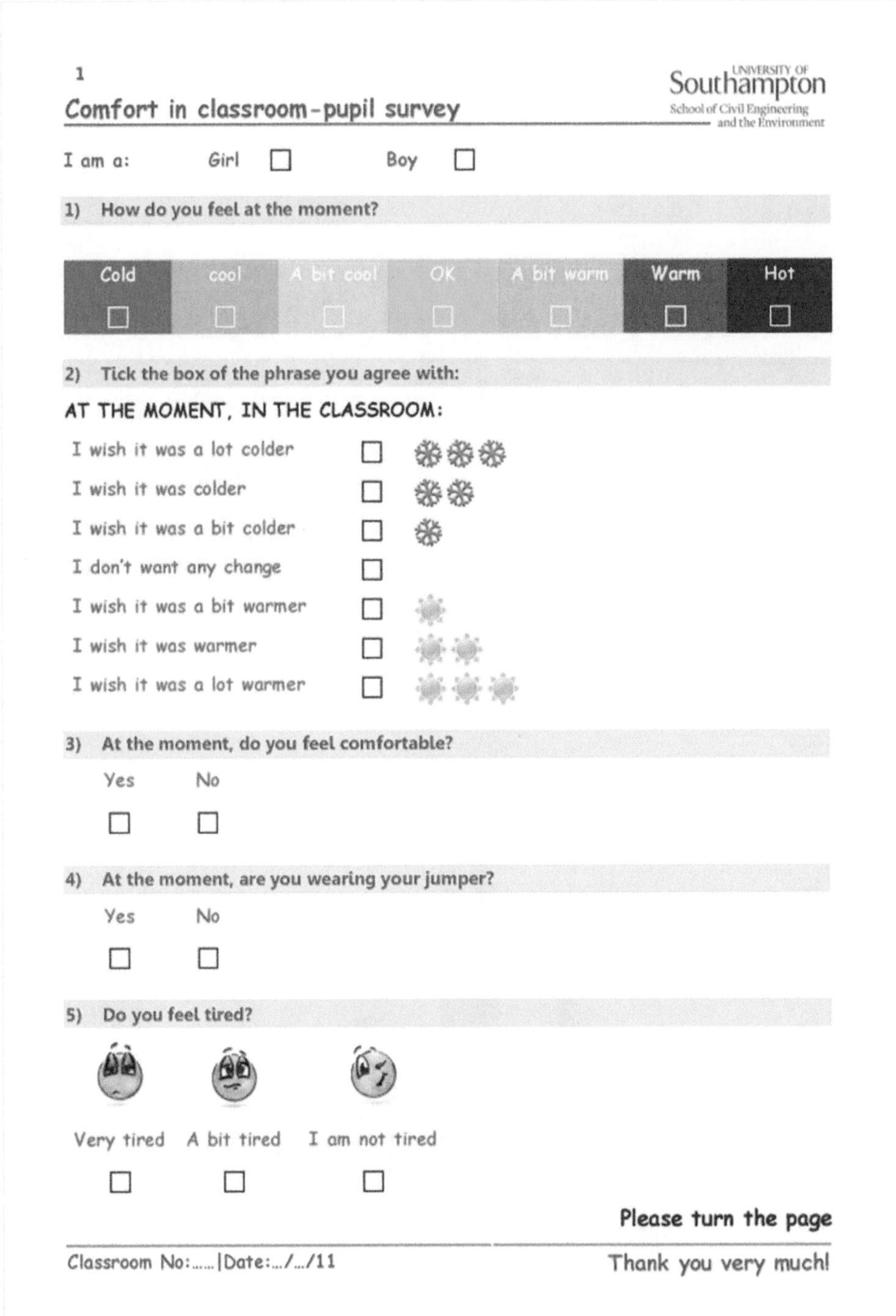

Comfort in classroom–pupil survey

UNIVERSITY OF
Southampton
School of Civil Engineering
and the Environment

I am a: Girl ☐   Boy ☐

1) How do you feel at the moment?

Cold	cool	A bit cool	OK	A bit warm	Warm	Hot
☐	☐	☐	☐	☐	☐	☐

2) Tick the box of the phrase you agree with:

AT THE MOMENT, IN THE CLASSROOM:

I wish it was a lot colder   ☐   ❄❄❄

I wish it was colder   ☐   ❄❄

I wish it was a bit colder   ☐   ❄

I don't want any change   ☐

I wish it was a bit warmer   ☐   ☀

I wish it was warmer   ☐   ☀☀

I wish it was a lot warmer   ☐   ☀☀☀

3) At the moment, do you feel comfortable?

Yes   No

☐   ☐

4) At the moment, are you wearing your jumper?

Yes   No

☐   ☐

5) Do you feel tired?

Very tired   A bit tired   I am not tired

☐   ☐   ☐

Please turn the page

Classroom No:......|Date:.../.../11

Thank you very much!

PLATE 19.1 The thermal questions in Teli *et al.*'s questionnaire.

Source: D. Teli.

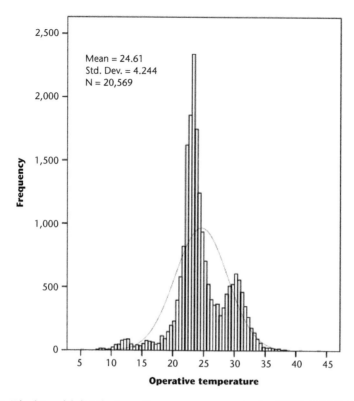

Mean = 24.61
Std. Dev. = 4.244
N = 20,569

FIGURE 21.8 The bimodal distribution of room temperature in the ASHRAE RP-844 database.

of. We first consider what happens to the regression line if the values on the temperature axis are rounded to the nearest degree. This forms groups (bins) of 1 K width in our data (Figure 21.9).

The regression line now has a gradient of 0.391 instead of the true value of 0.400, a 2 per cent fall. The difference is probably not large enough to matter. The R-squared value has reduced slightly from its true value of 0.437 to 0.427 – a similar percentage fall.

We consider the effect of a coarser grouping, allocating the data to groups (bins) of width 2 K (Figure 21.10). The regression gradient has fallen to 0.369, considerably below its true value of 0.400, and the R-squared value has again fallen (0.403). We conclude that this grouping is too coarse and has appreciably affected the regression gradient.

It is not obvious why binning the temperatures should affect the regression line at all, since it seems to be similar to selection on that axis. The explanation is that the temperature we allocate to a bin, the bin-centre temperature, does not in general equal the mean temperature of the data within the bin. Away from the centre of the distribution, there will be more observations in the half of the bin nearer to the centre of the distribution than in the half further from it. For example, the mean

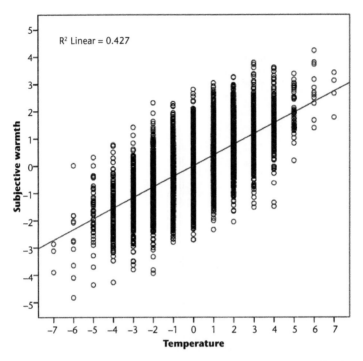

FIGURE 21.9 Effect of grouping (binning) the values of the temperatures (1 K bins).

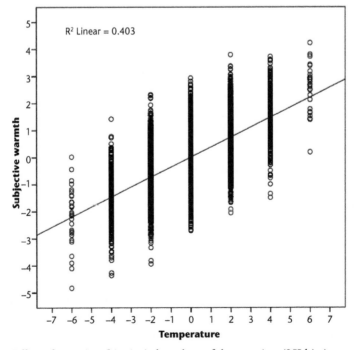

FIGURE 21.10 Effect of grouping (binning) the values of the x-variate (2 K bins).

temperature within the bin centred on +4 K is not 4 but 3.69 K. Replacing the bin-centre temperatures by the true mean temperatures in the bins restores the regression coefficient to its original value of 0.400. So provided that this procedure is followed, binning the temperatures does not alter the regression gradient.

It is not unusual to replace the individual values of the subjective warmth in each bin by the bin-mean value. Figure 21.11 shows the scatter plot after this has been done.

The regression line is unchanged, provided that the regression is weighted according to the numbers of observations in each of the bins, and so the gradient is the same as in Figure 21.10 (0.369). But the R-squared has increased to 0.999. It no longer indicates the correlation of the individual observations of the subjective warmth with the temperature, but of the bin-means of these variables. Its value is therefore much higher, and depends on the binning interval and on the number of observations in the dataset. Because of its systematic dependence on the number of observations, an R-squared based on bin-mean values is hardly a respectable statistic. Using the bin-means of the subjective warmth suppresses almost all the variation about the regression line, and is to that extent misleading.

It is certain that, in the analysis of thermal comfort field data, binning is used far more than is necessary or desirable. Originally the purpose of binning was to reduce the labour of statistical calculation. This was an important consideration before computers became available for statistical calculations.

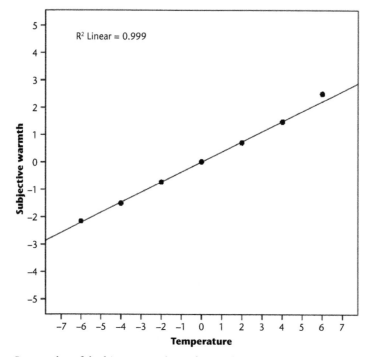

FIGURE 21.11 Scatter plot of the bin-mean values of warmth and temperature.

Nevertheless there are legitimate uses of binning. They can give a good visual portrayal of trends in the data, and are useful in exploratory analysis. Inspection of bin-means can be used to discover whether a linear fit is satisfactory, or whether a more complex model is needed. Because of these features, bins are also useful in the presentation of results, but care must be taken not to mislead either the researcher or the readers by hiding the extent of variation in the data.

Selection on the subjective warmth axis

We repeat the procedure, selecting on the subjective warmth axis instead of on the temperature axis. First we select only values that lie within one standard deviation of the overall distribution of the subjective warmth. The result is shown in Figure 21.12.

Curtailing the values of y has changed the regression line. Its gradient has fallen from 0.40 to 0.17, less than half its true value. The R-squared is low, too, at 0.185. So restricting the y-values has had a profound effect on the regression line.

Now we consider the case of the excluded middle, shown in Figure 21.13. The regression coefficient has increased to 0.60, considerably higher than its true value, and the R-squared has increased to 0.656.

Selection on the subjective warmth axis has again seriously affected the regression line. So, for example, one cannot leave out all the votes of 'neutral' in an analysis without risk of seriously biasing the regression coefficient. Selection on the warmth axis also changes the correlations.

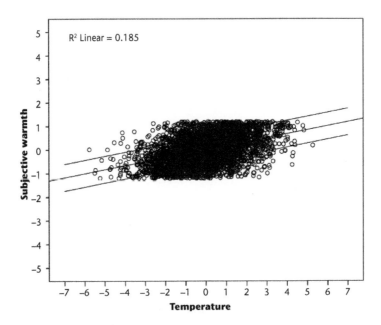

FIGURE 21.12 Restricting the subjective warmth axis (the y-variate).

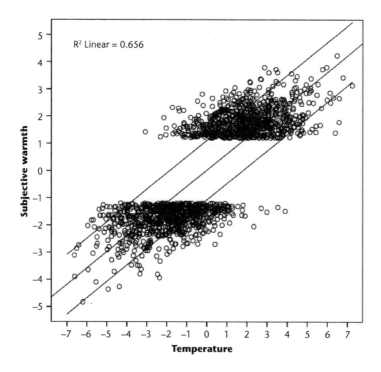

FIGURE 21.13 Scatter plot excluding the central values of the subjective warmth.

Grouping (binning) on the warmth axis

If selection on the warmth axis distorts the regression result, we might wonder if the same is true of grouping (binning) the data. In our computer-generated data, we have assumed that subjective warmth is a continuous variable – a common and reasonable assumption. But, in practice, most researchers use a scale with discrete categories, such as the ASHRAE scale or the Bedford scale. Some researchers use a continuous version of the ASHRAE scale, but even then most respondents place their marks only at the labelled points on the scale. So in survey-data we usually have seven discrete values rather than a continuum. Does this affect the regression line? We can check this by rounding our values to the nearest whole number and re-running the regression, as shown in Figure 21.14.

The value of the regression coefficient remains correct at 0.400, so this grouping of the points has not affected the regression line. The grouping does, however, produce other small effects. The R–squared value has fallen to 0.414, compared with the true value of 0.437 shown in Figure 21.5. The residual standard deviation of the subjective warmth values about the regression line has risen from 0.90 to 0.94. The overall variation of the subjective warmth has altered slightly too, increasing from a standard deviation of 1.20 to a value of 1.23.

Grouping or binning data slightly increases the variance of a set of data-points. It adds to the variance a quantity $(h^2/12)$, where h is the grouping interval, unity in our

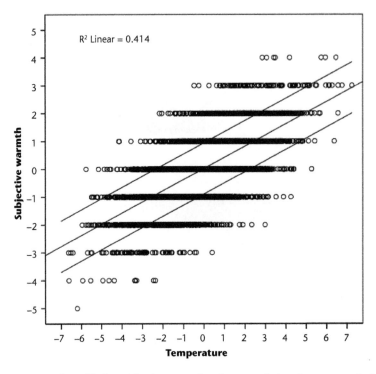

FIGURE 21.14 Scatter plot with the subjective warmth values rounded to the nearest whole number.

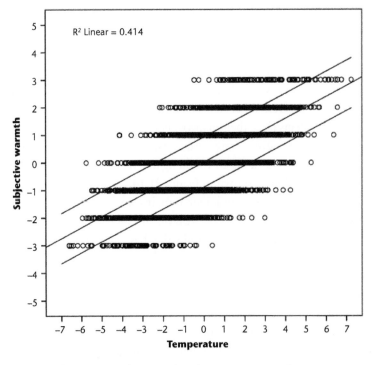

FIGURE 21.15 Effect of contracting the warmth scale to seven categories.

example. This quantity is known as Sheppard's correction. Remembering that the variance is the square of the standard deviation, we have:

Variance before grouping	1.435
Variance after grouping	1.517
Increase in variance	0.082
Sheppard's correction ($h^2/12$)	0.083

Sheppard's correction, while closely estimating the increase in the variance, does not make an important difference, being only 2.5 per cent of the total variance. So it seems that the grouping we have used is acceptable. It is not too coarse.

There is a further question to be considered. The ASHRAE scale has seven categories, ranging from −3 through to +3. Figure 21.14 shows that there are some votes of +4, some of −4 and one vote of −5. In practice in a thermal comfort field study, these votes would have been included in the categories +3 or −3. What happens to the regression line if we assign these numbers to the extreme values, so reducing our range to that of the ASHRAE scale? Such a re-allocation of the data changes the distribution on the subjective warmth axis, and would therefore be expected to affect the regression line.

Figure 21.15 shows the data and regression line after the re-allocation. The regression gradient has been reduced by this change. However, the difference is very small. It has fallen from its true value of 0.400 to a value of 0.395, a reduction of 1.3 per cent. The change is of no practical importance, so our result is not seriously biased by restricting the scale to seven categories. This is because of the very small proportion of votes (0.4 per cent) that lay outside the range ±3.

The effect can be further explored by supposing we had available in the survey only the five central scale values (cool, slightly cool, neutral, slightly warm, warm). All the values beyond this range would have been assigned either to the category 'cool' or to the category 'warm'. We make this change to our database and see what happens to the regression line (Figure 21.16).

The regression gradient has now fallen to 0.365, only about 90 per cent of its true value of 0.400. The bias in estimation is now becoming serious. It follows that the regression gradient will be biased downward if a considerable proportion of the votes fall in the extreme categories of the warmth scale. This shows the need for the categories of the verbal scale to cover fully the range of the respondents' subjective experiences. So, if the expected temperature range is large, it may be appropriate to use an extended version of the ASHRAE scale that adds the categories 'very hot' and 'very cold'. There is, however, a risk that the added categories will change the way the respondents interpret the existing categories.

Concluding note

In this chapter we have illustrated how a scatter plot of subjective warmth against room temperature is built up from its component parts: the distribution of the room

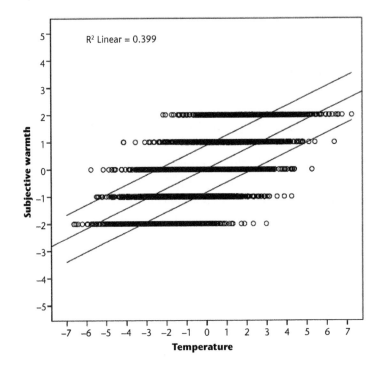

FIGURE 21.16 Effect of reducing the warmth scale to five categories.

temperature, the interpersonal dispersion of response to the room temperature, and the dependence of the subjective warmth on the room temperature (the regression coefficient). We have noticed that the regression coefficient is not affected by selection on the temperature axis (x). Selection does however markedly alter the correlation coefficient. Selection on the subjective warmth axis (y) alters the regression coefficient as well as the correlation coefficient.

We have also examined the effect of grouping (binning) the temperature data. This commonly used procedure is unnecessary, and if the binning is too coarse it alters the regression coefficient and the correlation coefficient.

It is acceptable to use a seven–category scale of subjective warmth to represent the underlying continuous variable, but trimming the scale to five categories would be likely to cause a substantial bias in the estimate of the regression coefficient.

Notes

1 Humphreys, M. A. *et al.* (2013) Updating the adaptive relation between climate and comfort indoors: new insights and an extended database, *Building and Environment* 63, 40–55, http://dx.doi.org/10.1016/j.buildenv.2013.01.024.
2 de Dear, R. *et al.* (1997) Developing an adaptive model of thermal comfort and preference, *Final Report ASHRAE on RP-884*, Macquarie Research Ltd, Sydney.

22

THE EFFECTS OF ERROR IN THE PREDICTOR VARIABLE

Introduction

Regression analysis is often used to predict the mean subjective warmth from the measurements of the thermal environment. In such a case, the thermal environment is the *predictor variable* and is by convention placed on the horizontal axis of a scatter plot. Again by convention, this axis is referred to as the x-axis, and the vertical as the y-axis. In this chapter, we explore some of the effects on regression analysis of the presence of error in the measurement of the thermal environment.

The simple regression model explored in the previous chapter applies when the values of the predictor variable, x, are free from error. Any error in x reduces the estimate of the regression coefficient. The reduction is small if the error is small compared with the range of x. The topic is not explained in the statistical textbooks which thermal environment researchers are likely to consult. In this chapter, we give a brief and largely visual account of the effects of the presence of error in the predictor variable. Some readers will need a more rigorous and complete account, and for them we suggest the book by Cheng and Van Ness.[1] It does, however, assume more facility with statistical theory and notation than most researchers in the architectural and building sciences have been able to acquire.

The history of the treatment of errors is convoluted. J. W. Gillard has given a good historical overview of the topic.[2] The current article in Wikipedia is also useful.[3] The usual advice given to the researcher to avoid the problems caused by error in the predictor variable is to increase the accuracy of measurement or increase its range. Increasing the range is not usually an option in thermal comfort fieldwork because the researcher does not control the thermal environment. This is not a matter only of practicality, but also of policy. The adaptive approach investigates people's responses to their normal conditions, so we do not wish to alter the room temperatures we encounter during a survey. To do so would risk changing the pattern of our respondents' subjective assessments, because putting people in unusual environments leads to unusual environmental behaviour.

The accuracy of the thermometry can of course be improved, but even were the thermometer perfectly accurate a problem would remain. The ideal would be to measure the temperature (and other environmental variables) at the place where the respondent is located and at the same time as she gives her responses to her environment. That is impossible, so either we must measure the temperature at a place nearby, so introducing an error because of the different location, or we must move the respondent away and measure the environment where she was sitting, so introducing an error caused by the displacement in time.

The problem may seem to be trivial, but in some common circumstances it is not. The temperature variation during a survey is often small, especially in well-constructed and well-controlled buildings. For example, analysis of the database from the SCATs project found that, during a working day, a typical standard deviation of room temperature in an office building was only about 0.8 K, while the estimated error in its measurement (a location error) had an estimated standard deviation of about 0.4 K.[4] The error was therefore a substantial proportion of the range.

Illustrating and explaining the effect

The effect can be modelled using the computer-generated dataset we constructed for the previous chapter. We add a new column to represent the error, drawing values from a Normal distribution with zero mean and standard deviation of 1.0. The computer-generated sample has a standard deviation of 1.015 K, quite close to the specified population value. We add these 'errors' to the original set of values of temperature (the x-values) to produce a column of values of temperature that includes an error term.

Consider again the figure showing the mean values of the subjective warmth predicted from the temperature, shown here as Figure 22.1. Each point on the line is the predicted mean warmth sensation for a particular temperature. The R-squared value is unity, for the observations are (in this ideal example) perfectly related.

To each point on the line we now add its temperature error. The y-value of each point remains the same, but its x-value now has its error attached to it. The point will move to the left or to the right depending on whether the value of its error term happens to be positive or negative. The effect is shown in Figure 22.2.

When y is regressed on x, the regression calculation procedure attributes all the errors to the y-variate, and therefore the calculated regression line (the solid line on the figure) does not truly represent the relation between the subjective warmth and the actual temperature. Its gradient is 0.318 compared with the true value of 0.400, a considerable reduction. The dashed line on the figure shows the true relation (before

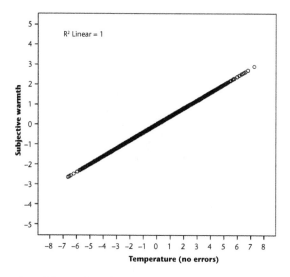

FIGURE 22.1 Scatter plot of the predicted warmth and the temperature (without error).

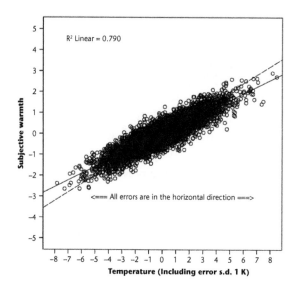

FIGURE 22.2 Scatter plot showing an error (s.d. = 1.015 K) introduced to the temperatures (x-values).

adding the error). The R-squared also has changed, falling from unity to 0.790. So introducing an error to the temperature has considerably reduced both the regression gradient and the estimate of the correlation.[5]

We now see what happens to the regression line if we use the individual values of the subjective warmth instead of its predicted mean values. The process introduces variation in the *vertical* direction. The scatter plot (Figure 22.3) is now more realistic.

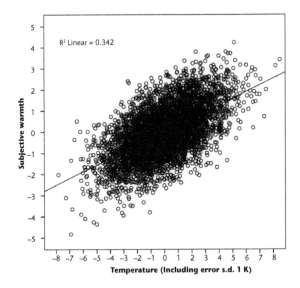

FIGURE 22.3 Scatter plot of thermal sensation (*y*) against temperature with an error (s.d. = 1.015 K) introduced to the temperature (*x*).

Adding scatter to the *y*-variate does not systematically alter the regression of *y* on *x*, as we noted in the previous chapter. The regression gradient therefore is unchanged from that in Figure 22.2, apart from a very small sampling variation (a gradient of 0.316, compared with 0.318). The R-squared is lower, at 0.342. This is because of the variation added to the subjective warmth.

Notice too that it is substantially lower than it was before the addition of error to the *x*-variate (0.437 – see Figure 21.5). The residual variation after regression is therefore correspondingly increased by the presence of the error in the temperature.

Adjusting a regression line for measurement error

It can be shown that the statistical estimate of the gradient, *b*, of a regression line is the ratio of the covariance[6] of the values of *x* and *y* to the variance of the values of *x*. The proof is given in statistical textbooks.

$$b = \text{covariance}(x,y)/\text{variance}(x)$$

In the presence of error (*e*) in the *x*-variate, the regression coefficient (b_{err}) becomes:

$$b_{err} = \text{covariance}\{ (x + e),y\}/\{\text{variance}(x) + \text{variance}(e) \}$$

Provided *e* is not correlated with *y*, the covariance$\{ (x + e),y\}$ is equal to the covariance (*x*,*y*), because the error is equally likely to be positive or negative and has zero mean. So the ratio of (b_{err}) to (*b*) becomes:

$$b_{err}/b = \{variance(x)\}/\{variance(x) + variance(e)\}$$

therefore:

$$b_{err} = b\{variance(x)\}/\{variance(x) + variance(e)\}$$

We can check this by reference to either Figure 22.2 or Figure 22.3. The true regression coefficient before introducing error was 0.400. The standard deviation of x is $2\,K$. Its variance is the square of this: $4\,K^2$. The error in the x-variate had a standard deviation of $1.015\,K$ and therefore a variance of $1.030\,K^2$. Substituting these values we obtain:

$$b_{err} = 0.400*4/(4 + 1.030) = 0.318$$

in agreement with the value we obtained above.

Usually we do not know the error-free values of the temperature, but only its values including the error. To obtain an estimate of what the regression coefficient would have been without the presence of error – the adjusted value (b_{adj}) for the regression coefficient – we re-write the equation as:

$$b_{adj} = b_{err}(variance(x_{meas})/\{variance(x_{meas}) - variance(e)\}$$

The room temperature as measured during the survey, including any error, is x_{meas}. The equation allows us to check how sensitive the estimate of the regression gradient is to some estimated value of its error, and if necessary adjust it accordingly.

Adjusting an R-squared value for measurement error

Similar logic applies to the R-squared value. R-squared is defined by the relation:

$$R\text{-squared} = (covariance(x,y))^2/\{(variance(x))(variance(y))\}$$

In the presence of error (e) in the x-variate, the R-squared value (R-squared$_{err}$) becomes:

$$R\text{-squared}_{err} = (covariance((x + e),y))^2/\{(variance(x + e))(variance(y))\}$$

and therefore, following the same method as with the regression coefficients, R-squared adjusted to remove the effect of the error in the predictor (R-squared$_{adj}$) becomes:[7]

$$R\text{-squared}_{adj} = \{R\text{-squared}_{err}(variance(x_{meas}))\}/\{variance(x_{meas}) - variance(e)\}$$

Adjusting the residual variation

In a later chapter we shall need to quantify the inflation of the residual variation that is caused by the presence of error in the predictor variable, and here we show how it may be estimated. The residual variance after regression (variance(y_{resid})) can be calculated from the variance in y and the regression coefficient. So before the introduction of error in the predictor we may write:

$$\text{variance}(y_{resid}) = \text{variance}(y) - b^2\{\text{variance}(x)\}$$

After the introduction of error in the predictor, the expression becomes

$$\text{variance}_{err}(y_{resid}) = \text{variance}(y) - b_{err}^2\{\text{variance}(x) + \text{variance}(e)\}$$

The variance of y is of course unchanged by introducing error in the predictor. The increase in the residual variance is given by subtracting the first expression from the second. So the inflation in the residual variance is:

$$b^2\{\text{variance}(x)\} - b_{err}^2\{\text{variance}(x) + \text{variance}(e)\}$$

Example1. Application to evaluating the sensitivity to temperature during the working day

A worked example of the correction for the presence of error in the predictor variable is given in Chapter 25, where the sensitivity of a population to change of room temperature during the working day is derived from multiple sources. The correction was needed because the variance of the room temperature was quite small, and the error in the temperature measurement could therefore not be ignored. The error in the temperature is not only the error attributable to the thermometer itself, but also that arising from its placing relative to the respondent, for the thermometer cannot be in the same location as the respondent at the time the measurement is made.

Example 2. Application to estimating the practical error in an estimate of PMV

It may occasionally happen that we have a known theoretical value for the true gradient, and wish to calculate from the apparent regression gradient the approximate size of the measurement errors in the predictor variate.

Figure 22.4 is a scatter plot of the values of PMV (the predictor variable) calculated for each of the 20,000 entries, against the corresponding actual votes recorded by the respondents. The values are from the ASHRAE RP-884 database. (The scatter plot looks unusual because some researchers used the seven-category version of the scale, while others allowed votes intermediate between the categories.)

Each of the 20,000 estimates of PMV has an error arising from the following sources:

- measurement of the environmental variables (air temperature, mean radiant temperature, air speed and humidity);
- the use of checklists to estimate the clothing insulation and the metabolic rate;
- location error or time displacement error, as noted above;
- equation error (error attributable to approximations made in the construction of the PMV equation).

All these combine to give an unknown practical error in the estimate of each individual value of PMV.

We have not previously mentioned 'equation error'. It arises because the PMV equation is a surrogate for a hypothetically perfect index for predicting the subjective warmth, as we noticed in Chapter 20. Such an index would be perfectly correlated with the mean thermal sensations of the population. Like all indices that combine a number of parameters, the approximations inherent in the formulation of PMV cause its values to be less than perfectly correlated with the values of the perfect index, even were the input data free from error. The discrepancy between the perfect index and the actual index is the equation error. Because the discrepancy is influenced by several variables, its distribution is likely to tend to Normal as the number of observations increases.

The solid line on Figure 22.4 is the regression-line of the actual votes on the ASHRAE scale, against the predicted votes (the calculated PMV values). The regression gradient is depressed by the presence of the error in the PMV estimates. In our terminology its gradient is b_{err}. The true (theoretical) gradient of the line is also known, because the mean of the actual responses ought by definition to equal PMV (the predicted mean vote). This is the dashed line on the figure, and has in our terminology gradient b_{adj}.

Let us suppose that the depression of the gradient is entirely attributable to the presence of error in the predictor. How big is this error? We can use the equation:

$$b_{adj} = b_{err}(\text{variance}(x_{meas})/\{\text{variance}(x_{meas}) - \text{variance}(e)\}$$

Rearranging the equation we have:

$$\text{variance}(e) = \text{variance}(x_{meas})(1 - b_{err}/b_{adj})$$

The standard deviation of the values of PMV, including its error, is 1.076 scale units. Its variance is therefore 1.158. The regression gradient (b_{err}) is 0.533. The true value (b_{adj}) is 1.00. From this we calculate the variance of the error to be 0.541 units on the PMV scale. The standard deviation of the error is therefore 0.735 units of the PMV scale. Thus the practical error associated with each individual estimate of PMV is some 0.74 of a unit, and leads to a large reduction in the gradient of the regression line.

We can also estimate what the R-squared value would have been in the absence of error in the predictor. The R-squared of the scatter plot in Figure 22.4 is 0.217, so the corrected value, R-squared$_{adj}$, equals:

$$R\text{-squared}_{err}(\text{variance}(x_{meas}))/\{\text{variance}(x_{meas}) - \text{variance}(e)\}$$
$$= 0.217(1.158)/(1.158 - 0.541)$$
$$= 0.407$$

We conclude that the error inherent in the estimate of a value of PMV can be a considerable nuisance when PMV is used as a predictor variable.

Notice that this example assumes that people use the ASHRAE scale in the same way in the laboratory as they do in normal life. There is evidence that people are slightly more sensitive to room temperature changes in daily life, as shown in Chapter 25. If this were so, it would imply that still larger errors were present in the formulation and use of the PMV index.

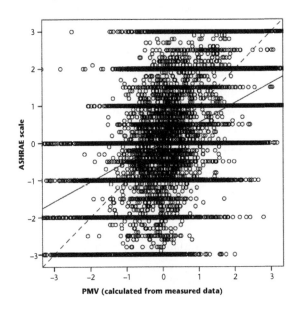

FIGURE 22.4 Scatter plot of the actual comfort votes in the ASHRAE RP-884 database and their measured PMV values (R-squared = 0.217).

Deciding whether an adjustment for error in the predictor is needed

Figure 22.5 shows how the ratio of the apparent gradient to the true gradient depends upon the ratio of the error standard deviation in the x-variate to the total standard deviation of the x-variate. It is useful when considering the accuracy that is likely to be needed in the predictor variable if the regression is to be substantially unaffected, and to show the seriousness of the effect on the regression gradient if the error in the predictor is unavoidable. It can be seen that if the error standard deviation is more than about 20 per cent of the total, the depression of the regression gradient becomes serious.

The result of example 2 can be located in Figure 22.5. On the horizontal axis, the ratio of the standard deviations is $0.735/1.076 = 0.68$. This leads to a value of 0.53 on the vertical axis, as shown in Figure 22.5.

Confidence intervals on the regression coefficient

The estimation of confidence limits on a regression line is affected by the presence of error in the predictor. This can be seen by considering an extreme case.

Suppose the x-variate is nothing but error; that is, there is no true variation in the predictor. If x is just error, the scatter plot can contain no information about the true regression gradient, so it could take any value. But, because the regression program assigns the error to the y-variate, it will calculate a regression line and attribute to it a gradient of zero. It will also calculate its confidence intervals, and, if there are numerous observations, the calculated confidence interval

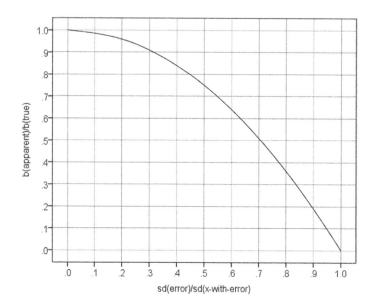

FIGURE 22.5 Effect of error in the predictor variable on the regression coefficient.

will be small. However, the calculated regression gradient and its confidence limits are meaningless.

The extreme example shows that the presence of error in the x-variate not only biases the regression gradient, but also upsets the estimation of its confidence limits. Thus a regression that appears to be estimated with considerable accuracy may not in fact be so. The behaviour of the confidence limits in the presence of error in the x-variate is complex, and no simple formula describes it. Some statistical packages now provide facilities for estimating regression gradients when there is error in the predictor variable, and the output includes the corrected estimate of the standard error of the regression coefficient, from which confidence intervals can be derived.[8]

Multivariate regression with measurement errors

The effect of errors in the predictor variables in multivariate regression (that is, when there are two or more predictor variables) is complex. The effects are not easy to foresee, especially if the predictor variables are correlated, as they usually are in thermal comfort field data. With a single predictor, the presence of error always depresses the regression coefficient, but this is not necessarily so in multiple regression.

While it is possible to measure the temperatures and the humidity with reasonable accuracy, the air speed presents more difficulty. Individual estimates of the clothing insulation and of the metabolic rate are probably accurate to no better than 20 per cent. If the error in a variable is proportional to its range, then no increase in its range will avoid the problem of error in that variable.

So if we were using multivariate regression to predict the subjective thermal response from the six principal thermal variables (air temperature, radiant temperature, humidity, air speed, clothing and metabolic rate) the prediction equation would be likely to be unreliable, and the effects of the errors difficult to disentangle.

Multivariate regression has often been used to derive multivariate indices of warmth, beginning with Thomas Bedford's pioneering attempt in the 1930s. The results have not been consistently reliable, and this method of obtaining indices of warmth has largely fallen into disuse. The sometimes strange behaviour of multivariate regression with errors in the predictors largely explains the difficulty that has been found in obtaining repeatable multivariate regression equations from different sets of data, where the errors and ranges are likely to be different.

In the face of this sometimes complex effect of errors in multiple regression, what can be done? An empirical method of checking for the effects of errors in the variables can be used. To do this we add to each variable a suitably chosen error term, and examine its effect on the regression coefficient. This can be done with each variable in turn, and then with all of them together. This will give a good 'feel' for the sensitivity of the model to error in the variables, and in propitious circumstances a corrected equation can be estimated.

Comment on binning the predictor variate

In the previous chapter, we showed that grouping or binning the data on the temperature (x) axis did not affect the regression gradient, provided the true bin–mean temperatures were used. Yet the binning process seemed to introduce error to the variate, so at first glance we would have expected the gradient to be reduced. The reason that it does not do so is as follows.

In the examples we have given above, the error in the predictor (x) was independent of the value of the predicted variate (y). This is not so when values of the predictor variable are grouped (binned). The 'error' introduced by the binning process is entirely correlated with the y-variate and consequently has no effect on the regression gradient. That is to say, had the value of x really been moved to the bin-mean value, the value of y would have moved correspondingly, the point 'sliding' along the regression line. No such change in y is associated with a genuine error in measuring x.[9]

We close this chapter on a note of caution. The handling of errors in the predictors is complex and it is easy to make mistakes. The identification of the error structure of a set of data requires careful and accurate thought. Modelling the behaviour of hypothetical errors by computer-simulation using a statistical package is an invaluable aid to thought, as too is the old-fashioned use of pencil-and-paper diagrams.

Notes

1 Cheng, C. L. and Van Ness, J. W. (1999) Statistical regression with measurement error, *Kendall's library of statistics*, Volume 6, Arnold, London.
2 Gillard, J. W. (2006) *An historical overview of linear regression with errors in both variables*. Technical Report, School of Mathematics, Cardiff University.
3 en.wikipedia.org/wiki/Errors-in-variables_models (accessed on 12 March 2013).
4 Humphreys, M. A. *et al.* (2013) Updating the adaptive relation between climate and comfort indoors; new insights and an extended database, *Building and Environment* 63, 40–55, http://dx.doi.org/10.1016/j.buildenv.2013.01.024.
5 In this particular example, where we have no residual variation of y, we could have obtained the original line by regressing x on y, for this regression would have assigned the errors to the temperature.
6 The sample covariance of x and y is $1/(n-1)*\Sigma(xy)$, where n is the number of observations and Σ the symbol for summation over all the observations. The variance of x is $1/(n-1)*\Sigma(x^2)$.
7 This R-squared$_{adj}$ should not be confused with the 'adjusted R-squared', where the value of R-squared in multivariate regression is adjusted to correct for the positive bias that is otherwise present.
8 The statistical package SAS offers this facility.
9 The matter is explained by Cheng and Van Ness (Op. cit.) when considering the 'Berkson model'.

23

MUTUAL RELATION BETWEEN ROOM TEMPERATURE AND SUBJECTIVE WARMTH

When we consider the relation between subjective warmth and room temperature, it is obvious that the room temperature, via the clothing insulation and via the characteristics of the body and the mind, causes the thermal sensation. This is a fairly straightforward example of the thermal environment being the predictor variable (or the 'independent' variable) when regression analysis is used, and therefore the room temperature can be used to predict the thermal sensation. Yet when people have easy control over their room temperature, a reverse path of causation is possible. If they feel too warm, they may turn the heating down, while if they feel too cool they may turn it up a bit. They adjust the room temperature to ensure that their warmth sensation is the one they desire at the time. The subjective warmth could then be regarded as *causing the room temperature*, and, if so, the subjective warmth becomes the predictor variable and the room temperature the dependent variable.

In this chapter, we are not so much concerned with establishing the actual paths of causation in any particular case as in considering the consequences for univariate regression analysis were the thermal sensation to 'cause' the room temperature. Again we use the computer-generated dataset of previous chapters, but look at the scatter diagram from a different perspective. We use for our example the data relating room temperature to subjective warmth (without the measurement error added to the temperature in the previous chapter).

If we assume that the room temperature causes the thermal sensation, we use the regression of the thermal sensation on the room temperature (the solid line on Figure 23.1). It gives the mean thermal sensation for any value of the room temperature. However, if the thermal sensation 'causes' the room temperature, then we would need the regression of room temperature on the thermal sensation. It would answer the question: granted that the thermal sensation has this value, what is the most likely room temperature? This regression equation has a steeper gradient (the dashed line).

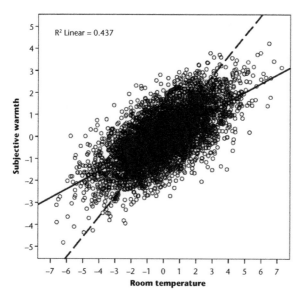

FIGURE 23.1 Scatter plot showing the two regression lines. The solid line is the regression of the subjective warmth on the temperature, the dashed line the regression of temperature upon the subjective warmth.

That is a reason why people often believe they are more sensitive to room temperature than they actually are. They tend to check the room temperature only when they feel too warm or too cold, and so they come to regard the temperatures of the steeper line as those causing warm and cold discomfort. They are not aware of all those occasions when they were perfectly comfortable at those same temperatures.

These two regressions are the extremes, where the causation is entirely one way or entirely the other way. If the causation is mutual, being partly one way and partly the other, then the line best representing the relation between room temperature and thermal sensation would lie somewhere between the two regression lines.

In terms of the standard regression model, the regression of y upon x allocates all the 'error' to the y-variate while the regression of x on y allocates all the 'error' to the x-variate. It would be better in this context to think of 'uncertainty' in the predicted value of x rather than 'error' – the values of x could be known with high precision.

Let us call the subjective warmth w and the room temperature t. The two regression lines cross at the joint means of t and w, which in our case is zero (ignoring the very small sampling variance in our data). The output from regression analysis gives us two equations.

The regression of w on t gives us the equation:

$$w = 0.400t$$

The equation gives us the estimate of the mean value of w for any particular value of t.

The regression of t on w gives the equation:

$$t = 1.093w$$

It gives the estimate of the mean value of t for any value of w. Rearranging this equation to make w its subject, it becomes

$$w = 0.915t$$

It can be shown that the ratio of the gradient in the first to that in the third equations is equal to the R-squared value. Thus in our example:

$$\text{R-squared} = 0.437 = (0.400/0.915)$$

The higher the value of R-squared, the closer the lines would be together, until, if R-squared is unity, the two lines coincide.

The geometric mean of the two gradients is equal to the ratio of the standard deviations of y and x. (The geometric mean of two quantities is the square root of their product.) So in our example:

$$\text{Sqrt}(0.400*0.915) = 0.605$$

Taking the values of the standard deviations from Chapter 21, the standard deviation of the thermal sensation was 1.199 scale units, and the standard deviation of the temperature 1.981 K:

$$\text{s.d.}(y)/\text{s.d.}(x) = 1.199/1.981 = 0.605$$

So the line whose gradient is the geometric mean of the two lines in Figure 23.1 is very simple to calculate. Its gradient is simply the ratio of the standard deviations of y and x.

We can superimpose on the data an ellipse, such that each point on the ellipse has equal probability density – that is to say, an observation is equally likely to occur at any place on the ellipse. The ellipse drawn in Figure 23.2 is at two standard deviations of the subjective warmth and of the room temperature. (We have reduced the size of the points to make the ellipse clearly visible.) The probability density at any point on this ellipse is 0.054, which is the ordinate of a Normal distribution at two standard deviations from the mean, as given in statistical tables of the ordinates of the Normal distribution. An ellipse drawn at one standard deviation of the joint distribution would have a probability density of 0.242, the ordinate at one standard deviation from the mean. Any number of such ellipses can be drawn, and so provide a 'contour map' of the probability density of a joint Normal distribution.

If R-squared were unity, the ellipse would collapse into a straight line whose gradient would be equal to the ratio of the standard deviation of y to the standard deviation of x, while if the correlation were zero the minor and major axes of the ellipse would lie along the vertical and horizontal axes of the scatter plot.

Figure 23.3 shows the relation between such an ellipse and the three lines. We see that the regression of y on x passes through the points where the vertical tangents touch the ellipse, while the regression of x on y passes through the points where the horizontal tangents touch the ellipse. The third (dotted-and-dashed) line has a gradient equal to the geometric mean of the two regression gradients, or the ratio of the standard deviations of y and x.

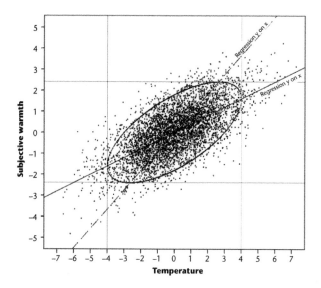

FIGURE 23.2 An elliptical 'contour line' of equal probability density on the joint distribution.

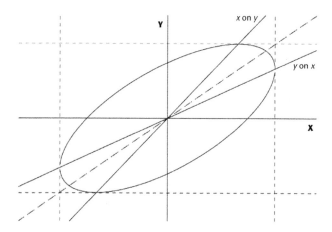

FIGURE 23.3 The relation between the two regression lines, the geometric mean line and an ellipse of equal probability.

Such a line might better represent the relation between the room temperature and the subjective warmth than would either of the two regression lines, in cases where the causation seems to be in both directions at once. An example might be a survey of thermal comfort in dwellings. People at home generally have control over their indoor temperature, so we are uncertain whether the temperature causes the thermal sensation, or the thermal sensation causes the temperature. In the absence of more precise information, neither regression line seems to have the logical priority, so the third (dashed) line might be chosen to represent the underlying relation between the subjective warmth and the temperature.

A word of caution is necessary. When the correlation between the variables is low, the gradient of the third line becomes very uncertain. Consider the extreme condition, where the correlation between the two variables is zero. In this case the third line has no definite location, although the ratio of the standard deviations of x and y remains definite. The geometric mean of the two regression gradients becomes indefinite, because it entails the product of infinity and zero [sqrt(∞*0)]. The practical implication is that the geometric mean line is ill defined if R-squared is very low.

When using the geometric mean to represent a relation, it is therefore not sufficient to calculate only the ratio of the standard deviations of the variables. It is always necessary also to calculate the R-squared value, or values of the two regression gradients, so as to obtain an indication of the strength of the relation. Showing the ellipse on the scatter plot can also help the visualisation of the relation between the two variables.

Examples

When inspecting the statistics from the first data-logging thermal comfort survey at BRS in the 1960s, Humphreys and Nicol were puzzled to find that the regression gradient of subjective warmth on room temperature depended on the standard deviation of the temperature. This was against all that they had understood of the properties of regression analysis. The contents of this chapter offer an explanation. Their respondents were mostly in individual offices, and each had some degree of control over their room temperature by adjusting radiator valves, internal venetian blinds, and opening or closing windows. The direction of causation was therefore ambivalent. Was the room temperature causing the thermal sensation? 'Yes'. Was the thermal sensation causing the room temperature? Again: 'Yes'. Humphreys found that a line with gradient equal to the ratio of the standard deviations of subjective warmth and room temperature gave more consistent results, across the various months and the several respondents, than did the regression of warmth on room temperature. Humphreys and Nicol made no reference to this in the subsequent publication, being unaware of a theoretical underpinning for the phenomenon. However, on one figure they represented the relation between subjective warmth and temperature by a line of gradient equal to the ratio of their standard deviations.[1]

We now give two examples where the use of the geometric mean line seems to be a suitable indication of the structural relation between the two variables.

Example 1. Scatter of neutral temperatures against mean indoor temperatures

In a recent analysis of field data from many sources, a scatter plot was made of the indoor temperature for thermal neutrality against the mean temperature in the accommodation during the survey period.[2] We use the original data to construct Figure 23.4 (see also Chapter 28).

In Figure 23.4 is the regression line of the neutral temperature (t_n) upon the mean operative temperature (t_{op}), together with the 95 per cent confidence limits for the individual observations. We consider whether this regression line correctly describes the relation between the variables.

By the adaptive principle the causation can be shown to be in both directions. Occupants will, over time, adjust their clothing so that they are comfortable at the mean temperature of their accommodation – thus the mean temperature causes the neutral temperature to be what it is. However, the occupants will, if they can, control the room temperature to make themselves comfortable – by the same adaptive principle – by adjusting blinds and window opening, as well as the heating or cooling if it is provided. The neutral temperature thus causes the mean temperature. Neither of the two regression lines adequately represents the relation between the variables. A line intermediate between the two would arguably better represent the relation between the mean temperature and the neutral temperature.

There is a further consideration. Some of the surveys were of short duration, maybe just a day or two. In this case, there is no guarantee that the mean temperature during the survey, even if accurately measured, truly represented the temperature experienced by the respondents, because the day-to-day fluctuations of temperature in a building can be considerable. So there may be an error in the values of this variable. It is not a measurement error but an equation error – we are using the mean temperature over the period of the survey as a surrogate for some more accurate representation of the temperature experienced by the respondents. In these circumstances, and until more is known of the relative strengths of the two directions of causation, it would be reasonable to use the geometric mean line to represent the relation between the variables.

From the underlying data we calculate that the regression of t_n upon t_{op} has gradient 0.785. The gradient of the regression of t_{op} on t_n has gradient 1.157. We take the inverse (0.864) as the gradient of this regression if it were to be drawn on the figure. The geometric mean of the two is a line of gradient:

sqrt(0.785*0.864) = 0.824

This line probably better represents the underlying relation than does either regression line.

In the absence of constraints on the processes of adaptation the t_n and t_{op} would become equal (the dashed line on the figure). The theoretical gradient would therefore be unity had there been complete adaptation without the hindrance of constraints. The geometric mean line is closer in gradient to this theoretical value than is the regression of t_n upon t_{op}, the regression line that would normally be assumed to apply. People have therefore adapted rather better than this line suggests.

The correlation between the two variables is high enough to make the geometric mean line useful. The R-squared value can be calculated from the gradients of the two regressions:

R-squared = 0.785/0.864 = 0.909

in agreement with the value given on the figure.

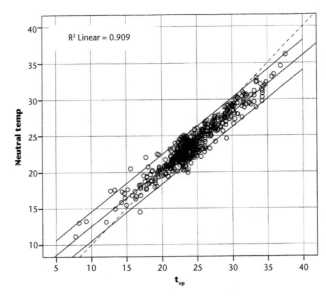

FIGURE 23.4 Neutral temperatures and the mean operative temperatures (the axis labelled 't_{op}').

Example 2. Differing neutral temperatures in buildings at the same prevailing outdoor temperature

This example is drawn from the same research paper (see also Chapter 28). The relation between the neutral temperature of a population and the corresponding prevailing mean outdoor temperature had been established for buildings operating in the free-running mode and for the heated-or-cooled mode of operation, using data from a large collection of thermal comfort field studies.

At any particular outdoor mean temperature there was a diversity of values for the indoor neutral temperatures among the various buildings, and it had already been shown that these differences could not be attributed to random errors of assessment of the neutral temperatures. What then might be their cause?

It was suggested that the differences were attributable to some buildings running warmer than others at the same prevailing outdoor temperature, and that their occupants had adapted to this elevation of their indoor temperature. Figure 23.5 is a scatter plot of the excess neutral temperature in the building (residual t_n) against the corresponding excess in the mean indoor temperature (residual t_{op}) for that outdoor temperature. The data are from buildings operating in the free-running mode, neither heating nor cooling being in use at the time of the survey.

A correlation between these two variables would suggest the presence of effective behavioural or psychological adaptation to the different mean indoor temperatures in the various buildings.

It was found that the two variables were indeed correlated, the R-squared value being 0.724, strongly suggesting that the occupants had adapted to the different indoor temperatures prevailing in their buildings. But we cannot say by inspecting the scatter plot in Figure 23.5 which is the dependent variable; that is to say, which is the direction of the causation. Both regression lines are shown on the figure, together with the line whose gradient is the geometric mean of their gradients.

The statistics for the relation between the residuals are as follows:

Regression of t_n upon t_{op}: $t_n = 0.668t_{op} - 0.02$
Regression of t_{op} upon t_n: $t_{op} = 1.085t_n + 0.05$

Rearranging to make t_n the subject of the equation:

$t_n = 0.922t_{op} - 0.05$

So the geometric mean gradient is: sqrt(0.668*0.922) = 0.785

A cross-check confirms that this is the ratio of the standard deviations of the residuals of t_n and t_{op}, as given in the regression output:

$$1.7075/2.1765 = 0.785$$

and that the ratio of the two gradients is equal to the R-squared value:

$$0.668/0.922 = 0.7245$$

agreeing with the value in the figure (0.724), apart from the small rounding error.

The geometric mean line perhaps better represents the underlying adaptive relation than does either of the regression lines, and the R-squared value of 0.724 assures us that the relation is reasonably robust.

The two examples we have given both arise from situations where we have reason to think that the causal relation between two variables is mutual, but we are uncertain of the relative strengths of the reciprocal causations. The use of the geometric mean line is therefore to some extent speculative; we surmise that it better represents the underlying relation than does either regression line, but no precision can be claimed for it. No doubt a wholly different kind of analysis would be superior. If we knew more about the way people control their thermal environment, a

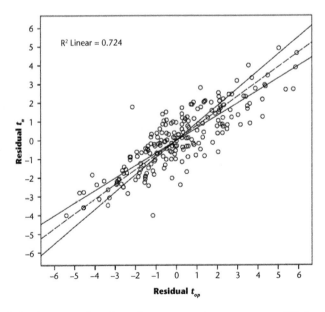

FIGURE 23.5 Relation between elevation of mean operative temperature (K) (horizontal axis, residual t_{op}) and the elevation of the neutral temperature (K) (vertical axis, residual t_n). (The horizontal axis was restricted to ±6 to exclude outliers.)

statistical feedback model could be built that would result in scatter plots such as those in Figures 23.4 and 23.5. Some of these behavioural modelling methods are discussed in later chapters.

A further example of the use of the geometric mean line will be found in Chapter 18, where it was used to verify the consistency of the behaviour of scales of subjective warmth and comfort across different batches of data. In this example, the two variables had equal status, and the geometric mean line described the expected relation between the two scales.

Alternatively the problem of mutual causation can be handled by the methods for regression with error in the predictor variable, as in the previous chapter. In favourable circumstances this approach permits an unbiased estimate to be made of the structural line relating the two variables by partitioning the 'error' between the two variables. The reader is referred to Cheng and Van Ness[3] for advanced treatment of such problems.

Concluding note

The use of the geometric mean line has been heavily criticised because it has some undesirable properties, the chief of which is that, when the correlation between the variables is low, the placing of the line has a large uncertainty, and may even be impossible to locate, as we noted earlier in this chapter.[4] However, the line necessarily lies between the two regression lines, and this places limits on the uncertainty of its location.

We again stress that it is unwise simply to calculate the gradient of the geometric mean line from the ratio of the standard deviations of y and x. If the correlation between the variables is low, the uncertainty in the gradient becomes very large, and the ratio of the standard deviations becomes useless as an estimate of the relation between two variables. Also, the ratio does not tell us whether the gradient is positive or negative. So it must be supplemented with a scatter diagram showing the two regression lines and the R-squared value. If this is done the method will not lead the user astray.

On the positive side, the geometric mean line has the virtue of a certain visual elegance and simplicity, as shown by Figure 23.3. Its gradient is invariant under changes of the scaling of the axes, and its value does not systematically depend on the sample size.

Notes

1 See Figure 9 in: Humphreys, M. A. and Nicol J. F. (1970) An investigation into thermal comfort of office workers, *J. Inst. Heat. & Vent. Eng.*, 38, 181–9.
2 Humphreys, M. A. *et al.* (2013) Updating the adaptive relation between climate and comfort indoors; new insights and an extended database. *Building and Environment*, 63, 40–55, http://dx.doi.org/10.1016/j.buildenv.2013.01.024.

3 Cheng, C. L. and Van Ness, J. W. (1999) Statistical regression with measurement error. *Kendall's library of statistics*, Volume 6, Arnold, London.

4 Compare the strictures of Cheng and Van Ness (Op. cit., p. 43) with the more sympathetic assessment given by J. W. Gillard (in: Gillard, J. W. (2006) *An historical overview of linear regression with errors in both variables*. Technical Report, School of Mathematics, Cardiff University, pp. 9–10). Much depends on the context.

24

THE RELATION BETWEEN REGRESSION ANALYSIS AND PROBIT ANALYSIS

Researchers have often used regression methods to analyse the data collected during a thermal comfort field study, following the example of Thomas Bedford's work in the 1930s, which used multiple regression analysis to form indices of thermal comfort from his extensive data. The application of Probit analysis to thermal comfort data is more recent. Charles Webb was, we believe, the first to apply it to field study data,[1] although F. A. Chrenko had applied it to thermal comfort data collected in laboratory experiments.[2] We briefly explained the application of both regression analysis and Probit analysis to thermal comfort field data in Volume 1, Chapter 10. Here we explore the close relation between the two methods.

Probit analysis was originally developed for quantifying the toxicity of pesticides, and used to determine what concentration of a pesticide is needed to kill a certain percentage of a particular agricultural pest. The response is binary – the insect is either dead or alive. The method can be extended to compare the effectiveness of the same pesticide on several different kinds of pest – perhaps insects of different kinds. The different kinds would in general respond differently to the insecticide; it would take a different dose to kill the same proportion of each kind of insect.

The classic presentation of the statistical method was set out by D. J. Finney in a book first published in 1947, a third edition appearing in 1971.[3] It is still well worth reading because it gives a clear step-by-step account of the logic of the method. Before the use of computers, Probit analysis was time consuming, because each calculation had to be iterated until the result converged on the final value, and much of Finney's book is directed towards reducing the labour and minimising the likelihood of making mistakes.[4]

Probit analysis can readily be applied to thermal comfort field data. The requisite binary response is obtained by separating the data into two groups. If the ASHRAE scale is used, we would first split the data so that group (1) consisted of all the votes of 'cold', and group (2) all the other votes – those warmer than 'cold'. This will yield a single Probit regression line showing how the proportion who are 'cold' diminishes as the temperature rises. The analysis continues with a further pair of groups: (1) all the votes of 'cold' and 'cool' and (2) all the votes warmer than 'cool'. The process continues until we have six pairs of groups. We notice at once that this is rather different from treating different kinds of insect. The insects of different kinds would be independent, while when we successively re-group the comfort votes, the new groups include members of the former groups. This 'nesting' of the groups needs to be remembered, because it affects any significance test that assumes the batches of data to be independent. Applying Probit analysis to a file of field study data will yield a set of Probit regression lines having a pattern like the lines in Figure 24.2, which may then be transformed into a set of sigmoid curves similar to those shown in Figure 24.3.

In this chapter we explore the relation between ordinary regression analysis and Probit regression analysis, drawing attention to some useful correspondences between the two methods. Certain correspondences are exact if the data fully conform to the conditions for the valid use of simple regression analysis. Practical thermal comfort datasets rarely, if ever, conform completely to these conditions, and so we consider the effects of various departures from them. We next explain the reason for expecting a set of Probit regression lines derived from a batch of thermal comfort field data to be parallel and consider what it means if they are not. We go on to explore the effect on the Probit regression lines of error in the predictor variable.

For these explorations we use the computer-generated database constructed for the previous chapters, and analyse it using ordinary regression analysis, and then by using Probit regression analysis. We begin by considering the regression of the thermal sensation on the temperature. Such a regression relation is shown in Figure 24.1.

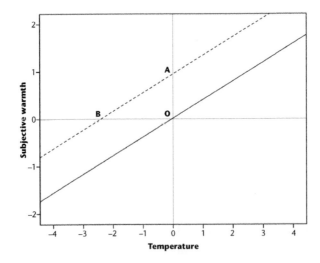

FIGURE 24.1 Detail of the central portion of the scatter plot of subjective warmth and temperature.

TABLE 24.1 Regression statistics for the different forms of the subjective warmth scale.

Form of subjective scale	Regression coefficient	Residual s.d. of subjective warmth
Continuous scale	0.400/K	0.899
Scale grouped to integer values	0.400/K	0.943
Scale restricted to seven integers	0.395/K	0.935

The solid line is the regression line of the subjective warmth on the temperature. We have also drawn a dashed line above it, displaced by one standard deviation of the scatter of the subjective warmth about the regression line. That is to say, the dashed line is elevated above the regression line by the residual standard deviation of the comfort votes. The segment OA, therefore, equals the residual standard deviation of the subjective warmth. OB is the shift in temperature necessary to shift the comfort vote by one (residual) standard deviation. That is to say, it is the standard deviation of the individual differences among the respondents, at a particular level of subjective warmth, expressed in terms of the room temperature. OA/OB is seen to be the gradient of either line, and is the regression coefficient.

We can write down the regression coefficient and the residual standard deviation for three closely related regression lines, and they are shown in Table 24.1, and are taken from Chapter 21. The first row is for the regression of the continuous (ungrouped) thermal sensation on the temperature. The second row applies to the regression when the thermal sensation has been rounded to the nearest integer, so forming discrete 'comfort votes'. It will be recalled that such grouping slightly inflated the residual standard deviation, but left the regression coefficient unchanged. The third row applies when the subjective warmth has been rounded to the nearest integer (comfort vote) *and* any extreme votes allocated to the categories +3 or −3, as shown in Figure 21.15.

We now Perform a Probit analysis on the data, using a discrete seven-category subjective warmth scale. The seven categories will produce a total of six Probit regression lines.

> Procedural note. It is convenient here to use the Ordinal Regression program available in SPSS to calculate the Probit regression statistics, because it produces the statistics for all six lines in a single run, without the need to re-arrange the data and bin the temperatures, in the manner that is usual for Probit regression programs. We set the 'Link Function' to 'Probit'.[5] The terminology of the output statistics differs from that of the usual Probit programs. The 'Location' is the Probit regression coefficient, but with its sign changed. The 'Threshold' is the intercept of the regression line on the y-axis. From the output we can write the equations of each of the six Probit regression lines. The program can test whether the lines are parallel, the null hypothesis being that they are.

The equations of the Probit regression lines are:

$$Z_{(\leq-3)} = -0.446t - 2.862$$
$$Z_{(\leq-2)} = -0.446t - 1.695$$
$$Z_{(\leq-1)} = -0.446t - 0.578$$
$$Z_{(\leq0)} = -0.446t + 0.543$$
$$Z_{(\leq1)} = -0.446t + 1.634$$
$$Z_{(\leq2)} = -0.446t + 2.749$$

$Z_{(\leq-3)}$ is the Probit of the proportion of the votes that are -3 and less, $Z_{(\leq-2)}$ is the Probit of the proportion that are -2 and less, and so on. The estimate of the Probit regression coefficient is $0.446/K$.

The set of parallel Probit regression lines is shown on Figure 24.2, the lowest line being that for $Z_{(\leq-3)}$, and t is the temperature. We see from the figure that the lines are almost equally spaced. Their horizontal (temperature) separations may be calculated from the set of equations, and are 2.62, 2.50, 2.51, 2.45 and 2.50 K respectively. The slight irregularities, none of which differs by more than 0.12 K from the true value of 2.5 K, are attributable to sampling variation. (With small samples this variation will be greater.) The mean horizontal separation of the Probit regression lines is 2.52 K. This is the temperature change required to move the median vote from one category to the next. Figure 24.3 shows the lines from Figure 24.2 with the Probits transformed into proportions. We now have a set of almost equally spaced parallel sigmoid curves, each of which represents a cumulative Normal distribution.

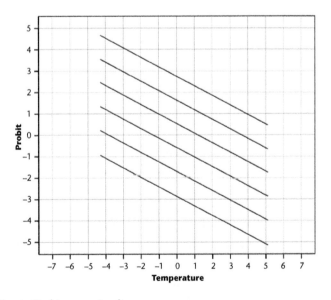

FIGURE 24.2 The six Probit regression lines.

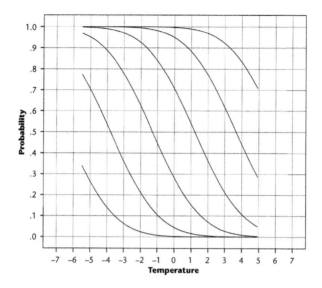

FIGURE 24.3 The six parallel sigmoid curves.

Two equivalences between Probit regression and simple regression

We now consider this from the perspective of simple regression. The simple regression equation of warmth on temperature had a gradient (regression coefficient) of 0.400/K (Table 24.1, row 1). The inverse of the regression coefficient, 2.50 K, is the temperature change needed to shift the mean vote by 1 scale unit. The small difference (0.02 K) between this value and that derived from the Probit analysis is attributable to sampling variations in the placing of the Probit lines. The result from Probit regression and ordinary regression are seen to be equivalent. We can therefore calculate the temperature separation of the Probit lines directly from the simple regression equation.

The regression gradient is slightly reduced when we allocate a value of 3 to those few votes that were above 3, and a value of −3 to those few that were below −3 (row 3 in Table 24.1), and is now 0.395/K. The inverse of 0.395/K is 2.53 K. The difference is too small to be of practical importance, being just 0.03 K more than the theoretical value of 2.50 K. So we may state the first equivalence between Probit regression and ordinary regression: *the temperature separation between the successive Probit regression lines is equal to the reciprocal of the simple regression coefficient.*

This relation between ordinary regression and Probit regression is perfectly general provided that two conditions are met: (a) the subjective warmth scale must be an equal interval scale, which the valid use of simple regression also requires; (b) the mean subjective warmth at any temperature is equal to the median response – that is to say, the distribution of the residuals of the subjective warmth is not skewed. But this is assumed both for ordinary regression analysis and for Probit regression analysis.

In practice neither condition is perfectly fulfilled in the sets of data obtained from a field survey of thermal comfort. We saw in Chapters 18 and 19 that scales of subjective warmth do not always have uniform category widths, and in this circumstance the Probit regression lines will not be quite equally spaced. The matter is further discussed in Chapter 26, where we use the results of a regression analysis to produce distributions of the incidence of thermal comfort despite the limitations of the scales of subjective warmth.

There is a second correspondence that is perhaps less obvious. It is the relation between the Probit regression coefficient and the residual standard deviation after simple regression. These residuals are shown in column 3 of Table 24.1.

The standard deviation of the cumulative Normal distribution used by Probit analysis is by definition the reciprocal of the Probit regression coefficient, because unit value on the Probit scale means one standard deviation from the centre of the distribution. So:

s.d.(Probit) = 1/(Probit regression coefficient)

In our example this is: $1/0.446 = 2.242\,K$. This corresponds in simple regression to the interval OB in Figure 24.1. It is the temperature difference corresponding to a change in the subjective warmth equal to the residual standard deviation. It follows that we can estimate the Probit regression coefficient directly from the statistics of the ordinary regression.

From Figure 24.1, the simple regression coefficient (b) is equal to OA/OB. Now OA is the residual standard deviation after regression. Hence:

b = (residual s.d.)/OB

We have already shown that OB is the inverse of the standard deviation of the Probit distribution, so:

b = (residual s.d.)*(Probit regression coefficient)

Rearranging we have:

Probit regression coefficient = b/(residual s.d.)

For the continuous scale of subjective warmth, as in the top row of the table, the simple regression coefficient is 0.400/K and the residual standard deviation after regression 0.899 scale units. From this we calculate the Probit regression coefficient to be 0.0400/0.899, which equals 0.445/K. The empirical value obtained from the Probit analysis is 0.446/K, very close indeed to the expected value. So the second equivalence is: *the Probit regression coefficient equals the simple regression coefficient divided by the residual standard deviation after simple regression.*

Again, the result is perfectly general, provided the conditions for regression analysis are fulfilled. Our experience shows that the equivalence is not particularly sensitive to the commonly present irregularities in the widths of the categories of the subjective scale.

Effect of grouping the subjective scale

The subjective warmth data are usually grouped to integers values. The process does not affect the estimate of the Probit regression gradient, but it inflates the residual standard deviation to some extent.

Using the statistics from the middle row of Table 24.1, we obtain for the estimate of the Probit regression coefficient:

Probit regression coefficient = 0.400/0.943 = 0.424/K

This is 5 per cent below the true value, but will often be sufficiently close for practical use.

When a more accurate estimate is needed, we can apply Sheppard's correction for grouping and correct the discrepancy. It will be recalled that grouping raises the variance of a set of data by $h^2/12$, where h is the grouping interval. Our data were grouped to the integer values, so $h = 1$. Sheppard's correction is therefore 1/12, or 0.0833. So the corrected residual standard deviation is:

Corrected residual s.d. = sqrt $(0.943^2 - 0.0833)$ = 0.898

We would then obtain for the Probit regression coefficient:

Probit regression coefficient = 0.400/0.898 = 0.445/K

which is equal to the true value apart from rounding error. If we apply the same procedure to the bottom line of the table, the seven-category scale, we obtain a value of 0.444/K. We conclude that applying Sheppard's correction for grouping will yield a value for the Probit regression gradient that is scarcely to be distinguished from the true value.

Usefulness of the equivalences

The usefulness of these two equivalences is considerable. If a published paper or research report gives only the results of a Probit analysis, the simple regression coefficient and the residual standard deviation can be estimated. The reverse is also true. If a publication supplies only the statistics from a regression analysis of the data, the Probit statistics can be calculated from them and the usual bell curve for the proportion in thermal comfort can be constructed. In Chapter 26, there is an example of using the equivalences to obtain from regression statistics of a meta-analysis the curve relating the proportion of people comfortable to the departure of the room temperature from its current optimal value.

Why the Probit regression lines should be parallel

The Probit regression lines in Figure 24.2 are parallel because our data have the same residual standard deviation of subjective warmth right across the temperature range. (That is to say, they fulfil the requirement for regression analysis of homoscedasticity of the residual standard deviation.) If the data conform to this pattern, the Probit regression lines are necessarily parallel. There is therefore good reason when applying Probit analysis to such data to force the lines parallel. This is done automatically when using the SPSS Ordinal Regression program, and is an option often available in ordinary Probit regression programs.

Should the lines be found not to be parallel it is possible that something is wrong with the underlying model that is being used. But this is unlikely with thermal comfort field study data, in view of the long experience of the applicability of the ordinary regression model, and hence of the Probit regression model. Irregularities in the *widths* of the subjective scale categories cannot cause the Probit regression lines to differ in gradient. Such irregularity alters only the temperature displacement between the lines – they remain parallel. A more likely reason for the lines not to be parallel is that there is one or more *poor category label* in the subjective scale – a category whose description is not clear and distinct to the respondents. This can take the form of a blurred distinction between two adjacent categories, such as can arise from an unhappy translation of the ASHRAE scale from one language to another, a topic discussed in Chapter 18.

We can explore the effects of this. Let us suppose (just for the purpose of illustration) that people can make no distinction between the 'neutral' and the 'slightly warm' categories. To model this in the database, we allocate the responses in these two categories at random to either category. The randomisation process is equivalent to saying a person voting 'neutral' might equally well have voted 'slightly warm', and vice-versa. The modified histogram of the 'comfort votes' is shown in Figure 24.4, beside the histogram of the data before the randomisation. Notice the now nearly equal numbers of votes in categories zero and one, and that the distribution of the votes is no longer Normal. The mean and standard deviation show surprisingly little change. We then redo the Probit regressions, this time not forcing the lines to be parallel, and the result is shown in Figure 24.5.

The dashed line is the Probit regression of the new boundary between 'neutral' and 'slightly warm'. It has now a much lower gradient, and the program output indicates that the lines do not now constitute a parallel set. Notice that none of the other lines in the set has been changed in either gradient or intercept. It is these steeper lines that represent the true sensitivity of the population to temperature. Notice too

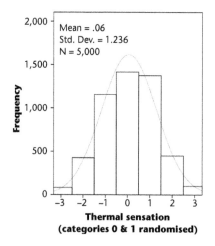

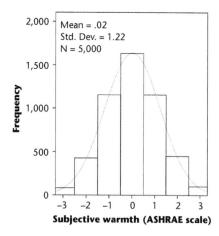

FIGURE 24.4 Histograms of comfort votes. Left: after randomising votes of 0 and 1. Right: before randomisation.

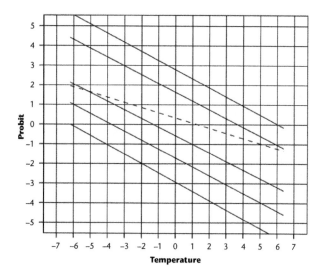

FIGURE 24.5 Probit regression lines after randomising votes of 0 and 1.

that the dashed line, if extended, would cross the other lines. A crossover represents an unreal situation, for it implies a negative number of votes in some category. This is another reason for supposing that the true lines must be parallel.

We see from this rather extreme example that a Probit regression having a lower than expected gradient could indicate a 'fuzzy' boundary between two category labels on the subjective warmth scale.

This property of the scale is unlikely to be picked up until after all the data have been collected and the analysis is in progress. The scale will need to be amended and validated before it is used again (see Chapter 19), but what is to be done at this late stage of analysis? For if the researcher is unfortunate enough to find a fuzzy boundary between categories that define the change from comfort to discomfort, the data will yield a misleading and asymmetric bell-curve for the proportion of respondents who are comfortable. No corrective procedure is entirely satisfactory. We could force the lines parallel, but doing so lowers the Probit regression coefficient from 0.446/K to 0.380/K, a considerable reduction. The reduction makes a substantial increase to the temperature separation of the successive parallel lines, and has a consequent effect on the bell-curve indicating the proportion of people in thermal comfort. It becomes broader and the peak, which shows the maximum proportion of people who are comfortable, becomes lower. If, alternatively, we apply regression analysis and reconstruct the Probit regression lines from it, we find that the regression coefficient has fallen from 0.395 to 0.362, and the residual standard deviation of the comfort vote has increased from 0.935 to 1.01 scale units, making the estimates of the Probit regressions inexact, with a similar consequent effect on the bell-curve.

Perhaps the best rescue procedure would be to take the average gradient of the three steepest Probit regression lines, together with the intercepts of the set of six lines at the mean temperature of the data. This would be an improvement, because there are a number of reasons why a Probit regression coefficient might be depressed, but no reasons, apart from sampling variation, why it might be artificially raised. Such a procedure, if applied to Figure 24.5, would restore it to the situation portrayed in Figure 24.2, before the introduction of the 'fuzzy' boundary.

> Practical Note. Sometimes the Probit regression lines must be derived from a rather small dataset. When this is done, the lines may appear far from parallel. This can be because the gradients of the regression lines are imprecisely determined (they have a large standard error). This will affect most seriously the lines derived from the more extreme categories, where data are sparse. In these circumstances the wisest course of action is to force the lines parallel before deriving the usual bell-shaped curves from them.

Effect on Probit regression of error in the predictor variable

It has sometimes been thought that Probit analysis is, unlike ordinary regression, unaffected by the presence of error in the predictor variable. This seems unlikely if the two procedures are as equivalent as we have seen them to be. Figure 24.6 shows the result of a Probit regression after the introduction of an error of standard deviation 1 K into the temperature. Comparison with Figure 24.2 shows that the gradient of the Probit regression lines has been reduced, and that consequently the temperature separation of the lines has increased. The effect is exactly equivalent to the effect on ordinary regression that we quantified in Chapter 22.

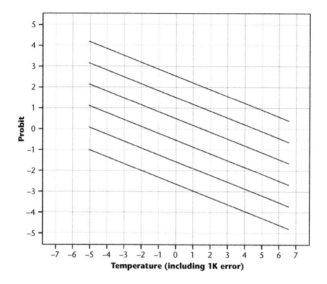

FIGURE 24.6 The effect on the Probit regression lines of error in the predictor variable.

Note on Ballantyne's method of obtaining the optimum temperature using Probit analysis

Ballantyne published a method of applying Probit analysis to obtain the optimum temperature from a set of responses to the McIntyre three-category scale of thermal preference,[6] and the method has not infrequently been used to analyse thermal comfort survey data. Ballantyne's procedure was to randomly allocate all the 'no change' votes into either the 'prefer warmer' category or the 'prefer cooler' category. He then calculated two Probit regression lines. One showed the increase of 'prefer cooler' votes, together with the extra votes allocated to the category, as the temperature increased. The other showed the increase of 'prefer warmer' votes, together with the extra votes allocated to the category, as the temperature decreased. The intersection of the curves was taken to be the optimum condition. The method is open to objection:

- The two Probit regressions produced by the procedure are mirror images of each other. Either line contains all the information that is available.
- The random allocation of the 'no change' votes into the other categories introduces scatter. This scatter depresses the gradient of the Probit regression, as we saw earlier in this chapter.
- The method produces a biased estimate of the optimum temperature if the numbers of votes in the 'prefer warmer' and the 'prefer cooler' categories are unequal. This can be seen by considering the extreme condition where no 'prefer warmer' votes were cast, a result sometimes obtained from surveys in hot weather. In this situation, it is evident that no estimate of an optimum

may legitimately be extracted from the data. However, applying Ballantyne's randomisation procedure will produce a pair of Probit regression lines that intersect to yield a value for an optimum temperature. With less extreme asymmetry in the voting, the procedure will produce a biased estimate of the optimum temperature.

For these reasons, despite its place in the progress of adaptive thermal comfort, we advise against the use of Ballantyne's method.

Concluding comment

Probit analysis has been widely used in the interpretation of field study data. In this chapter we have explored the close relation that exists between regression analysis and Probit analysis, and shown that in certain important respects they are equivalent. This equivalence enables the construction of distributions of comfort directly from the statistics of a regression equation, a feature that is very convenient when analysing the pooled results of many individual surveys, as we show in later chapters.

Traditional Probit analysis now has a rather old-fashioned feel and is rather inconvenient to use for thermal comfort data when the subjective warmth scale has several categories, requiring the repeated application of the analysis to achieve the desired curves. Its modern equivalents, Logistic Regression and Ordinal Regression, may be less familiar to thermal comfort researchers but are easier to use and lead to the same results. They can also (in some statistical packages) handle a linear combination of predictor variables, rather than accepting just a single predictor, a limitation of traditional Probit analysis.

Notes

1 Webb, C. G. (1959) An analysis of some observations of thermal comfort in an equatorial climate. *British Journal of Industrial Medicine* 16(3), 297–310.
2 Chrenko, F. A. (1953) Probit analysis of subjective reactions to thermal stimuli – a study of radiant panel heating in buildings, *British Journal of Psychology*, General Section 44(3), 248–56.
3 Finney, D. J. (1947, 1952, 1971) *Probit analysis*. Cambridge University Press, Cambridge.
4 Since the advent of computers and statistical packages alternatives to Probit analysis have become available, and can readily be applied to thermal comfort data. They include ordinal regression and simple or multiple logistic regression. We refer to them from time to time in this chapter.
5 The default link function is the logistic distribution. We use the Probit link in this chapter because it makes the comparison with simple regression easier. Analysis using Ordinal Logistic Regression would give cumulative sigmoid curves (see Figure 24.3) almost indistinguishable from those obtained by Ordinal Probit Regression. The methods are interchangeable.
6 Ballantyne, E. R. *et al.* (1977) Probit analysis of thermal sensation assessments, *International Journal of Biometeorology* 21(1), 29–43.

25

THE DEPENDENCE OF SUBJECTIVE WARMTH ON WITHIN-DAY CHANGES IN ROOM TEMPERATURE

Introduction

Chapters 18 to 24 concerned methodology: the testing and construction of subjective scales, and the statistical analysis of thermal comfort survey data. In this and subsequent chapters we examine the principal databases of thermal comfort from an adaptive perspective and see what general trends emerge from them.

In this chapter we look at how people respond to a varied room temperature during a single day. During any day, the temperature in a room is likely to vary because it is influenced by such factors as the outdoor temperature, solar radiation and ventilation rate, together with the action of any heating or cooling system that may be operating. How sensitive are people to such changes of temperature during the daytime? Does the sensitivity depend on whether the building is air conditioned? Do men and women differ in their sensitivity to temperature? Does the sensitivity depend on the mean temperature of the accommodation, and thus on the season of the year?

We begin by explaining what we mean by 'sensitivity'. As the temperature of a room changes, the average thermal sensation of its occupants changes too. The sensitivity to such temperature variation is expressed in scale units per degree of room temperature, usually using a seven-point scale of subjective warmth. It can be estimated from a set of experimental data by applying regression analysis, with the thermal sensation (the 'comfort vote') as the dependent variable and the room temperature as the predictor variable. The regression line gives the relation between the room temperature and the thermal sensation of the average occupant, and the estimate of the sensitivity is the regression coefficient.

Background considerations

Before focussing on within-day temperature changes, we consider the question more broadly.

Studies over the years in climate laboratories, where groups of people have been exposed to different room temperatures for periods of usually three hours, find the sensitivity of subjects to thermal changes to be in the region of 0.33/K. This value applies to people in a standard uniform of thermal insulation of 0.6 clo, engaged in sedentary activity (1 met) such as reading or watching television, and having reached steady-state thermal equilibrium. The classic large-scale experiment of this kind was conducted by R. G. Nevins and his team at Kansas State University in the 1960s.[1] This was a cross-sectional experiment, each person being exposed to just one of a wide range of experimental conditions. The comfort votes from the last part of the exposure only were used for analysis, because by then steady-state thermal equilibrium had been reached. From such experiments men have been found to be somewhat less sensitive to temperature difference than are women, but it is uncertain whether the difference should be attributed to a different interpretation of the scale, to the physiological differences between the sexes, or to social and cultural factors.

Experiments of this kind provide us with a value of the temperature sensitivity with which to compare values obtained in more usual conditions, where people are to some extent free to choose their clothing and their activity, and where the room temperature is varying during the day, and steady-state thermal equilibrium is never established. It is also possible that the thermal sensation scales may be interpreted differently in the context of a laboratory test, where the subjects might reasonably expect to encounter some less-than-comfortable experimental conditions, or perhaps be less forgiving of thermal experiences with which they were unfamiliar.

The ability to adapt to the thermal environment, by choosing suitable clothes or by other means, has a profound effect on a person's apparent sensitivity to the temperature of a room. For example, in summertime a person may be thermally neutral when wearing light summer clothes in a room at 27°C, but they could also be thermally neutral at 20°C when wearing warmer clothing in winter. The apparent sensitivity to temperature change would be zero because, by means of selecting and wearing suitable clothing, adaptation is complete. So the estimate of the sensitivity depends on how complete is the process of adaptation.

In general, it is easier for people to adapt to gradual seasonal changes in indoor thermal environments than to more sudden changes that may occur within a day or from day to day. This is because the processes of adjustment take time to complete, unless one is already prepared for the change by having suitable clothing immediately available. So surveys conducted over a period of several months will, if people are free to adapt, find a smaller sensitivity to room temperature than will surveys conducted over a few days.

For example, Hindmarsh and MacPherson, studying office workers in Sydney, Australia, obtained responses from their experimental subjects on a seven-point thermal sensation scale at about 11 am every day for a whole year.[2] They obtained a regression coefficient of just 0.16/K, or about half the value found in laboratory studies. The result suggests that the effects of day-to-day temperature variations were 'diluted' by the adaptation of their subjects to the seasonal trends in their temperature during the year of the experiment. Davies obtained a similar result for secondary-school children.[3] She sampled the response of a class at weekly intervals for more than a year. The school was of heavyweight construction with an unusually high 'thermal mass', and so the temperature varied little during the school day. The children's thermal responses were therefore attributable chiefly to the week-to-week variations in room temperature over a year and more. She obtained a regression coefficient of just 0.10/K. The children's clothing varied with the season, which was how they adapted to the seasonal variation in their classroom temperature during the school year.

If we wish to obtain a value for the underlying sensitivity to temperature change in everyday life, without resorting to the climate laboratory, it follows that it is to be obtained from surveys of the shortest duration, or from longer surveys whose data can be separated into batches of short duration.

There is therefore sense in taking the *single working day* as the unit from which to derive people's everyday sensitivity to temperature variation. It is often inconvenient to change clothing during a day at work, and there can also be social considerations that can limit the choice of clothing, so changing clothing during the working day is quite rare. Field studies in offices and schools commonly report very little systematic change of clothing during a working day. Both the ASHRAE RP-884 database and the SCATs databases find systematic changes in clothing insulation within the working day at the office to average about 0.01 clo/K. Full compensation for temperature change would need about 0.2 clo/K. The clothing change within the working day is therefore only about one twentieth of that needed for complete adaptation. So within-day batches of data are virtually free from the effects of clothing change, and can therefore be expected to yield an estimate of sensitivity to temperature variation that approaches its underlying value.

Using field databases to derive the within-day sensitivity to temperature change

Databases of thermal comfort field studies commonly consist of numerous blocks of data obtained from buildings in various seasons and climates. A block of data may be identified in a manner such as: Building A, Aberdeen, UK, 7 April, 2007. Each row of data in the block contains the data from a single interview and gives the thermal sensation of the respondent together with the corresponding environmental data. The best-known databases of this kind are the ASHRAE RP-884 database and the SCATs database, but there exist other such databases not currently in the public domain.

It is possible to complete the data collection in a small building in one day, while a survey in a large office block may take several days. The data can be separated into daily batches if the date of each record was recorded. We shall call such a batch of data a *day survey*. It can be regarded as a quasi-random sample of the temperatures occurring throughout that building during that day, together with the occupants' subjective responses to them. We say 'quasi-random' because, in a field survey, it is hardly practicable to strictly randomise the selection of respondents, their locations in the building, or the timing of the visits to their workstations.

When estimating the sensitivity, we use data only from surveys that are 'transverse' in design: that is to say, surveys in which each respondent gave just one response during the day. An alternative research design often used to supplement the transverse data is the 'longitudinal' design. In this design, each person in a small cohort of respondents gives several assessments each day over a period of weeks or months. These surveys are useful for tracking the responses of individuals over longer periods, but are not well suited to the day survey approach. This is because the several thermal responses the individual gives during a single day lack the degree of independence desired for regression analysis, and this leads to spuriously low estimates of the sensitivity to temperature change.

Such low estimates can be demonstrated from the SCATs database. In addition to the monthly transverse surveys, quasi-random sub-samples of respondents in each building gave their comfort votes up to four times a day, for periods that varied from a week to several months. The within-day sensitivity derived from these sub-samples was found to be less than a third of that derived from the day surveys of transverse design. But a sub-sample cannot yield a genuine statistic significantly different from that of its parent population. The reason for the low value seems to be that the first 'comfort vote' of the day acted as an 'anchor' for subsequent votes. Such 'anchoring effects' are well documented in experimental psychology. Data from longitudinal surveys have not therefore been used in the sensitivity estimates obtained below.

Figure 25.1 is a scatter diagram of the data from a day survey. The 'thermal feeling' (tf) is the ASHRAE vote of each of the 26 respondents and is plotted against the globe temperature (tg) at their workstation. Shown on the diagram is the linear regression line. Its gradient indicates the group-sensitivity of the respondents to temperature. Also shown are the means of both the variables (the dashed lines). The regression line necessarily passes through the joint mean, as shown in the figure. The curved lines are the 95 per cent confidence limits of the mean vote, and therefore indicate limits of the placing of the regression line. The estimate of the regression coefficient obtained from this small set of data is evidently not precise enough to be of practical use. To raise the precision we can combine the information from all the day surveys in the database, or from deliberately selected subsets of them.

We explain the method with the help of a worked example that uses the SCATs database. For each day survey in the database, we first subtract from the globe temperature its mean value to form a new variable (δtg) and from the thermal sensation (tf) its mean to form the variable δtf.[4] We next perform a regression analysis of δtf on δtg, thereby pooling the information from all day surveys in the database.

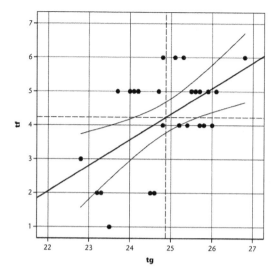

FIGURE 25.1 Example of data from a day survey (UK, building U1, 6 August, 1998). The vertical axis, tf (thermal feeling), is the vote on the ASHRAE scale (1 = cold, 7 = hot). The horizontal axis, tg (globe temperature), is the room operative temperature.

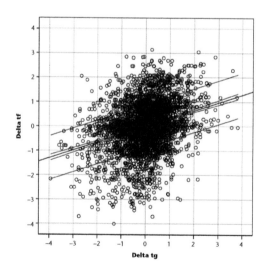

FIGURE 25.2 Pooled regression of the day surveys in the SCATs database. The range of delta tg has been restricted to ±4 K to remove outliers. The outer lines show the residual standard deviation of the ASHRAE scale votes (delta tf). The inner pair of lines shows the 95 per cent limits on the placing of the regression line.

The regression now gives a correctly weighted value of the day survey regression gradient. Figure 25.2 shows the scatter diagram of the pooled day surveys in the SCATs database. Table 25.1 shows the regression coefficients and other statistics derived from the database, and Table 25.2 shows the same statistics derived from the ASHRAE RP–884 database. The regression coefficients are now usefully precise.

TABLE 25.1 Regression coefficients from the SCATs data (349 day surveys).

Type of climate control	No. of observations	Variance of δtg	Estimate of error variance in δtg	Regression coefficient	s.e. of coefficient	Adjusted regression coefficient
AC	1339	0.6631	0.158	0.385	0.031	0.513
MM	372	0.8093	0.158	0.360	0.057	0.447
MV	167	1.3428	0.158	0.352	0.058	0.400
NV	1440	0.7444	0.158	0.361	0.030	0.458
PP*	1236	0.9650	0.158	0.231	0.022	
All (PP omitted)	3318	0.7362	0.158	0.368	0.019	0.469

* 'PP' (part-and-part) indicates buildings with AC in some rooms only.

TABLE 25.2 Regression coefficients from the ASHRAE data (483 day surveys).

Type of climate control	No. of observations	Variance of δtop	Estimate of error variance in δtop*	Regression coefficient	s.e. of coefficient	Adjusted regression coefficient
AC	5731	0.7389	0.158	0.400	0.014	0.508
MM	162	1.1679	0.158	0.367	0.079	0.424
NV	2585	0.5550	0.158	0.308	0.024	0.432
All	8478	0.6892	0.158	0.377	0.012	0.489

* The ASHRAE database uses operative temperature (top) rather the globe temperature. They are virtually indistinguishable.

The tables also give the statistics for subgroups within the data: air-conditioned accommodation (AC), naturally ventilated accommodation (NV), mixed-mode accommodation (MM), where cooling is used just to peak temperatures, and buildings where some rooms are AC and others not (PP, part-and-part).

The overall regression coefficients of 0.37/K and 0.38/K, seen in the last line of each table, are a little higher than the value of 0.33/K typical of laboratory experiments. Yet there is reason to think that even these values are too low, as we now explain.

The standard regression model that we have used takes no account of the presence of error in the predictor variable. If error forms a substantial part of the total variance of the predictor variable, its effect is to bias downwards the estimate of the regression coefficient (see Chapter 22). The variances of the room temperature from the pooled within-day surveys in either database are quite small, as can be seen from column 3 of Tables 25.1 and 25.2. It is therefore advisable to adjust the regression for the presence of error in the measurement of the room temperature. The adjusted regression coefficient (b_{adj}) can be estimated as:

$$b_{adj} = b(\sigma^2_{stop})/(\sigma^2_{stop} - \sigma^2_{err})$$

σ^2_{stop} is the variance of the operative temperature and σ^2_{err} its error variance. The relevant error is not just that arising from limitations of the thermometer, but also from its placing, for it cannot be in the same place as the respondent and at the same time.

A useful estimate of the magnitude of this error can be obtained from the ASHRAE database by using data from surveys where the instruments were mounted on a measurement-cart. Globe temperatures were measured simultaneously at heights of 0.1 m, 0.6 m and 1.1 m. For some 5000 observations, the variance of the difference between the temperatures at 0.1 m and 1.1 m (eliminating their mean difference) was $0.316\,K^2$. Halving this variance will give the error variance attributable to one of the thermometers. This gives $0.158\,K^2$ for the error variance attributable to a globe thermometer, with a good-quality resistance thermometer at its centre, and placed about 1 m from the desired location in a typical room.

The final column in the tables shows the adjusted regression coefficients, assuming that this error is present in the globe temperature measurements in all the data batches. It is evident that the presence of this error-variance has made a considerable difference – as much as 25 per cent in the AC buildings – to the estimate of the regression coefficient. The estimates are quite sensitive to the error-variance in the predictor variable, which we have been able only to approximate, and our estimate of the sensitivity is to that extent uncertain. There is room for further research on quantifying such errors.

In the 'PP' buildings of the SCATs data, some rooms were air conditioned while others were not. The regression coefficient was much lower in these buildings (0.23 ± 0.02). In summer and winter, these buildings show bimodal distributions of operative temperature, because the different types of accommodation have different characteristic mean indoor temperatures. People would have become to some extent adapted to these different mean temperatures of their rooms, and pooling the responses of these two groups therefore artificially lowers the regression coefficient. These data have been excluded from the overall estimate of sensitivity.

The overall estimates of sensitivity (disregarding the type of plant and adjusted for the presence of error in the predictor variable) are $0.47 \pm 0.02/K$ in the SCATs data and $0.49 \pm 0.01/K$ in the ASHRAE data. The agreement between the two independent databases is impressive. The difference between the two values is not statistically significant, and a combined value of $0.48/K$ is therefore appropriate.

Nicol and Humphreys,[5] using different reasoning, arrived at a value of $0.5/K$, which is quite close to the more rigorous estimate of $0.48/K$ we have obtained from the day surveys in the two databases. A sensitivity of $0.48/K$ means that a 2.1 K change in operative temperature is associated with a change of 1 scale unit in the average thermal sensation on the ASHRAE scale. The corresponding value if the sensitivity were $0.5/K$ would be 2.0 K. The difference is not of practical importance.

It is interesting to note that the value of 0.48/K is some 50 per cent greater than that expected on the basis of steady-state experiments in the climate chamber. The difference shows that sensitivities obtained in the climate laboratory under steady-state conditions do not necessarily apply to people in everyday life. The difference is large enough to be of practical importance.

Does the sensitivity to temperature depend on the type of climate control in the building?

During their analysis of the ASHRAE database, de Dear, Brager and Cooper found that the sensitivity depended on whether the building was centrally conditioned (AC) or naturally conditioned (NV), with people being much more tolerant in the NV buildings. This finding perhaps resulted in part from discarding all non-significant regressions before estimating the average gradient, a procedure that is likely to artificially increase the estimates.[6] For the day surveys in the SCATs data, Table 25.1 shows that that the regression coefficient is slightly higher in the AC buildings than in the NV buildings, but the difference is not statistically significant. When the regressions are derived from the ASHRAE data (Table 25.2), a difference is again present, but it is small. Taken together, these results suggest that people in AC buildings are a little more sensitive to temperature variation than people in NV buildings. The effect is in the region of 10 per cent.

Does the sensitivity depend on the sex of the respondents?

The ASHRAE database indicates the sex of the respondents, and the day surveys have 3938 responses from men and 4125 from women. The numbers are sufficient to enable us to test whether men and women were on average equally affected by changes in temperature during the working day. Their responses are shown in a scatter plot (Figure 25.3). The open points are the data from the men and the filled points from the women. The dashed line shows the sensitivity of the men; the solid line shows the sensitivity of the women. The figure suggests that the women were on average more affected by temperature change than were the men. The regression statistics of the two lines are given in Table 25.3.

The regression coefficient for the women was about 20 per cent greater than that for the men. The difference between the sexes is statistically significant ($p = 0.012$). So we can say with some confidence that the women were on average more sensitive to temperature changes during a working day than were the men. Since the database includes data-files from a variety of climates, cultures and types of building, and the surveys were conducted by many different research teams, we may perhaps generalise this finding: women are on average more sensitive to temperature change than are men. The difference is substantial. We will explore its practical importance for the incidence of discomfort in shared accommodation in Chapter 26.

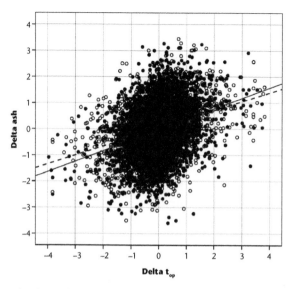

FIGURE 25.3 Scatter plot from all the day surveys in the ASHRAE database, showing a difference in sensitivity between men and women. Delta ash is the ASHRAE scale value with its day-mean subtracted. Delta top is the operative temperature minus its day-mean value. The range of delta top has been restricted to ±4 K to remove outliers. The open points are the men, the closed points the women.

TABLE 25.3 Regression statistics for the sensitivities of men and of women.

	Regression coefficient (s.e.)	Intercept (s.e.)	Mean delta ash (s.e.)	Mean delta top (s.e.)
Men	0.341 (0.016)	0.029 (0.014)	0.026 (0.015)	−0.011 (0.013)
Women	0.401 (0.018)	−0.027 (0.015)	−0.024 (0.016)	0.008 (0.012)

We have also explored the differences in sensitivity of women and men more generally in the database. Briefly the findings are as follows. In most files in the database, the women were found to be more sensitive than the men. The same was true when the files were split by building, so that the data from each building were separated. When all the some 20,000 observations were pooled, and the thermal environment expressed in terms of PMV, the women were again found to be more sensitive than the men. The same is true if the environment is expressed in terms of operative temperature. The conclusion is that, although an individual survey may not pick up the difference, it pervades the database. All this supports the generalisation that women are on average more sensitive to a change in the room temperature. The generalisation is further supported by the findings of experiments in climate-controlled laboratories over many years, as we noted above.

The day surveys regression lines shown in Figure 25.3 can also tell us whether men and women sharing the same thermal environments differed in their average sensation of warmth. The very small, non–significant difference between the mean values of the *delta top* (Table 25.3, column 5) confirms that they occupied virtually identical thermal environments. The overall means of *delta ash* (column 4) show that the men were on average just 0.05 of a scale unit warmer than the women. This difference of one–twentieth of an ASHRAE scale division is negligibly small. The same conclusion can be drawn from the respective intercepts for men and women (column 3). The difference of 0.06 of a scale unit is statistically significant ($p = 0.007$) but of no practical importance. So for practical purposes the men and the women overall felt equally warm when sharing the same accommodation. We conclude that they had adapted equally to their mean thermal environment during a working day, presumably by their day–to–day choice of clothing. This conclusion is to be expected from the adaptive principle, and will hold whenever men and women are free to choose thermally appropriate clothing, and use that freedom.[7]

Summary of values of the within-day average sensitivity to temperature change

At this point it is useful to summarise the findings on sensitivity that arise from the analysis of the day surveys in the ASHRAE database and the SCATs database. In Table 25.4, the sensitivities, pooled for men and women, are the mean values from the databases. To arrive at values for women we have raised the sensitivity by 10 per cent and lowered it by 10 per cent for men. The lowest sensitivity is 0.41/K, for men in naturally ventilated accommodation. This means that a change of 2.4 K in the room temperature changes their thermal sensation by 1 scale unit. The highest sensitivity is 0.56/K, for women in air–conditioned accommodation. It would need a change of 1.8 K to change their thermal sensation by 1 scale unit.

We now consider the implication of a temperature sensitivity in the region of 0.5/K. It is usual to assume that the three central categories of a seven–point scale (such as the ASHRAE or of the Bedford scale) indicate thermal comfort. So to take a group of people across this zone would require a room temperature shift of some

TABLE 25.4 Average sensitivities to temperature change during the working day.

	Women and men	*Women*	*Men*
All buildings	0.48	0.53	0.43
AC buildings	0.51	0.56	0.46
NV buildings	0.45	0.50	0.41

Notes: The values are group averages. The units are scale-points per degree (/K), and assume a seven–point scale. The values are corrected for the presence of error in the predictor variable. Values apply to lightly active duties typical of office work, and assume that the people are free to choose their clothing.

6 K. This seems a surprisingly wide temperature range. We would not be content with such a wide range of room temperature if we were sitting at home of an evening reading a book or watching a film. This is because, at any particular temperature, there is a large scatter of 'comfort votes', as is evident from Figure 25.2. At the lower end of the 6 K range, the average vote would be −1.5. This would mean that half the people in the group would be uncomfortable, or that an average person would be uncomfortable for about half the time. The room temperature would need to stay within a degree or two of the optimal value to minimise the incidence of discomfort. It is therefore important to distinguish between the temperature range that corresponds to three units of the subjective scale and the range over which the probability of discomfort is small.

Does the sensitivity depend on the mean temperature during the day survey?

We tested whether the regression coefficient varied systematically with the mean indoor temperature on the day of the survey. This was done by separating the SCATs data into ten percentile groups (deciles) of day-mean indoor temperature, the groups containing approximately equal numbers of observations. The same procedure was applied to the ASHRAE RP-884 database. Figure 25.4 shows the result as a scatter plot. Splitting the data in each database into these smaller groups of day-mean temperature has the consequence of increasing the uncertainties in the estimates of the sensitivity. The standard error in each estimate of the sensitivity is about 0.04/K in the ASHRAE data and about 0.06/K in the SCATs data. The gradient of the regression line on

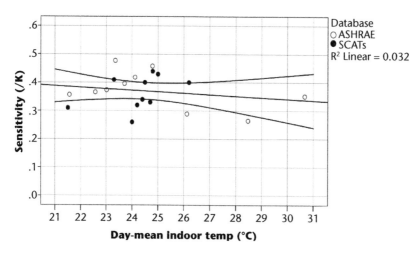

FIGURE 25.4 Scatter plot of the sensitivity to a change in room temperature, related to the mean temperature during the day survey.

Note: The sensitivities (regression coefficients) are given before adjustment for error in the predictor variable. Data from the PP buildings in the SCATs database have been excluded.

the figure (−0.005/K) is consequently not very precisely determined. Its 95 per cent confidence intervals are −0.019/K and +0.009/K. The regression line gives us no reason to suspect a greater sensitivity at higher room temperatures – indeed the trend line suggests that the opposite may be so – but neither is the possibility of a slight dependence in the direction predicted by the PMV equation excluded.

Does the sensitivity depend on the mean clothing insulation during the day survey?

We used the same procedure to check for a dependence on the clothing insulation. For this test, the databases were divided into deciles of day-mean clothing insulation. The data are shown as a scatter plot in Figure 25.5.

The trend line is in the direction indicated by the PMV equation, and its gradient differs significantly from zero (p = 0.03). The 95 per cent confidence intervals on its gradient are −0.04/clo and −0.71/K. That is to say, the effect could be almost absent (0.04/clo) or implausibly large (0.71/clo). The data are insufficient to tell us more.[8]

The results for room temperature and for clothing appear to conflict. One would expect, because temperature and clothing are related, that Figures 25.4 and 25.5 would lead to the same conclusion. However, the conflict may be more apparent than real, for both are consistent with a very small dependence of the sensitivity on the day-mean room temperature and on the day-mean clothing insulation. This would mean that more clothing and lower room temperatures were associated with a very slightly lower sensitivity to temperature change. Any dependence of the sensitivity on the mean temperature and the mean clothing is therefore small, and may be

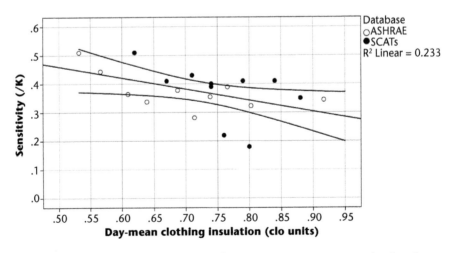

FIGURE 25.5 Scatter plot of the sensitivity to a change in room temperature, related to the mean clothing insulation during the day survey.

Note: The sensitivities (regression coefficients) are given before adjustment for error in the predictor variable. Data from the PP buildings in the SCATs database have been excluded.

disregarded for practical purposes. The values for the sensitivity given in Table 25.4 can therefore be used with little error if the day-mean temperature prevailing in the accommodation lies within the range 20–30°C. Allowing for within-day variations of up to ±4K (Figure 25.2) extends the range to 16–34°C, which well covers the practical range of indoor temperature.

Do people differ from one another in their sensitivity to temperature change?

We have seen that on average women are more sensitive to temperature change than are men. But how much do people differ from one another in their sensitivity, apart from this sex difference? A survey of transverse design cannot answer the question, because each person gives just one response. Longitudinal surveys, where each respondent gives repeated responses over an extended period, are needed to answer this question. In the large year-round BRS survey described in Chapter 3 the data from each respondent were separated into monthly batches.[9] Fifteen respondents provided more than one batch of monthly data. Together these 15 respondents provided 85 person-months of data. From each of these monthly records, the regression coefficient of thermal sensation (Bedford scale) on the globe temperature was obtained. The regression coefficient is the sensitivity of the person to temperature change within a month. The mean sensitivity was 0.21/K, the rather low value being attributable to the adaptation that would occur during a month, and also to the 'anchor effect' previously noted in longitudinal surveys.

Analysis of variance of the monthly records showed that the between-respondent variation of sensitivity had a standard deviation of 0.19/K, while the within-subject variation had a standard deviation of 0.11/K. The differences between the respondents in their sensitivity to room temperature were statistically significant ($p < 0.01$). The result therefore showed that there were systematic differences among the respondents in their sensitivity to temperature, and that these differences were substantial.[10] However, the sensitivity of the same respondent also varied substantially from month to month, and this variation was not entirely attributable to chance variations in the estimates of sensitivity ($p < 0.03$).[11] Humphreys and Nicol put it like this: 'Over a long period one person's responses vary in much the same way as do those of other people.'

As far as we know there are no other field study reports that have explored this question. Fishman and Pimbert's near-replication of the BRS survey in a commercial setting did not divide the data into monthly batches, and gave no information on individual differences in sensitivity.[12] Williamson's Australian study of thermal comfort in houses obtained long-term records from individuals, but information on differential sensitivity has not been given.[13,14] The longitudinal data collected during the SCATs project could provide further information on the question, but the requisite analysis has yet to be performed.

There are, then, systematic differences in sensitivity between people, and the sensitivity of an individual is not constant but varies from month to month. If the room temperature moves away from its current optimum, there will be a tendency for some people to be more aware of the change than others. However, a person's sensitivity varies over time, so it will not always be the same people who become aware of the change in room temperature. This variation over time reduces the practical importance of the differences in sensitivity between people.

The differences among individuals in their sensitivity far exceed the difference in sensitivity between the sexes. So although on average women are more sensitive than men, this does not apply to individual men and women. A particular man could be more sensitive than a particular woman, or vice-versa. Again, this much reduces the practical importance of the difference between the sexes.

Notes

1 Nevins, R. G. et al. (1966) A temperature-humidity chart for thermal comfort of seated persons, *ASHRAE Transactions* 72(2), 283–91.
2 Hindmarsh, M. E. and MacPherson, R. K. (1962) Thermal comfort in Australia, *Australian Journal of Science* 24, 335–9.
3 Davies, A. D. M. (1972) *Subjective ratings of the classroom environment: a sixty-two week study of St George's School, Wallasey.* University of Liverpool, Liverpool.
4 This sounds laborious but is quick to perform within a statistical package such as SPSS, using the aggregate facility and the 'append data' option.
5 Nicol, F. and Humphreys, M. (2010) Derivation of the adaptive equations for thermal comfort in free-running buildings in European standard EN15251, *Building and Environment* 45, 11–7.
6 de Dear, R. et al. (1997) *Developing an adaptive model of thermal comfort and preference.* Macquarie University, Sydney.
7 The SCATs database does not identify the sex of the respondents. Background surveys asked the sex of the respondent, but only about half returned the completed questionnaire, and so the information was not entered in the database.
8 The question could be resolved in the climate laboratory, where a wider range of thermal insulation could be used.
9 Humphreys, M. A. and Nicol, J. F. (1970) An investigation into thermal comfort of office workers, *J. Inst. Heat. & Vent. Eng.* 38, 181–9.
10 Among the possibly numerous reasons for such individual differences are differences in body build and in characteristic levels of metabolic rate.
11 Some of the statistics in this paragraph are not in Humphreys and Nicol's original 1970 publication. We have calculated them from Figure 10 in the paper and from the accompanying table.
12 Fishman, D. S. and Pimbert, S. L. (1982) The thermal environment in offices, *Energy and Building* 5(2), 109–16.
13 Williamson, T. J. et al. (1989) Thermal preferences in housing in the humid tropics, *ERDC No 1*, Energy Research and Development Corporation, Canberra.
14 Williamson, T. J. (1989) *Thermal comfort and preferences in housing: south and central Australia.* University of Adelaide, Adelaide.

26

CONSTRUCTING BELL-SHAPED CURVES FOR COMFORT AND TEMPERATURE

Curves relating the proportion of the population likely to be in thermal comfort with the departure of the room temperature from its optimal value, as illustrated in Figure 26.1, are usually derived from a comfort field survey by applying Probit analysis. But because of the equivalences that exist between Probit analysis and ordinary regression analysis, as shown in Chapter 24, they can equally well be derived from the statistics of an ordinary regression equation. In this chapter, we show how this is done, and proceed to apply the method to derive bell-shaped comfort curves for short-term variations in temperature in the workplace. By this we mean temperature changes that occur during a single day at work, where people adapt very little to the temperature variations, as shown in Chapter 25. By extension, the results will apply to any circumstances where for any reason people cannot or do not adapt to temperature changes.

There are reasons to prefer the derivation from regression statistics:

- A small irregularity in the interpretation of the scale will lead to large changes in the bell curve if it is derived from the Probit statistics, but if the curve is derived from regression statistics it is much less affected by such irregularities. When the curve is derived from the regression statistics, the bell curve depends very little on the idiosyncrasies of the particular subjective scale that is used. The ASHRAE scale, for example, does not behave in exactly the same way when translated into different languages, as we have shown in Chapter 19. The irregularities found in the behaviour of a particular version of a subjective scale when used in a particular setting are attributable to the exact meanings and semantic ranges of the particular words rather than to the underlying sensation of warmth. It is unreasonable to link the incidence of discomfort to these irregularities in the interpretation of the scale. Derivation of the bell curve from the regression statistics is therefore to be preferred.

- If Probit statistics are used to derive the bell curve, it is necessary to take certain scale-descriptors as indicating comfort. It is usual to take the three central categories of a seven-point scale as definitive of comfort; in the ASHRAE scale, these are 'slightly warm', 'neutral' and 'slightly cool'. This procedure is rather inflexible and arbitrary. When statistics from an ordinary regression are used to construct the curve, any length of the subjective continuum can be chosen to describe comfort, as may be appropriate. This flexibility enables us to see how the proportion of the population considered to be comfortable depends upon the interval of the subjective warmth scale that is taken to indicate thermal comfort. The examples given later in the chapter will clarify this statement.

Figure 26.1 shows how such a bell-shaped graph is derived in the usual way from the results of a Probit analysis. The right-hand cumulative normal curve (dashed line) is derived from a Probit regression and represents the proportion of people who vote either 'hot' or 'warm' on the ASHRAE scale. The cumulative normal curve on the left is the number who vote 'hot', 'warm', 'slightly warm', 'neutral', or 'slightly cool'. Subtracting the right-hand curve from the left-hand curve leaves those who vote 'slightly warm', 'neutral' or 'slightly cool'; that is, those who are conventionally taken to be comfortable on the ASHRAE scale. This is the bell-shaped curve on the figure. The horizontal axis is the chosen thermal index, and may be expressed in terms of the departure from its optimal value for the population.

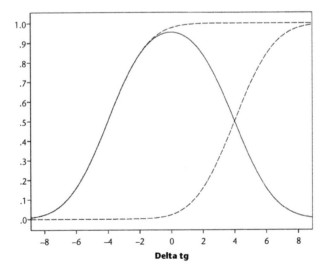

FIGURE 26.1 Constructing the bell-shaped curve showing the probable proportion in comfort against the departure (delta tg) from the day-mean operative temperature.

Using regression statistics to derive the bell curve

The shape and positions of the two cumulative curves in Figure 26.1 can be cal-culated directly from the statistics of the regression of the thermal sensation on the room temperature. The curves are cumulative distributions whose standard deviation is the between-persons standard deviation of room temperature for exceeding a particular value on the ASHRAE scale. This standard deviation of temperature is, as shown in Chapter 24, equal to the residual standard deviation σ_{rv} of the ASHRAE vote divided by the regression coefficient (b). It is therefore equal to σ_{rv}/b. The temperature separation of the curves in the figure is three categories of the ASHRAE scale, and is therefore equal to 3/b. The curve on the right is calculated using the normal cumulative distribution function (Cdf. Normal) available in a statistical package or from statistical tables, with the value 3/(2b) for the mean and σ_{rv}/b for the standard deviation. The curve on the left has a mean of $-3/(2b)$ and the same standard deviation. Subtracting them gives the required bell.

Thus the proportion in thermal comfort is therefore given by:

$$P_{(comf)} = P_{(left\ hand\ curve)} - P_{(right\ hand\ curve)}$$

The following example might help:

In SPSS we used the 'transform variable' menu to create the bell (bell9) and to output its shape as a scatter plot on the temperature axis (delta_tg') for the particular circumstance that the residual standard deviation is 0.9 scale units, the regression coefficient 0.5/K and the length of the thermal sensation scale reck-oned as comfortable is 3 scale units.

The commands are as follows:

```
COMPUTE     bell9=CDF.NORMAL(delta_top,-3/2/.5,0.9/0.5)
-CDF.NORMAL(delta_top,3/2/.5,0.9/0.5).

EXECUTE.

GRAPH

/SCATTERPLOT(BIVAR)=delta_top WITH bell9

/MISSING=LISTWISE.
```

Using this procedure we shall first explore some general features of these curves, and then go on to produce a family of standard curves applicable to different populations.

Effect of the residual standard deviation on the bell curve

The maximum of the bell curve depends only on the residual standard deviation of the subjective warmth after regression on the temperature, and on the length of the scale included in the comfort zone. We illustrate this with a family of bell curves, all of which are for a regression coefficient of 0.5/K and a scale-length of 3 scale units as indicating comfort, as shown in Figure 26.2. The curves differ only because of the different values of the residual standard deviation. The curve with the highest peak, where everyone is comfortable at the optimum, has a residual standard deviation after regression of 0.5 scale units. It is possible for everyone to be comfortable, because the more similar are people's responses the more of them can be satisfied by a single well-chosen room temperature. The curve with the lowest peak (0.86) applies to a residual standard deviation of 1.0 scale units.

Effect of the scale-length taken to indicate thermal comfort

Figure 26.3 shows the effect, on the proportion in comfort at the optimum temperature, of different assumptions about the length of the subjective scale that is taken to indicate comfort. Figure 26.3 shows the effect of the scale-interval on the shape of the bell curve. The family of curves are for a residual standard deviation of 0.8 scale units and regression coefficient 0.5/K. The upper curve takes a scale-length of three units as indicating comfort. Then in succession the curves are for scale-lengths 2.5, 2.0, 1.5 and 1.0 scale units. Three units is the conventional assumption, and on

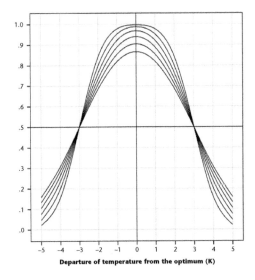

Departure of temperature from the optimum (K)

FIGURE 26.2 Illustrating the effect of the residual standard deviation of the thermal sensation on the proportion of people in comfort. The values of the residual standard deviation are: 1.0, 0.9, 0.8, 0.7, 0.6, 0.5 scale units, the highest peak applying to 0.5 scale units. The regression coefficient is 0.5/K and a scale-length of three units is taken as comfortable.

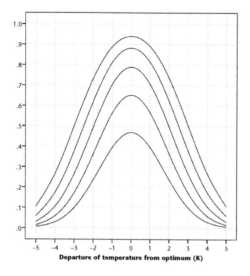

FIGURE 26.3 The effect of altering the length of the subjective scale considered to indicate comfort.

the ASHRAE scale this would include 'slightly warm', 'neutral' and 'slightly cool'. An interval of 1 scale unit is equivalent to the more stringent standard of assuming that only those who are 'neutral' are truly comfortable. It is evident from the figures that the decision about the length of the scale to include within the definition of comfort has a large effect on the bell curve.

The combined effect of the residual standard deviation and the number of scale units (n) indicating comfort is illustrated in Figure 26.4. A little consideration will show that it is independent of the regression coefficient and is given by the relation:

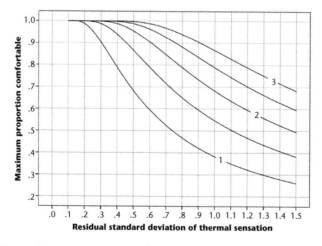

FIGURE 26.4 The maximum proportion comfortable against the residual standard deviation of the comfort scale after regression. The lower curve includes a scale-length of 1 scale unit within the comfort zone. The remaining curves are for 1.5, 2.0, 2.5 and 3 scale units.

$$P_{comf} = 2\{Cdf.normal(n/2\sigma_{rv})\} - 1$$

The distribution of the quantity $n/2\sigma_{rv}$ has zero mean and unit standard deviation. The figure shows this relation for n = 1 and for n = 3. The predicted proportion dissatisfied (PPD) in the Fanger comfort equation has a minimum of 5 per cent. This corresponds to 95 per cent of the population being comfortable, and implies a residual standard deviation (the between–subject differences) of 0.76 scale units. Fanger's model takes three units of the subjective scale as representing comfort, so n = 3.

Effect of the regression coefficient on the shape of the bell curve

Figure 26.5 illustrates the effect of the regression coefficient on the bell curve. The family of curves on the figure all have a residual standard deviation of 0.8 scale units, and a scale-length of 3 scale units to indicate comfort. The innermost bell curve is for a regression coefficient of 0.5 scale units per degree. The other three bell curves, in sequence, are for regression coefficients of 0.4, 0.3 and 0.2 scale units per degree. The effect of a lower regression coefficient is to flatten the bell curve. The effect is large, so it is important that the regression coefficient be estimated with a good degree of accuracy. Typically, in small datasets this is not possible, and hence the need for the procedures explained in the previous chapter to arrive at sound estimates of the day survey coefficients.

The importance of the statistical model

Figures 26.2 through to 26.5 illustrate not only the manner in which the bell curve depends on the regression statistics, but also show the sensitivity of the curve to their

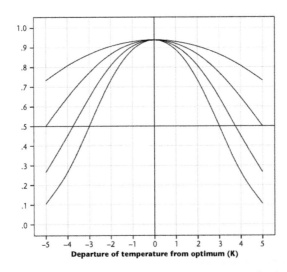

FIGURE 26.5 The effect of the regression coefficent on the shape of the bell curve. In succession from the innermost curve, b = 0.05, 0.4, 0.3, 0.2/K. The residual standard deviation is 0.8 scale units.

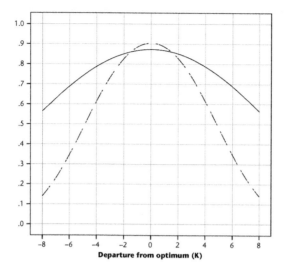

FIGURE 26.6 SCATs data: the effect of the statistical model on the bell curve. The solid line is for the data taken all together. The broken line is for the data broken down into day surveys.

values. The statistics that are obtained from a database may depend strongly on the statistical model that is assumed in the analysis. The importance of choosing a suitable model can be illustrated from the SCATs data. The naïve method would be to estimate the relation between the thermal sensation and the room temperature from the data as a whole, making no distinction between the countries from which the data come or the time at which they were collected. This simple model gives a regression coefficient of 0.167/K and a residual standard deviation of 0.985 scale units. A more appropriate model breaks the data down into the results of day surveys, as described in the previous chapter, and this model gives a regression coefficient of 0.31/K and a residual standard deviation of 0.90 scale units.[1] Figure 26.6 compares the bell curves derived from the regression statistics of the two models. The flatter of the two curves is for the simple model. It under-estimates the proportion in comfort at the optimum temperature, but very seriously over-estimates it when the temperature deviates far from its optimum.

Fitting the data from the statistics of the day surveys

We can now draw the bell curves for the regression coefficients of the day surveys that we established in Chapter 25 (see Table 25.4).

The other statistic we need is the residual standard deviation of the thermal sensation scale. This was estimated as follows: the residuals have been pooled from the two databases and a correction made to the degrees of freedom for the number of day surveys. This gave a value of 0.917. From the variance, we subtracted Sheppard's correction for grouping, and the small amount by which the error in the predictor inflated the residual standard deviation. This gave a value of 0.854 scale units for

TABLE 26.1 Estimating the residual standard deviation of the subjective warmth.

	Standard deviation	*Variance*
Residual after regression analysis	0.917	0.841
Correction for error-in-predictor		0.028
Sheppard's correction		0.083
Adjusted estimate of the residual	0.854	0.730

the residual (see Table 26.1) for men and women together. From the ASHRAE database, we found that the residual for women was 8 per cent greater than for men. So for women we increased the value by 4 per cent and for men we reduced it by 4 per cent, giving the residual standard deviation for women as 0.89 scale units, and for men 0.82.[2]

From these statistics we draw a number of curves. The first (Figure 26.7) makes no distinction between men and women or between the modes of operation of the heating or cooling system. The second (Figure 26.8) introduces a distinction between men and women. Inspection of this figure suggests that the distinction is large enough to be of some practical importance. For any equal reported likelihood of comfort, the women need a narrower band of room temperature than do the men. The third (Figure 26.9) retains this distinction but applies to the AC mode only, while Figure 26.10 applies to the NV mode. In practice, it is the upper parts of the curves that are important, because of the need to keep a high proportion of

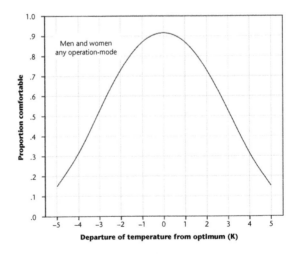

FIGURE 26.7 The proportion of a population that is comfortable. The curve applies to the following circumstances. AC and NV modes combined. No distinction between men and women. A scale-length of 3 units is considered comfortable. Sensitivity 0.48/K. Residual standard deviation of the subjective warmth response: 0.870 scale units.

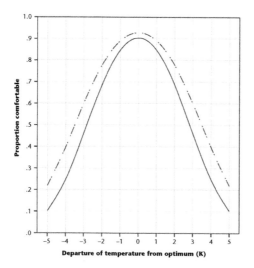

FIGURE 26.8 Proportion comfortable, AC and NV modes pooled. Upper curve: men; lower curve: women. The curves apply to the following circumstances. A scale–length of 3 units is considered comfortable. Sensitivities: men 0.43/K; women 0.53/K. Residual standard deviation of the subjective warmth response: men 0.835 scale units; women 0.904 scale units.

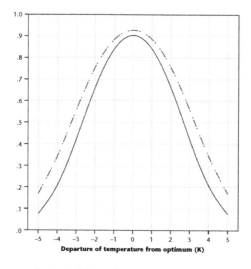

FIGURE 26.9 Proportion comfortable, AC mode. Upper curve: men; lower curve: women. The curves apply to the following circumstances. A scale–length of 3 units is considered comfortable. Sensitivities: men 0.46/K; women 0.56/K. Residual standard deviation of the subjective warmth response: men 0.835 scale units; women 0.904 scale units.

the people comfortable. At the 80 per cent level (p = 0.8), the differences between the curves for the two modes are small, and so it is reasonable to use Figure 26.8 and make no allowance for the mode of heating or cooling.

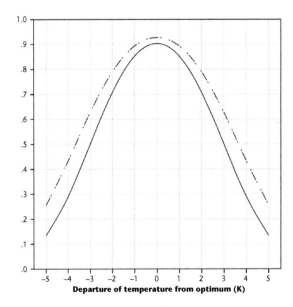

FIGURE 26.10 Proportion comfortable, NV mode. Upper curve: men; lower curve: women. The curves apply to the following circumstances. A scale-length of 3 units is considered comfortable. Sensitivities: men 0.41/K; women 0.50/K. Residual standard deviation of the subjective warmth response: men 0.835 scale units; women 0.904 scale units.

In conclusion

In this chapter we have shown how bell curves may be constructed using the statistics obtained by regression analysis. We have explored the separate effects of deriving the bell curve using different regression statistics. We have used the method to construct practical bell curves for the results of the day surveys in the ASHRAE RP-884 database and the SCATs database. We conclude that:

- The bell curve for women is distinctly different from that for the men.
- The bell curves for the NV and AC modes are so similar that no distinction between them is needed.

The curves may be used to estimate the proportion of people likely to be comfortable, and apply when people adapt little to the temperature fluctuations they encounter. The curves strongly suggest that temperature fluctuations about the optimal temperature within a working day should be kept small. This may be achieved either with a well-controlled heating or cooling system, or, in the free-running mode, by having a building of sufficiently high thermal inertia (a heavy-weight building with a high 'thermal mass').

Notes

1 The analysis pooled all the modes of operation and is without the correction for error in the predictor.
2 These estimated values of the residuals are still slightly high. This is because they are derived from databases where the ASHRAE scale, or the Bedford scale, was presented in a variety of languages. This means that the pooled regressions used to derive the statistics apply to a 'pooled' version of the seven-point scale. This will tend slightly to inflate the residual variance after regression. Estimates of the size of this effect are not yet available. The curves therefore slightly underestimate the proportion comfortable at the optimal temperature.

27

TOLERANCE OF SEASONAL DRIFT OF INDOOR TEMPERATURE

In Chapter 25, we examined the day survey, where there was very little adaptation by means of clothing change, and showed that an increase in room temperature of 2 K raised the comfort vote by about 1 scale unit – a sensitivity in the region of 0.5/K. In this chapter, we explore the sensitivity when people have had more opportunity to adapt to their thermal environment. We touched on this topic in Part I of this volume, while recounting the story of the development of the adaptive approach.

We introduce our more extensive consideration of the topic by considering the oldest survey that compares winter and summer comfort. It comes from Brazil.[1] Sà, in a longitudinal thermal comfort survey in Rio de Janeiro, collected some 1,000 responses from just eight respondents, winter and summer, using a seven-point scale of warmth. He reported that his respondents had a mean comfort vote in winter of +0.05 scale units, when the mean indoor temperature was 24.7°C, while in summer they had a mean vote of 1.00 at a mean indoor temperature of 28.8°C. The winter–summer sensitivity is therefore 0.23/K. The wintertime mean comfort vote of 0.05 shows that they were well adapted to a mean indoor temperature of 25°C. However, the summertime mean comfort vote of +1.00 scale units, while not excessively high, indicates that Sà's respondents had not completely adapted to a summer indoor mean temperature of 29°C.

The adaptive model predicts that, given time and opportunity, people will adapt so that they are comfortable in *the prevailing thermal environment they are experiencing*. So we would expect a near-zero sensitivity to gradual seasonal indoor temperature changes experienced in a building. We do not so much ask why the sensitivity to seasonal changes in indoor temperature was so much lower than the values obtained from day surveys (0.23/K for the seasonal change, compared with around 0.45/K found from the analysis of within-day variations of temperature in Chapter 25). Rather, the adaptive model asks why it was not zero. Answers are of three kinds.

First, it should be noticed that *the mean indoor temperature during a survey* is only a surrogate variable for *the prevailing thermal environment experienced by the respondents*, and may not be a very good one. That is to say, the mean indoor temperature during the survey hours may differ from the normal temperature experience of the respondents while up and about during the day. We notice that:

- The workplace provides just part – perhaps about half – of respondents' waking hours thermal experience. Away from work, they may experience environments systematically warmer or cooler than those recorded during the survey at the workplace.
- The survey period is sometimes just a day or two. The workplace may have been unusually warm or cool on those days, if for example there was a brief hot spell or cold spell, or if the heating or cooling plant was not providing its usual room temperatures.
- Measurements and subjective responses are sometimes recorded during the warmest period of the day, rather than being distributed uniformly throughout the day. A bias from this source is likely in summertime surveys in buildings operating in the free-running mode. The effect will be greatest in buildings subject to large diurnal swings of indoor temperature.

In such examples there is no failure of the adaptive process. It is just that the mean temperature recorded during the survey did not correctly represent *the prevailing thermal environment experienced by the respondents*.

Second, it may be that there is some constraint on the adaptive process. For example, social custom or institutional policy may prevent the adoption of clothing that would be sensible from a thermal point of view, for clothing has important social functions besides providing thermal insulation. People 'say things' by what they are wearing. The social requirement may clash with the thermal requirement. Adaptation may then be incomplete for social reasons.

Third, it may be that the thermal environment is outside the range to which people can adapt completely. Adaptation is then incomplete because of the limitations of our thermal physiology or the adaptive opportunities afforded by the building. Here we move beyond the realm of thermal comfort and into the realm of thermal stress.

Nearly 80 years have passed since Sà conducted his survey, and we cannot now discover to what the apparently incomplete summer adaptation should be attributed. We do not know how typical recorded temperatures were of the experience of the respondents, nor what social constraints they had on their clothing. With the high summer humidity that prevails in Rio, an indoor temperature of 29°C is quite close to the physiological limit for comfort. Thermal comfort would be possible with light, loose clothing and an elevated air speed, but we do not know about the clothing or the air movement in Sà's Rio. So the apparently incomplete adaptation to the summer condition could be for any or all of the three kinds of reason.

We turn to more recent surveys. There are a number of studies from the 1960s and 1970s that touch on the effects of seasonal temperature changes. We have mentioned these in earlier chapters, but here consider them in more detail.

Adaptation to day-to-day variations in room temperature

Hindmarsh and MacPherson studied office workers in Sydney, Australia. They obtained responses from their 14 experimental subjects on a seven-point thermal sensation scale at about 11 am *daily* for a whole year.[2] The design of the experiment therefore excluded the effect of temperature change during a single day. The comfort votes would reflect any day-to-day changes in room temperature, together with the seasonal variations. Analysis of their data gives a regression coefficient of just 0.16/K, or about a third of the value found in day surveys. The result suggests that there was considerable adaptation to day-to-day and longer-term temperature trends. The warmest indoor temperatures (mean 24.2°C) were in January–March, and the coolest in October–December (21.4°C). The seasonal variation was therefore quite small (2.8 K). The overall range of the 11 am indoor temperature was much greater (about 15–30°C). The result therefore tells us more about the response to day-to-day temperature variations than it does to seasonal variations. The value of 0.16/K indicates considerable adaptation to the day-on-day temperature variations in their offices.

Adaptation to week-to-week room temperature variations

Davies conducted a similar experiment for secondary school children in the UK.[3,4] She sampled the response of a class at *weekly* intervals for more than a year. The school was of heavyweight construction with an unusually high 'thermal mass', and so the temperature varied little during the school day. There was no heating system, the classroom being designed to use solar heat and the children's metabolic heat. We do not know the exact extent of the seasonal variation in indoor temperature, but it was substantial. The design of the experiment excluded the effects of temperature variations during a week. The children's thermal responses were therefore attributable chiefly to the week-to-week variations in room temperature over a year and more. She obtained a regression coefficient of just 0.10/K. The result shows that there was considerable but not quite complete adaptation from week to week throughout the school year. Davies also recorded the children's clothing. It varied systematically with the week-to-week changes in classroom temperature, which explains how the children had adapted to these changes.

Adaptation to month-to-month variations in room temperature

The analysis of the BRS data-logging project[5] described in Part I of this volume, where the data were separated by individual respondent and by month, informs us of the extent of adaptation from month to month. The statistical analysis showed that the monthly mean comfort votes of each respondent depended little on the monthly mean temperature of their office during the working day. The experimental range of these monthly means was from 18.5°C to 25°C. None of the respondents individually could be shown to be sensitive to the month-to-month variation of their office temperature, though the hypothesis that the sensitivity of an individual was 0.03/K best fitted the data. This very low value demonstrated that their adaptation was almost complete. Humphreys and Nicol also analysed the dependence of the mean vote on the corresponding mean monthly indoor temperature, pooling the mean values from all the respondents. This gave a regression coefficient of just $0.049 \pm 0.025/K$. The very small coefficient marginally failed to reach the $p = 0.05$ level of confidence. So the respondents, whether considered as individuals or as a group, had adapted almost completely to the month-to-month variation in the mean temperature of their offices, and, consequently, their sensitivity to such changes was close to zero. The seasonal drift of 6.5 K in the monthly mean indoor temperature did not therefore produce any thermal discomfort. This leads us to explore the extent of adaptation to seasonal variations of indoor temperature.

Adaptation to seasonal changes in the indoor temperature

Auliciems studied the thermal comfort of UK secondary school children, and we may compare his winter and summer results.[6–8] The mean classroom temperature in winter was 19.9°C and in summer 21.8°C, quite a small seasonal difference. In winter the mean warmth sensation was 0.59 scale units (Bedford scale) while for the summer survey it was 0.29 scale units. The children had, it seems, over-compensated for the seasonal difference in their classroom temperature (or perhaps the setting for the school heating system was higher than they were experiencing at home). However, his summer survey was a quarter of the size of his winter survey, and the data came from fewer schools. The populations were thus not precisely the same, so the result is not very secure.

Seasonal adaptation and the ASHRAE database

We turn now to more recent and more specific evidence. Within the ASHRAE RP-884 database, there are a number of surveys that have investigated similar populations in the same buildings winter and summer (Table 27.1). We omit from the table our own surveys in Pakistan, as they will be examined later. Seven winter/summer-paired comparisons are to be found in the database. For all but one of these, the mean indoor temperature hardly differed between winter and summer (1 K or less), and therefore contains no useful information on adaptation to a seasonal variation of indoor temperature. The differences in mean comfort vote that were found remind us only of the somewhat rough nature of subjective data obtained in the field.

TABLE 27.1 Seasonal comparisons for populations from the ASHRAE RP-884 database.

File no:	Country	Region or town	Summer				Winter				Differences	
			Mean vote	Mean temp	N	hc	Mean vote	Mean temp	N	hc	Mean vote	Mean temp
1,2	Wales	Glamorgan	1.45	22.8	80	1	.42	20.3	38	1	1.03	2.6
9,10	Canada	Montreal	−.28	23.6	443	1	−.28	22.6	426	1	.00	1.0
29,30	Australia	Sydney	.10	24.0	137		.63	23.3	170		−.53	0.7
32,34	USA	San Francisco	.25	23.2	673	1	.17	22.9	923	1	.08	0.3
33,35	USA	San Francisco	.32	23.9	360	0	.20	23.0	390	1	.12	0.9
38,39	England	Merseyside	.43	21.9	167	0	.70	21.9	209	1	−.27	−0.1
44,45	USA	San Ramon	.15	22.8	96	1	−.20	22.1	285	1	.35	0.8

Note: The database is made up of numerous files. The first column is the file number in the database. hc = 1 indicates heating/cooling plant was operating during the survey.

Small winter–summer differences in indoor temperature can occur for various reasons:

- A building in a temperate climate is likely to be heated in winter and unheated in summer. If the summer happens to be cool during the survey period the mean indoor temperature can even be below the winter value.
- A building that has a central HVAC system might operate with the same indoor temperature set point all year round.
- A building may be in an equatorial climate, where the seasonal range in the outdoor temperature is small. There will be little if any seasonal drift in indoor temperature.

There remains one comparison in the database that has a substantial seasonal variation of indoor temperature. It comes from Brown's study of the thermal comfort of workers engaged in industrial processes in factories in South Wales. It has a seasonal difference of 2.6 K. The mean thermal sensation in summer was 1.45 scale units and in winter 0.42. The figures suggest that there was almost no adaptation to the seasonal change, the sensitivity being similar to that found from day surveys. There are reasons to doubt the reliability of this finding. The winter data were few (just 38 observations) and came from fewer factories than did the summer data, and so the populations were different in the two seasons. It is likely that the industrial processes differed too. They ranged from light industrial to heavy industrial tasks. Modest differences in activity level can cause big changes in the thermal sensation. Nevertheless, it should be noted that the workers failed to adapt successfully to a very moderate summer mean temperature of 23°C – and we do not know why this was so.

Seasonal adaptation and the Pakistan surveys

We have the results of two research projects that provide useful information on adaptation to larger seasonal temperature changes. The first project gives a simple winter–summer comparison. The experiments were of longitudinal design. Five respondents, who were native residents, gave hourly responses for a period of a week, winter and summer, in each of five cities.[9] Their thermal environment was logged by a personal portable monitor set. The mean votes and mean globe temperatures are shown in Table 27.2. Within this project a small supplementary survey of transverse design was conducted among the employees of a bank in Mingora.[10] The result is shown as the last row of the table.

The seasonal temperature differences are large, ranging from 7 K to 17 K. The sensitivity to seasonal change of indoor temperature is given in the final column. The low values all indicate substantial adaptation even to large seasonal changes. The adaptation was very strong in Karachi, Peshawar and Quetta. Our respondents in Mingora did not adapt successfully to the low winter indoor mean temperature of 12.7°C in their dwellings, but the bank employees in the same town adapted well to their winter workplace temperature of 15.9°C. The larger seasonal drift found in Multan (11.4 K) caused some discomfort winter and summer. Seasonal differences of 10 K or less seemed to be quite well accommodated, giving a value for the sensitivity of 0.1/K or less. So it would need some 10 K of seasonal change in temperature to change the comfort vote by 1 scale unit.

This project was followed up a few years later by monthly transverse surveys in the same regions of Pakistan, but with different respondents.[11] This year-round project had the advantage of sampling the entire seasonal ranges rather than just the extremes. Figure 27.1 shows the mean indoor temperatures and the mean comfort vote for each survey in each of the five regions. The expected pattern of adaptation is that there would be a temperature range over which people could easily adapt to

TABLE 27.2 Summer–winter comparisons in five cities in Pakistan.

City	Summer			Winter			Seasonal difference		
	Mean vote	Mean temp (°C)	N	Mean vote	Mean temp (°C)	N	Mean vote	Mean temp (K)	Sensitivity (/K)
Karachi	.49	31.4	190	−.19	24.6	470	.68	6.8	.10
Peshawar	.60	30.2	556	−.77	18.7	513	1.37	11.5	.12
Quetta	.77	29.5	492	−.29	19.4	425	1.06	10.2	.10
Multan	0.56	30.2	437	−1.25	18.8	582	−1.25	11.4	.16
Mingora	.90	29.3	568	−2.36	12.7	548	3.26	16.6	.20
Mingora(t)	−0.05	26.2	42	0.11	15.9	19	−0.16	10.3	−.02

Note: Mingora(t) gives the result from the small transverse survey of bank employees in Mingora.

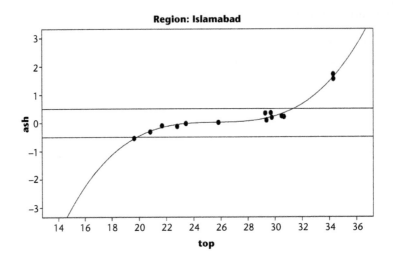

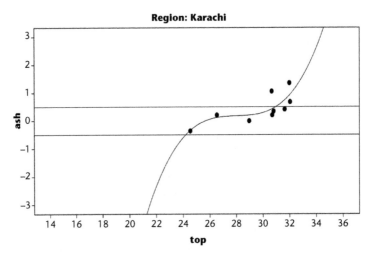

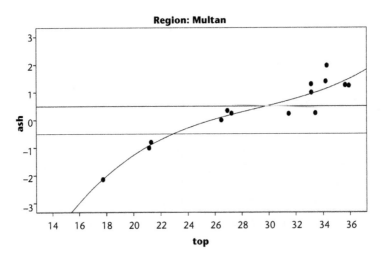

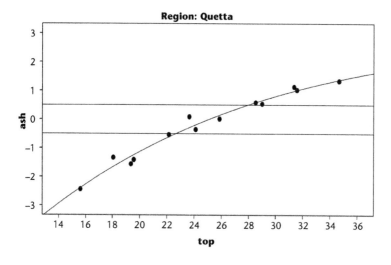

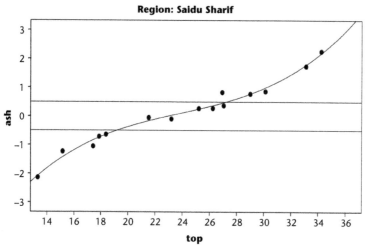

FIGURE 27.1 (a–e) The extent of adaptation to seasonal indoor temperature variations in five cities of Pakistan. (Saidu Sharif and Mingora are adjacent cities). 'ash' is the month mean response on the ASHRAE scale, and 'top' the month mean indoor globe temperature.

their seasonal indoor temperature. Towards the ends of this range, failure of adaptation would begin to show. At the hot end people would become increasingly hot, while at the cold end they would become increasingly cold. This expected pattern can be approximately modelled by the use of a cubic fit to the data, and we have drawn such lines in Figure 27.1 (a–e).

The lines fitted to the data from Islamabad and Karachi show this expected pattern, while the lines fitted to the data from Multan and Saidu Sharif show the pattern to some extent. The line fitted to the data from Quetta fails to reveal it; the upturn in the line at high temperatures fails to occur within the range of our data. We have

TABLE 27.3 Approximate adaptive limits for five cities in Pakistan.

City	Lower limit (°C)	Upper limit (°C)	Seasonal range (K)
Islamabad	20	31	11
Karachi	<24*	31	>7
Multan	23	30	7
Quetta	19	27	8
Saidu Sharif	17	27	10

*The lower limit for Karachi is not defined, there being no data below 24°C.

drawn a band across the figures to indicate a zone where the level of adaptation can be considered to be very good. This horizontal band is set at ±0.5 units of the seven-point scale, a band in which a very high proportion of people would be comfortable. It is approximately equivalent to the room temperature being within 1 K of its current optimum value during a working day – a high standard.

Using this quite stringent standard, we can quantify approximately the limits of adaptation in the five regions (Table 27.3). Despite the wide ranges over which the people adapted successfully to their mean indoor temperatures, the figures reveal considerable discomfort at more extreme temperatures. At these extremes, the adaptive strategies available to our respondents had become inadequate to secure a high standard of thermal comfort.

Seasonal adaptation and the SCATs data

The SCATs project followed the same monthly pattern of data collection, and we have data from five European countries. The indoor temperature was but one aspect of the environment that was monitored in the project, but it is with this that we are currently concerned. We examine the month-by-month sets of data to see what they tell us about the completeness or otherwise of seasonal adaptation in Europe. The results from the five countries are shown in Figure 27.2 (a–e) Each observation on the graphs represents the mean comfort vote (ASHRAE scale, numbered 1–7) for the month and the corresponding mean of the globe temperature.

Figure 27.2 (a–e) shows that the adaptation to the seasonal variations in the indoor temperature is very good. Few points are outside the band that indicates a high standard of comfort. Because the limits of seasonal adaptation are rarely if ever reached we have not fitted cubic lines to the observations.

Two observations fall above the band in the French data, and the higher of these, at 28°C, is probably attributable to the high indoor temperature, and suggests the upper limit to their ability to adapt is about 27°C. The other observation, at 25°C, falls only marginally above the band. The lack of full adaptation does not appear to arise from the room temperature itself, as there are higher month mean temperatures to which the respondents successfully adapted.

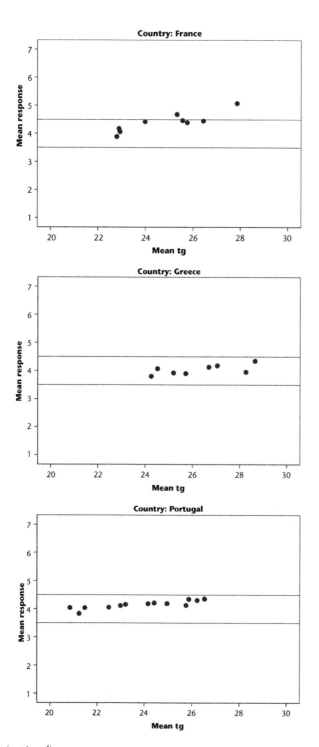

FIGURE 27.2 (a–e) *(continued)*

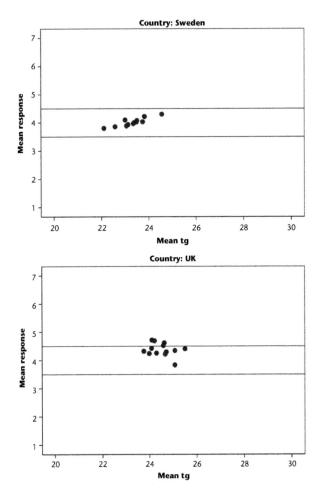

FIGURE 27.2 (a–e) The extent of adaptation to seasonal indoor temperature variations in five European countries. 'Mean response' is the month mean response on the ASHRAE scale, numbered 1–7, and 'Mean tg' is the month mean indoor globe temperature.

None of the observations from Greece, Portugal or Sweden fall outside the band. The respondents adapted entirely successfully to the seasonal ranges of their month mean indoor temperatures.

The UK data show three observations that fall marginally above the band, but they do not appear to be related to the month mean temperature. Both in the French data and the UK data the overall mean comfort vote is noticeably on the warm side of thermal neutrality. This is probably attributable not to the thermal environment itself, but to the way the ASHRAE scale is interpreted in French and English. The matter is discussed in Chapter 19.

The SCATs data therefore do not indicate, with the possible exception of the French data, any upper or lower limits to the seasonal variations of room temperature. Table 27.4 sums up the information.

In the data from all countries except the UK, there appears to be a slight upward trend of comfort vote as the month mean temperature increases. This trend may be quantified. Figure 27.3 summarises the effect of pooling the data from the five countries, and thus provides an overall estimate of the sensitivity to seasonal room temperature variations. For this purpose the data are expressed in terms of departures from the annual mean comfort vote and the annual mean indoor temperature for the country.

The linear trend line (the dashed line) has a gradient of 0.09 ± 0.02 scale units per degree. The sensitivity to seasonal variations in mean indoor temperature is therefore about one-fifth of the sensitivity to within-day changes of temperature. The observations on the figure suggest that seasonal drifts of six degrees (the range of our data) need not be a source of thermal discomfort. We have also fitted a cubic trend line (the solid line). It has the expected shape, and suggests that a high standard of comfort is compatible with a seven-degree seasonal drift. However, this estimate depends on an extrapolation beyond the range of the data, and may be unreliable.

TABLE 27.4 Information on the extent of seasonal adaptation from the SCATs data.

Country	Lower limit (°C)	Upper limit (°C)	Range (K)
France	<23	27	>4
Greece	<24	>29	>5
Portugal	<21	>27	>6
Sweden	<22	>25	>3
UK	<24	>26	>2

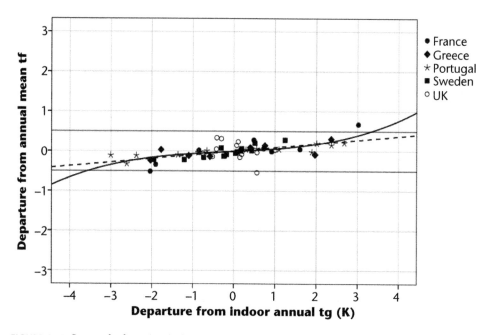

FIGURE 27.3 Seasonal adaptation in European countries – pooled data.

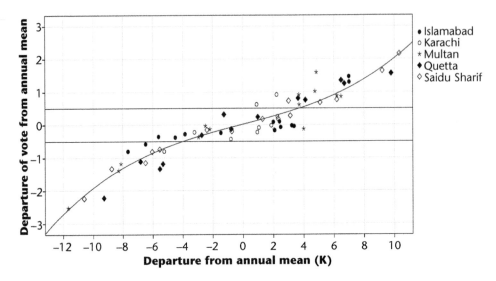

FIGURE 27.4 Seasonal adaptation in five cities in Pakistan – pooled data.

This result may be compared with a similar figure we have derived from the Pakistan data (Figure 27.4). The form of the figure is remarkably similar. A cubic fit for the trend line is needed to describe the Pakistan data because of its much greater range of indoor temperature. The cubic trend line suggests that a seasonal range of indoor temperature of 8 K is consistent with a very high standard of thermal comfort. The gradient at the centre of the figure is 0.12 ± 0.02 scale units per degree, a figure not significantly different from that obtained from the European data.

It seems then that the range of seasonal variation of indoor temperature that can be accommodated in Europe and in Pakistan is similar. This would be expected if the means of adaptation were also similar. The chief means of adapting to a changed room temperature are changes in the thermal insulation of the clothing and the use of fans to increase air movement when the room is hot. There may also be contributions from changes in activity and posture: people in colder rooms tend to increase their metabolic rate slightly, and to adopt a more 'closed' posture.

Summary comment

In this chapter, we have seen that the opportunity to adapt to a changing environment is very powerful in maintaining thermal comfort. Within a population, seasonal drifts in the indoor temperature of less than seven or eight degrees are well tolerated, and are consistent with a very high level of thermal comfort. In these circumstances, the apparent sensitivity to seasonal temperature drifts is about one-fifth of that found in data from day surveys. It follows that adaptation – chiefly by wearing warmer clothing in winter and by using fans in summer – compensates for some 80 per cent of the seasonal variation in room temperature. The advantage of permitting free choice of clothing and the use of efficient fans is evident, both for the thermal comfort of people in buildings, and perhaps on a consequent maximisation of their mental performance. The data strongly suggest that a considerable winter–summer drift in indoor temperature is entirely permissible. Such drifts can save energy for heating or for cooling, or for both.

Notes

1 Sà, P. (1938) *Conforto termico*. Departmento de Estatistica e Publicidade, Rio de Janeiro.
2 Hindmarsh, M. E. and MacPherson, R. K. (1962) Thermal comfort in Australia, *Australian Journal of Science* 24, 335–9.
3 Davies, A. D. M. (1972) *Subjective ratings of the classroom environment: a sixty-two week study of St George's School, Wallasey*. University of Liverpool, Liverpool.
4 Davies, M. G. and Davies, A. D. M. (1973) Warmth ratings and temperature ratings in a classroom, in: *Thermal comfort and moderate heat stress*, Eds: Langdon, F. J. *et al.*, *Proceedings of the CIB Commission W45 (Human Requirements) Symposium*, Building Research Station, 13–15 September 1972, HMSO, London, pp. 79–85.
5 Humphreys, M. A. and Nicol, J. F. (1970), An investigation into thermal comfort of office workers, *J. Inst. Heat. & Vent. Eng.* 38, 181–9.
6 Auliciems, A. (1969) Thermal requirements of secondary school children in winter, *Journal of Hygiene* 67(1), 59–65.
7 Auliciems, A. (1969) Some group differences in thermal comfort, *Heating and Ventilating Engineer and Journal of Air-Conditioning* 71, 562–4.
8 Auliciems, A. (1973) Thermal sensations of secondary school children in summertime. *Journal of Hygiene* 71, 453–8.
9 Nicol, F. *et al.* (1994) *A survey of thermal comfort in Pakistan*. Oxford Brookes University, School of Architecture.
10 Humphreys, M. A. (1994) An adaptive approach to the thermal comfort of office workers in North West Pakistan, *Renewable Energy* 5(5–8), 985–92.
11 Nicol, F. *et al.* (1996) *Thermal comfort in Pakistan II*. Report to the Overseas Development Administration. Oxford Brookes University, School of Architecture.

28

ESTIMATING NEUTRAL TEMPERATURES FROM SURVEY DATA

A regression equation of thermal sensation (TS) on operative temperature (T_{op}) allows the temperature for thermal neutrality (the neutral temperature T_n) to be estimated for the population represented by the data. It is the temperature at which the average thermal sensation would be neutral.

The regression equation is normally written in the gradient–intercept form:

$$TS = bT_{op} + c$$

where b is the regression coefficient and c the intercept on the thermal sensation axis.

The thermal sensation scale is numbered here so that 'neutral' is assigned the value zero. So, setting TS = 0 for thermal neutrality we have:

$$T_n = -c/b$$

There is, however, an alternative way of calculating the neutral temperature from the regression equation. A regression equation always passes through the joint means of its two variables. It can therefore be re-written in terms of the regression coefficient and the joint means of the thermal sensation and the room temperature. Thus, if for any block of data we know the mean operative temperature ($T_{op-mean}$) and the mean thermal sensation (TS_{mean}), we may calculate the neutral temperature (T_n), provided we have a value for the regression coefficient:

$$T_n = T_{op-mean} - TS_{mean}/b$$

The neutral temperature can therefore be calculated from the regression coefficient and the joint means. This formulation is useful if we wish to use a standard value of the sensitivity to temperature variation in place of a value of the regression coefficient derived from a particular survey. The procedure is appropriate when the regression coefficient is either unavailable or thought for some reason to be unreliable, as is frequently so.

The reliability of the estimate of the neutral temperature, if the mean thermal sensation differs from thermal neutrality, depends on the reliability of the estimate of the regression coefficient. We saw in Chapters 25 to 27 that the regression coefficient, indicating the apparent sensitivity to temperature change, depends on the extent to which adaptation occurred during the period over which the data were collected. The value of a regression coefficient derived from a single survey is therefore apt to be an uncertain estimate of the sensitivity to temperature change, and can lead to an unreliable estimate of the neutral temperature.

The extent of the problem can be seen by inspecting any large thermal comfort database. We use the ASHRAE RP-884 database to explore the matter. Figure 28.1 is a funnel plot of the regression coefficients obtained from each of the 160 buildings included in the database. The vertical axis is the number of observations upon which the regression is based. Some wild values occur when observations are very few, and with fewer than 50 observations little credence can be placed on the value obtained. Even with much larger surveys, the regression coefficients differ markedly from survey to survey. For building surveys that have over 300 observations – quite large surveys – the coefficients range from zero to 0.7.

Figure 28.2 is a histogram of the regression coefficients excluding those buildings where fewer than 50 observations were made. Although the exclusion of the small

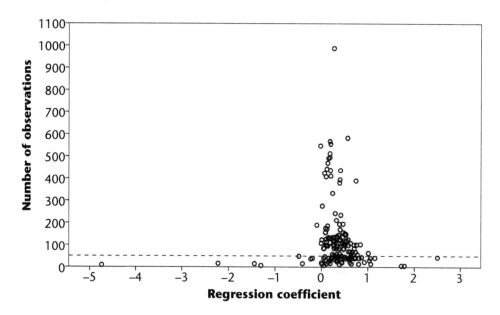

FIGURE 28.1 Funnel plot of the regression coefficients of thermal sensation on operative temperature from the buildings in the ASHRAE RP-884 database. The dashed line is drawn at 50 observations. The vertical axis is the number of observations in the dataset.

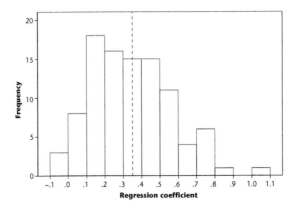

FIGURE 28.2 Histogram of the regression coefficients of thermal sensation on operative temperature from the buildings in the ASHRAE RP-884 database (N > 50). The dashed line is the mean value.

surveys eliminates the wildest estimates, the estimates still range from −0.1 to 1.0 scale units per degree. It is evident that such diversity cannot be attributed to the random scatter that arises from the numbers of observations in the surveys. Rather, it is attributable to the pattern of data acquisition in the survey, and perhaps to unknown factors that apply to particular sets of data. The diversity is so great that it cannot represent real differences of thermal sensitivity between people in different buildings. It is apparent, then, that no single estimate of the coefficient can be relied upon, no matter how large the survey from which it was obtained.

Such uncertainty in the estimate of the regression coefficient leads to a corresponding uncertainty in the estimate of the neutral temperature if the mean thermal sensation differs from neutral. It is therefore necessary, when estimating a neutral temperature from a set of data, to use a regression coefficient that is more soundly based. The use of an appropriate regression coefficient derived from the pooled day surveys (Chapter 25) meets this need, enabling the neutral temperature for that day to be estimated without bias.

The procedure of calculating the neutral temperature from the joint mean is sometimes called the 'Griffiths method'. It is usually associated with the use of a value for the regression coefficient drawn from laboratory studies, and has been criticised because the value may differ from that found in everyday life. The objection is justified, as we have shown that sensitivity to temperature change in daily life differs considerably from that obtained in the climate laboratory. The use of a value estimated from the best available field-survey databases entirely escapes this objection.

At first glance, the use of a regression coefficient derived from day surveys would seem to restrict the usefulness of the method to such surveys. The method will, however, give a valid neutral temperature even if the survey took place over a protracted period of days, weeks or months. In such a case, we use the mean values of the thermal sensation (TS) and of the operative temperature (T_{op}) for the entire survey. The resulting neutral temperature will therefore be for a day whose mean operative temperature equalled the mean operative temperature during the survey, and whose mean thermal sensation equalled the mean vote during the survey. This day will necessarily lie within the range of real days encountered during the survey – a typical survey day rather than any actual day – and its neutral temperature will similarly be for this same typical day within the survey period.

Alternatively, should line-by-line data be available, a neutral temperature can be obtained from each individual 'comfort vote'. The value obtained from an individual comfort vote is of very low precision but is free from systematic bias.

The accuracy of the estimate of the neutral temperature: The error of the estimate of the neutral temperature is attributable to the scatter of the thermal sensation, and can be estimated from the residual standard deviation (σ_{resid}) of the thermal sensation, the regression coefficient (b), and the number of observations (n). It is given by the relation

$$(\sigma_{resid})/(b\sqrt{n})$$

Figure 28.3 shows this relation for a regression coefficient of 0.48/K, the value derived from the day surveys, disregarding the difference between the sexes and the type of heating or cooling. Lines are shown for values of the residual standard deviation of 0.7, 0.8, 0.9 and 1.0 scale units. With as few as 20 observations, the standard error of the estimate is between 0.3 and 0.5 K, so even a small survey can now provide a usefully reliable estimate of the neutral temperature. If the neutral temperature were estimated from a single vote, as for example was done in the analysis of the SCATs data,[1] its standard error is simply the residual standard deviation divided by the regression coefficient [$(\sigma_{resid})/b$].

It will be noticed that the method of estimating the neutral temperature just described avoids the need to perform a regression analysis, because, provided we have a reliable and appropriate value for the sensitivity, the neutral temperature can be estimated simply from the means of the thermal sensation and the room temperature. This is advantageous because there are numerous published thermal comfort field surveys for which we have only summary statistics, and not the line-by-line data that would be needed for regression analysis. Provided only that the mean operative temperature and the mean comfort vote are sound, such summary statistics can be used to obtain sound estimates of the neutral temperatures.

We have over a number of years been assembling a database of such summary statistics (DBSS), and it can now be drawn upon to estimate neutral temperatures from numerous surveys of thermal comfort and to compare them with the corresponding

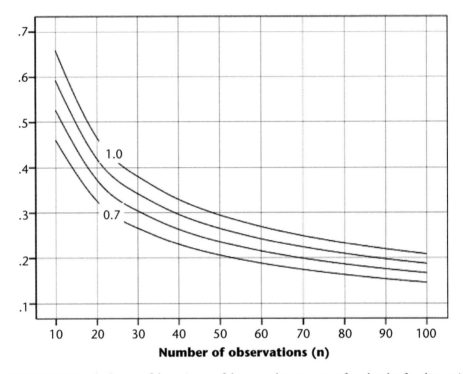

FIGURE 28.3 Standard error of the estimate of the neutral temperature for a batch of n observations, for various values of the residual standard deviation of the thermal sensation. The vertical axis is the standard error of the estimate of the neutral temperature (K).

mean temperatures within the accommodation. The ability validly to use neutral temperatures from surveys for which we have only summary statistics vastly extends the body of available data from which to construct scatter plots relating the climate and the neutral temperature.

The data used by Humphreys in his 1970s meta-analyses are not included in DBSS. The DBSS includes the ASHRAE RP-884 data. They are included in the following way. The RP-884 database consists of numerous files, containing survey data from one or more buildings in a single town or district. There are separate files for buildings with central air conditioning (HVAC), mixed-mode (MM) and natural ventilation (NV). For the DBSS, each file is separated by building, and the summary statistics for each building are entered as a row in the DBSS.

The SCATs data are also included in DBSS. They are entered in the following way. Each row in DBSS consists of the summary statistics for a single month in a single building, the data being collected usually over a few days in that month. The buildings had a variety of heating and cooling systems (air-conditioned (AC), naturally ventilated (NV), mixed mode (MM), mechanically ventilated (MV), part AC part not (PP)).

The many other sources from which we have drawn summary statistics are listed at the end of this chapter. For these sources, unless we hold the raw data, we have necessarily drawn the summary statistics from the data-blocks as given in the publications cited. The database currently has 741 rows of data, the data coming from 30 countries. It summarises over 185,000 sets of observations, or about 90 times as many rows as in the ASHRAE RP-884 database.

The following countries are represented in the DBSS. The number of rows of summary statistics is given in parentheses:

Australia (74), Bangladesh (1), Brazil (1), Canada (29), China (5), France (35), Germany (4), Greece (45), India (13), Indonesia (8), Iran (15), Italy (5), Japan (43), Malaysia (1), Nepal (11), Nigeria (2), Pakistan (76), Papua New Guinea (2), Poland (4), Portugal (51), Saudi Arabia (1), Singapore (2), Slovakia (2), Solomon Islands (1), Sweden (50), Thailand (6), Tunisia (48), UK (168), USA (36), Zambia (2).

Many further surveys await inclusion in the database. In particular we note the absence from the database of most of the numerous studies that underlie the adaptive relation in the recently published Chinese adaptive standard. The database will be available from the authors on request when documentation is completed. The ability to use the summary statistics from so many surveys, and to do so in a logical and standard way, considerably improves the robustness of our knowledge of thermal comfort worldwide.

The neutral temperatures derived from the database are operative temperatures or globe temperatures. There are 732 rows in the database for which both the mean temperature and the mean thermal sensation are available. Figure 28.4 shows the distribution of the neutral temperatures from the surveys in the database.

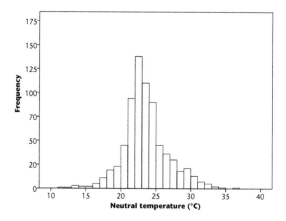

FIGURE 28.4 The distribution of the neutral temperatures in the database of summary statistics.

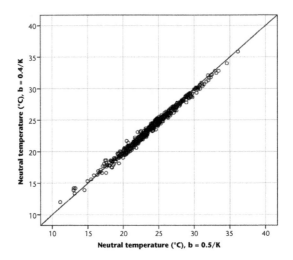

FIGURE 28.5 Comparison of the neutral temperatures calculated using sensitivities of 0.4/K and 0.5/K.

It is evident that the range of neutral temperatures is very large. There are some surveys in which the population reported thermally neutrality below 15°C while populations in many surveys reported neutrality at temperatures above 30°C. (This may seem surprising, but at these temperatures fans are usually in use, and have raised the neutral temperature. The absolute humidity has little effect on the temperatures found to be thermally neutral, even at these room temperatures, but a high humidity reduces people's tolerance of temperatures elevated above their neutral temperature.) The excess of neutral temperatures at 21–25°C is attributable to the great number of surveys conducted in offices in Europe and the USA, where such room temperatures are commonly maintained or encountered, and people have adapted to them.

Figure 28.5 explores the sensitivity of the neutral temperature to the exact value chosen for the sensitivity to temperature variation, as indicated by the regression coefficient. Analysis of the ASHRAE RP-884 database and the SCATs database showed that the value of the regression coefficient for day surveys is unlikely to be less than 0.4/K or more than 0.5/K. (The value applies to surveys that used the ASHRAE or the Bedford scale, or translations of these scales into other language, such that the category intervals were reasonably uniform in psychological width.) The figure compares neutral temperatures calculated for regression coefficients of 0.4/K and 0.5/K, and shows that the neutral temperatures are very little affected by the precise value used for the sensitivity to temperature variation.

The relation between the neutral temperature and the mean indoor temperature

According to the adaptive principle, people tend to adjust their clothing, posture, etc. to suit their local conditions and so become comfortable at the temperatures they

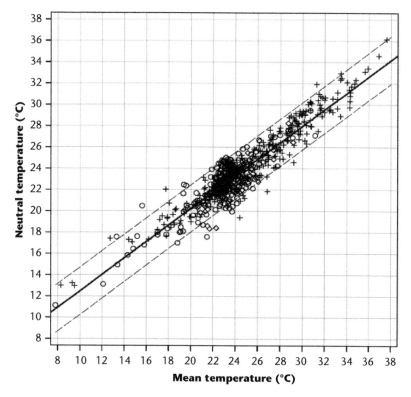

FIGURE 28.6 Neutral temperatures and the mean room temperatures for buildings operating in the FR mode (crosses), HC mode (circles) and indeterminate modes (diamonds). The dashed lines include 95 per cent of the points. r = 0.94, N = 732.

Note: FR = free-running mode, HC = heated-or-cooled mode.

typically encounter, while at the same time adjusting the room temperature for comfort. This adaptive process tends to make their neutral temperature close to the mean temperature they experience. We conclude this chapter by illustrating that this is so.

A scatter plot of the relation between the neutral temperature and the mean temperature, obtained from the DBSS, is shown in Figure 28.6. Each point on the plot represents a single survey of thermal comfort. Inspection of the graph shows that the relation applies, irrespective of the mode of operation of the building.

The coefficient of correlation between the neutral temperature and the mean room temperature is 0.94. The equation of the regression line is:

$$T_n = 0.783(\pm 0.011)T_{op} + 4.5$$

with the standard deviation of the regression coefficient shown in parentheses. The residual standard deviation (the standard error of predicting a survey's neutral temperature) is 1.1 K.

If adaptation were complete across the entire range of room temperature shown in Figure 28.6, the line, experimental error apart, would have unit gradient and unit correlation. A number of factors contribute to make the adaptation, although very strong, a little less than complete:

- If the survey is short, and the room temperature varies from day to day, people are unlikely to adapt completely to the day-to-day changes.
- The data include instances where the temperature approaches a cultural limit for adaptation. That is to say, though adaptation is possible, it would require a change in social customs that would take a long time to occur. We discussed this in Chapter 23 when considering the limits of seasonal changes of room temperature in Pakistan.
- We have already noted that the mean indoor temperature is a surrogate variable for the thermal environment to which the respondent is currently adapted. How good a surrogate it is depends on the duration of the survey, and upon the time sequence of indoor temperature experienced by the respondents prior to and during the survey period. Little research has been done on this topic.
- Finally we might question whether it is correct to use the mean temperature as the predictor variable, as to do so attributes all errors to the estimate of the neutral temperature. Partitioning the error equally between the two axes would have slightly increased the gradient of the line, raising it to 0.834.

The data amply demonstrate both the presence and the power of the adaptive process, and that it applies irrespective of the mode of operation of the heating or cooling system. This adaptive relation between the neutral temperature and the prevailing indoor temperature experienced by the occupants is fundamental, and is the basis for the better-known relations between the outdoor climate and comfortable indoor temperatures that are included in guidelines and standards. These are the topic of Chapter 29.

The database of summary statistics

In addition to the ASHRAE RP-884 database and the SCATs database, summary statistics have been obtained from the following sources:

Abdelrahman, M. A. (1990) Field evaluation of a comfort meter, *ASHRAE Transactions* 96(2), 212–15.

Auliciems, A. (1975) Warmth and comfort in the subtropical winter: a study in Brisbane schools, *J. Hygiene* 74, 339–43.

Auliciems, A. (1977) Thermal comfort criteria for indoor design in the Australian winter, *Arch. Sci. Rev.* 20, 86–90.

Bailie, A. P. *et al.* (1987) *Thermal comfort assessment: a new approach to comfort criteria in buildings.* Dept of Psychology, ETSU S-1177, University of Surrey, Guildford.

Ballantyne, E. R. *et al.* (1977) Probit analysis of thermal sensation assessments, *Int. J. Biometeorology* 21(1), 29–43.

Bouden, C. and Ghrab, N. (1999) *Thermal comfort in Tunisian buildings: results of an enquiry.* Final Report, Solar Energy Laboratory, ENIT (in French).

Bouden, C. and Ghrab, N. (2001) Thermal comfort in Tunisia: results of a one year survey, *Moving thermal comfort standards into the 21st century*. Conference Proceedings compiled by Kate McCartney. OCSD Architecture Group, Oxford Brookes University, Oxford, pp. 197–207.

Cao, B. *et al*. (2009) Field study of human thermal comfort and thermal adaptability during summer and winter in Beijing, *The 6th International Symposium on Heating, Ventilating and Air Conditioning*, Nov. 6–9, 2009, Nanjing, China (data received from Dr. Bin Cao), pp. 1207–16.

Cena, K. and de Dear, R. (1998) *Field study of occupant comfort and office thermal environments in a hot-arid climate*. Final Report on ASHRAE RP-921, American Society of Heating, Refridgeration and Air-Conditioning Engineers, Atlanta.

Cena, K. *et al*. (1986) Thermal comfort of the elderly is affected by clothing, activity and psychological adjustment, *ASHRAE Transactions* 92(2), 329–42.

Cena, K. M. *et al*. (1990) A practical approach to thermal comfort surveys in homes and offices: discussion of methods and concerns, *ASHRAE Transactions* 96(1), 853–8.

Cena, K. M. *et al*. (1988) Effect of behavioral strategies and activity on thermal comfort of the elderly, *ASHRAE Transactions* 94(1), 83–103.

Chan, D. W. T. *et al*. (1998) A large scale survey of thermal comfort in office premises in Hong Kong, *ASHRAE Technical Data Bulletin* 14(1), 76–84 (see also *ASHRAE Transactions* 104(1)).

Coldicutt, S. *et al*. (1991) Attitudes and compromises affecting design for thermal performance of housing in Australia. *Environment International* 17, 251–61.

Feriadi H. and Wong, N. H. (2004) Thermal comfort for naturally ventilated houses in Indonesia, *Energy and Buildings* 36(7), 611–13.

Fishman, D. S. and Pimbert, S. L. (1982) The thermal environment in offices, *Energy and Building* 5, 109–16.

Griffiths, I. D. (Undated, circa 1990) *Thermal comfort in buildings with passive solar features*. Report to the Commission of the European Communities ENS-090-UK.

Grivel, F. and Barth, M. (1981) Thermal comfort in office spaces: predictions and observations, in: *Building Energy Management*, Eds: de Fernandes, O., Woods, J. E., and Faist, A. P. Pergamon, Oxford.

Howell, W. C. and Kennedy, P. A. (1979) Field validation of the Fanger thermal comfort model, *Human Factors* 21(2), 229–39.

Hunt, D. R. G. and Gidman, M. I. (1982) A national survey of house temperatures, *Building and Environment* 17(2), 107–24 (also information by personal communication).

Kato, T. *et al*. (1996) Difference between winter and summer of the indoor thermal environment and residents' thinking of detached houses in Nagano city, *J. Archit. Plann. Environ. Eng*. 481, 23–31 (in Japanese with English abstract).

Kwok, A. G. (1998) Thermal comfort in tropical classrooms, *ASHRAE Technical Data Bulletin* 14(1), 85–101.

Li, H. *et al*. (2009) Measurement and field survey of indoor thermal comfort in rural housing of northern China in winter, in: *The 6th International Symposium on Heating, Ventilating and Air Conditioning*, Nov. 6–9, Nanjing, China (data are received from Dr. Bin Cao) pp. 245–51.

Malama, A. *et al*. (1998) An investigation of the thermal comfort adaptive model in a tropical upland climate, *ASHRAE Technical Data Bulletin* 14(1), 102–11.

Mallick, F. H. (1992) Thermal comfort in tropical climates: an investigation of comfort criteria for Bangladeshi subjects, *PLEA Conference Proceedings*, pp. 47–52.

Matthews, J. (1993) *An investigation into the thermal comfort of factory workers in North India using the Probit method of analysis*. MSc Dissertation, University of East London, London.

Matthews, J. and Nicol, J. F. (1994) Thermal comfort of factory workers in Northern India, in: *Standards for thermal comfort*, Eds: Nicol, F. *et al*., E. & F. N. Spon (Chapman & Hall), London, pp. 227–33. (See also: Ramasodi, L. (1993) *Thermal comfort and the Tropical Summer Index*. MSc Dissertation, University of East London, London.)

McCartney, K. J. and Nicol, J. F. (2001) Developing an adaptive control algorithm for Europe: results of the SCATs project, in: *Moving thermal comfort standards into the 21st century*, Ed.: McCartney, K., Oxford Brookes University, Oxford, pp. 176–97.

Nakamura, Y. *et al.* (2008) Method for simultaneous measurement of the Occupied Environment Temperature in various areas for grasp of adaptation to climate in daily life, *Journal of Human and Living Environment* 15(1), 5–14 (in Japanese with English abstract). (Data received from Prof. Nakamura.)

Nakaya, T. *et al.* (2005) A field study of thermal environment and thermal comfort in Kansai region, Japan: neutral temperature and acceptable range in summer, *J. Environ. Eng.* 597, 51–6 (in Japanese with English abstract).

Nicol, J. F. *et al.* (1997) *Thermal comfort survey in Pakistan, II.* School of Architecture, Oxford Brookes University, Oxford.

Oseland, N. A. (1994) Predicted and reported thermal sensation votes in UK homes, in: *Banking on design?* Ed.: Seidal, A., *Proceedings of 25th Annual Conference of the Environmental Design Association*, St Antonio. EDRA, pp. 175–9.

Oseland, N. and Raw, G. (1990) Thermal comfort in starter homes in the UK. *BRE Note PD186/90.* Building Research Establishment, Watford.

Oseland, N. A. (1994) A comparison of the predicted and reported thermal sensation vote in homes during winter and summer, *Energy and Buildings* 21, 45–54.

Oseland, N. A. (1998) Acceptable temperature ranges in naturally ventilated and air conditioned offices, *ASHRAE Technical Data Bulletin* 14(1), 50–62.

Pirsel, L. (1989) Which temperature is being felt comfortable in dwellings? *Proc. CIB 1989* 2, pp. 167–72.

Rijal, H. B. and Yoshida, H. (2006) Winter thermal comfort of residents in the Himalaya region of Nepal, *Proceedings of International Conference on Comfort and Energy Use in Buildings – Getting Them Right*, Windsor. Organised by the Network for Comfort and Energy Use in Buildings.

Rijal, H. B. *et al.* (2002) Investigation of the thermal comfort in Nepal, *Proceedings of International Symposium on Building Research and the Sustainability of the Built Environment in the Tropics*, Indonesia, pp. 243–62.

Saito, M. (2009) Study on occupants' cognitive temperature scale for their environmental controls behaviors: in the case of University laboratories in summer in Sapporo, *J. Environ. Eng.* 74(646), 1291–97 (in Japanese with English abstract). (Data are received from Dr. Saito.)

Sawachi, T. and Matsuo, Y. (1989) Daily cycles of activities in dwellings in the case of housewives: Study on residents' behavior contributing to formation of indoor climate (part 2), *J. Archit. Plann. Environ. Eng.* 398, 35–46 (in Japanese with English abstract).

Sharma, M. R. and Ali, S. (1986) Tropical Summer Index – a study of thermal comfort in Indian subjects, *Building and Environment* 21(1), 11–24.

Sliwowski, L. Z. *et al.* (1983) Indoor climate problems in Polish apartment blocks, in: *PLEA Conference*, Ed.: Yanas, S., Pergamon, Oxford.

Stevenson, F. and Rijal, H. B. (2008) *Post-occupancy evaluation of the Stewart Milne Group's Sigma® House.* Stewart Milne Group (Final report).

Suzuki, N. and Shukuya, M. (2009) Field and semi-field surveys on thermal-environment experience and its associated acquired cognition by family members, parents and children, *26th Conference on Passive and Low Energy Architecture*, Quebec City, Canada, 22–24 June. (Data received from Ms N. Suzuki.)

Tobita, K. *et al.* (2009) Difference of the thermal sensation votes by the scales in the field study of houses during winter, *J. Environ. Eng.* 74(646), 1291–97 (in Japanese with English abstract).

Tobita, K. *et al.* (2007) Calculation of neutral temperature and acceptable range by the field study of houses in Kansai area, Japan, in winter, *J. Environ. Eng.* 614, 71–7 (in Japanese with English abstract).

Turnquist, R. O. and Volmer, R. P. (1980) Assessing environmental conditions in apartments of the elderly, *ASHRAE Transactions* 86(1), 536–40.

Umemiya, N. *et al.* (2008) Setting temperature and clothing insulation in student rooms in summer, *Proceedings of the 29th Air Infiltration and Ventilation Centre Conference 3*, pp. 79–84. (Data received from Prof. N. Umemiya.)

Williamson, T. J. *et al.* (1989) *Thermal comfort and preferences in housing: south and central Australia.* University of Adelaide, Adelaide.

Williamson, T. J. *et al.* (1989) Thermal preferences in housing in the humid tropics, *ERDC No 1,* Energy Research and Development Corporation, Canberra.

Woolard, D. S. (1981) Thermal sensations of Solomon Islanders at home, *Architectural Science Review* 24(4), 94–7.

Yamashita, H. *et al.* (2009) Setting temperature, air temperature and thermal comfort in student rooms of university, *Proceedings of the Kinki Chapter of the Society of Heating, Air-Conditioning and Sanitary Engineers of Japan (SHASE)* 38 (in Japanese with English abstract), pp. 25–8. (Data received from Prof. N. Umemiya.)

Note

1 McCartney, K. J. and Nicol, J. F. (2002) Developing an adaptive control algorithm for Europe: results of the SCATs project, *Energy and Buildings* 34(6), 623–35 (see Figure 3).

29

THE ADAPTIVE RELATION BETWEEN INDOOR NEUTRAL TEMPERATURES AND THE OUTDOOR CLIMATE[1]

When people speak of the adaptive model of thermal comfort they usually mean the graph or chart that relates the prevailing outdoor temperature to the temperature required for comfort indoors. The graph has practical importance because it indicates what temperatures are likely to be acceptable in a building at a particular outdoor temperature. The relation is strongest when a building is operating without either heating or cooling, that is to say, in its 'free-running' mode (FR). This relation is particularly useful because it enables building engineers to check, using building thermal simulation software, whether a proposed design would be likely to provide thermal comfort during hot weather without mechanical cooling, and during cooler weather without heating. It also gives a very general idea of how much heating and cooling will be required in that climate. This relation is a sufficiently useful design tool to be included in some of the current standards and guidelines for thermal comfort.[2-4]

The principal models

We begin with short critical reviews of the principal graphs that have related thermal comfort indoors to the outdoor climate. The first was set out by Humphreys, who derived it from all the field studies of thermal comfort then available.[5] The idea was taken up in Australia by Auliciems,[6] who also worked with Richard de Dear on the subject in the mid-1980s. In the 1990s, ASHRAE became interested in the relation, and commissioned field studies of comfort from buildings in various climates and seasons. De Dear, Brager and Cooper collected the results of these and other studies into a database and produced a new adaptive relation.[7,8] Then in the 2000s, a team headed by Nicol collected and analysed extensive data from Europe, so producing further adaptive relations between climate and comfort indoors, applicable in Europe, and subsequently included in the European Standard EN 15251.[9,10]

Humphreys's original model

The original meta-analysis using the summary statistics of some 30 field surveys of thermal comfort conducted between 1930 and 1975, representing more than 200,000 records of thermal sensation, led to the first quantitative relation between indoor comfort and the climate.[11-13] Humphreys quantified two relations, one for buildings where heating or cooling was in use at the time of the comfort survey, the 'Heated-or-Cooled' mode (HC), and another for buildings where no heating or cooling was in use, the 'Free-Running' mode (FR) (Figure 29.1). The same building could be FR at one time of the year and HC at another. It was a classification of the *mode of operation* rather than of the *type of building*, although Humphreys sometimes spoke more loosely of 'free-running buildings'. The relation between the temperature for comfort indoors and the monthly mean of the outdoor temperature was found to be linear for the buildings in the FR mode, and surprisingly strong ($r = 0.97$):

$$T_n = 11.9 + 0.534T_o$$

T_n is the neutral or preferred indoor temperature (°C), no distinction being made between the two, and T_o the outdoor monthly mean air temperature (°C) for that region, taken from the world meteorological tables published by the UK government. The variation about the line had a standard deviation of 1.0 K.

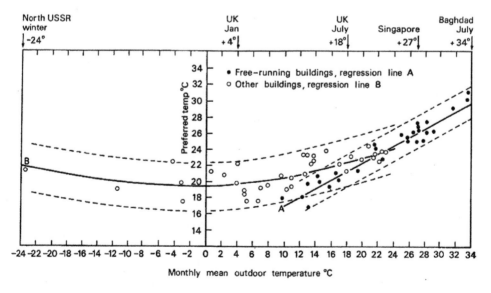

FIGURE 29.1 Humphreys's scatter diagram for neutral/preferred temperature and the mean outdoor temperature.

Source: M. A. Humphreys (BRE slide).

For buildings operating in the HC mode, the relation was weaker ($r = 0.72$) and curvilinear, so the data were fitted to a mathematical function whose shape was consistent with that hypothesised, having a gradient positive in the central zone and becoming horizontal as either extreme is approached. The chosen function was of the form:

$$f(x) = x.\exp(-x^2)$$

When fitted to the data it yielded the complicated-looking equation:

$$T_n = 23.9 + 0.295(T_o - 22).\exp(-[(T_o-22)/(24\sqrt{2})]^2)$$

The variation about the curve had a standard deviation of 1.5 K. There were few survey results from cold climates, so the relation was rather uncertain for cold outdoor temperatures. Both relations were highly significant statistically.

The difference between the two relations can be understood by considering the effect of the close relation between mean indoor temperature and the neutral temperature, explored in Chapter 28 and illustrated by Figure 28.6. The indoor temperature of a building in the free-running mode will be strongly related to the outdoor temperature, but modified by the fabric and layout of the building. We would therefore expect the neutral temperature and the outdoor temperature also to be strongly related, though the precision of the relation will be reduced by the variety of the buildings included in the data. In buildings that are being heated or cooled, the indoor temperature is largely decoupled from the outdoor temperature by the action of the heating or cooling system. So a smaller range of indoor temperature is likely to occur, and therefore a smaller range of neutral temperatures is likely to be found.

It had been difficult to discover what aspects of the climate influenced the comfort temperatures. Only the data from world meteorological tables were available in the 1970s, except for those few surveys where researchers had obtained local weather data for the duration of the survey period. The highest correlations between neutral temperature and climate for the FR mode were found when using the monthly mean outdoor temperature as the predictor variable. This was the mean of the daily maximum and daily minimum temperatures for the month. The correlation could not be improved significantly by including the annual average temperatures, or summer or winter maxima and minima, or by altering the relative weights of the average daily maximum and minimum temperatures from their simple average. The same was found for the HC mode, except that the mean maximum temperature of the hottest month of the year also had a statistically significant effect. This effect had not been hypothesised, had no obvious cause, raised the correlation but slightly, and applied only to the HC mode, so it was not pursued.

The tabulated average climate data were not ideal as predictors, for two reasons. First, people tend to adapt to the conditions they are experiencing, and so it seemed probable that concurrent temperature data would be more relevant than data expressed as the averages over many years. Second, work on the rapidity of adaptive actions had shown that clothing adaptations in response to temperature drifts were likely to be complete within about a week.[14] A month would therefore be too coarse a division of the time.

Auliciems' development of Humphreys' model

Auliciems revised and updated the relation, adding data from some good quality newly completed surveys and deleting data from a few surveys he considered to be of doubtful quality.[15,16] This had the incidental effect of restricting the range to monthly mean outdoor temperatures above $0°C$. He pooled the data for the FR and HC modes and performed a new linear regression analysis, using as predictors both the mean indoor temperature during the survey and the outdoor monthly mean temperature. He obtained the relation:

$$T_n = 9.2 + 0.48T_m + 0.14T_o$$

where T_m is the indoor mean temperature during the period of the survey.

However, using the indoor mean temperature as a predictor is of doubtful value, since for optimum comfort the mean indoor temperature (T_m) would necessarily be equal to the neutral or preferred temperature (T_n). If we set T_m equal to T_n, Auliciems's equation becomes:

$$T_n = 9.2 + 0.48T_n + 0.14T_o$$

which reduces to:

$$T_n = 17.7 + 0.27T_o$$

This is a line of lower gradient than Humphreys had obtained for the FR mode but higher than for the HC mode, reflecting the inclusion of surveys in both the FR and the HC mode in the same regression equation. The disadvantage of using a single line to represent the two modes of operation is that its regression gradient would depend upon the relative number of surveys in the database in either mode of operation. So it is necessary to retain the distinction between the modes of operation. Also the mean indoor temperature cannot be used as a predictor if the purpose of the relation is to suggest appropriate indoor temperatures from the prevailing outdoor conditions alone.

Auliciems observed that indoor temperatures that are constant in all seasons were neither necessary nor desirable, and that indoor temperature control could be by means of a 'thermobile' whose set point varied in relation to the outdoor temperature, rather than a 'thermostat' with a fixed set point. The use of such a 'thermobile' to control indoor temperature could, he argued, both save energy and improve thermal comfort, as has indeed been found in subsequent studies.[17]

The ASHRAE model

A wholly new database of thermal comfort surveys (or field experiments) was assembled by de Dear and colleagues in the 1990s. It had some 21,000 observations of thermal sensation from field studies in 160 buildings from 9 countries.[18,19] It was collated for the ASHRAE research project RP-884 to develop an adaptive model of thermal comfort and preference. The type, accuracy and range of related instrumentation had advanced significantly over the intervening decades. Computerised data logging and rapid response instruments had speeded the acquisition of the indoor environmental data. More complete measurement had also become practical, and many surveys included temperatures at three heights above the floor, measurements of radiant asymmetry and of air turbulence in addition to the classic comfort measurements of air temperature, mean radiant temperature, air speed and humidity. The new data generated included estimates of clothing insulation and metabolic rate, so enabling the calculation of thermal indices such as Fanger's PMV and Gagge's SET.[20,21]

De Dear and Brager performed an extensive analysis of these data. Only those sections of the analysis that underlie the relation between climate and comfort indoors concern us here. They divided their data according to building type – naturally ventilated (NV), centrally heated, ventilated and air-conditioned buildings (HVAC), or mixed-mode buildings (MM). This division rested on the hypothesis that people responded differently to their thermal environment in buildings having different types of climate control (HVAC or 'naturally conditioned') – that they would have different 'expectations'. They further separated the data from each building. This was a sound procedure because the different buildings were likely to be run at different indoor temperatures and would in consequence be expected to yield different neutral temperatures, as people tend to adapt to the conditions they are given. Typically, each building survey took a few days to complete. In many buildings the surveys had few observations.

De Dear and Brager used regression analysis to quantify in each building the dependence of the thermal sensation (the 'comfort vote' on a seven-point scale, usually the ASHRAE scale) upon the indoor operative temperature, retaining only those regressions that were significant at the 95 per cent confidence level. Table 29.1 is an extract from their report. The resulting regression equations were solved to

TABLE 29.1 Summary of weighted linear regression of bin-mean thermal sensation on operative temperature (after de Dear *et al.* 1997).

	HVAC	NV	MM
Number of buildings	109	44	4
	2 missing	1 missing	None missing
Number of buildings included in the analysis	63	36	3
Mean gradient (s.d.)*	0.51 (0.25)	0.27 (0.13)	0.39 (0.11)

* Based on those models (y = a + b*T$_{op}$) achieving 95% confidence or better.

find the operative temperature that would be thermally neutral – the neutral temperature. There was a wide scatter among the estimates of the regression gradients (s.d. of ±0.25 for the HVAC buildings, ±0.13 for the NV buildings and ±0.11 for the MM buildings). Regressions whose coefficients were non-significant at the $p < 0.05$ level were discarded because they could not yield a finite confidence interval on the estimate of the neutral temperature. However, this procedure biases upwards the estimate of the mean of the regression gradients, for it is usually those batches with low gradients that fail to reach statistical significance. The effect is greatest for the HVAC buildings, where data from 44 out of 109 buildings had to be discarded. The more recently developed procedure we set out in Chapter 28 avoids this problem.

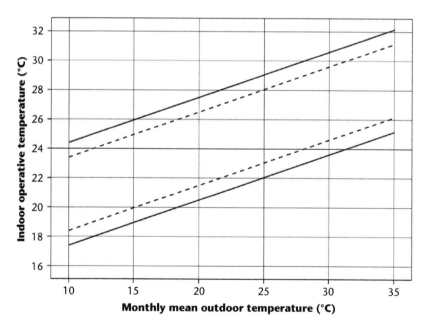

FIGURE 29.2 Acceptable operative temperature ranges for naturally conditioned spaces (after ASHRAE Standard 55-2013). The limits for 90 per cent in comfort are the dashed lines, for 80 per cent in comfort are the solid lines.

Many of the surveys included in the ASHRAE database contained contemporary local meteorological data, but, if not, de Dear and Brager used the historic monthly mean values from the nearest meteorological station. The relation obtained between climate and the indoor operative temperature for comfort (Figure 29.2) was included in ASHRAE Standard 55-2004.

In de Dear's research report, a more complex 'outdoor effective temperature' had been used as the index of the climate, so as to include effects of humidity. It was not decisively superior to the air temperature, and it would be more convenient to incorporate the humidity as a modifier of the neutral temperature rather than include it in the climate metric. The similarity to Humphreys's relation for the FR mode is immediately apparent, giving general confirmation from an entirely independent database, but the gradient of the relation is considerably lower. The different classification should be remembered when comparing the relations. NV buildings are free-running (FR mode) in warm weather but heated in cool weather (HC mode). A new feature, shown in Figure 29.2, was the estimation of the limits for the acceptability of the indoor environment.

The SCATs project (Smart Controls and Thermal Comfort)

A third independent database of thermal comfort responses in offices comes from the SCATs project, and is the basis of the adaptive relation included in European Standard EN 15251.[22] The SCATs project was a year-round study of the indoor environment in European offices, with particular emphasis on the thermal environment. The project consisted of a study of selected office buildings in France, Greece, Portugal, Sweden and the UK. We describe only those parts of this extensive study that are relevant to the relation between climate and comfort indoors. Each month a survey was conducted in each participating office building. The surveys yielded some 5000 subjective thermal responses on the ASHRAE scale, each accompanied by measurements of air temperature, globe temperature, air speed and relative humidity, together with estimates of clothing insulation and metabolic rate. Contemporary hourly meteorological data were obtained from nearby weather stations.

A 'comfort temperature' (T_{comf}) was derived from each individual interview. It was taken to be the operative temperature at which the respondent would have recorded thermal neutrality had the divisions on the ASHRAE scale corresponded to a change in operative temperature of 2K.[23] Each 'spot-value' of the comfort temperature was based on a single thermal sensation on a single occasion and therefore subject to considerable random uncertainty, but not to systematic error if the chosen value of 2K was correct. The value 2K had been adopted after examination of the regressions of thermal sensation on operative temperature from each building in each month, and then making an approximate adjustment to allow for the effect on the regression gradient of the presence of error in the predictor variable. Our subsequent more rigorous evaluation, described in Chapter 27, confirms that the value 2 K was indeed suitable. A scatter plot was made of these spot estimates of comfort temperature against the outdoor temperature. The metric chosen for the outdoor temperature was an exponentially weighted running mean (T_{rm}) of the daily mean outdoor temperature. Exponentially weighted running means give greatest weight to temperatures in the most recent past, and progressively less weight to those in the more remote past. The best results were found using the equation:

$$T_{rm(tomorrow)} = 0.8*T_{rm(yesterday)} + 0.2*T_{m(today)}$$

The exact value of the relative weighting (0.8, 0.2) was not critical. The regression equation for buildings operating in the FR mode was:

$$T_{comf} = 0.33T_{rm} + 18.8$$

The calculation procedures behind the adaptive comfort chart in EN 15251 (Figure 29.3) have been fully set out by Nicol and Humphreys.[24] The same data and

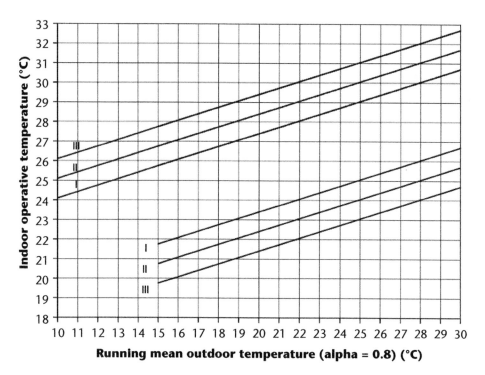

FIGURE 29.3 Design values (vertical axis) for the operative temperature for buildings operating in the FR mode as a function of the exponentially weighted running mean of the external temperature (horizontal axis). (after EN Standard 15251).

Note: The comfort temperature in type I buildings (close control) should lie between the pair of lines I, etc.

the same method of calculation underlie the adaptive chart in the CIBSE Guide A Chapter 1 (Comfort).

The similarity to Humphreys' original chart, and to the chart used in the ASHRAE Standard, is immediately evident, giving general support to both. However, the charts are not precisely comparable, this time because the metric on the climate axis is different – a running mean instead of the monthly mean temperature – and because of the different methods of calculating the neutral temperatures.

Our brief review has shown two matters that needed to be considered if the relation between indoor comfort and outdoor climate is to be further developed:

- the formulation of a suitable metric for the outdoor temperature;
- a method of derivation for the neutral temperature from batches of data.

The second of these was the topic of Chapter 28.

The metric for the climate

We now consider what climate metric best predicts the indoor comfort temperature. Three principal kinds of metric have been used:

- the monthly mean maxima and minima outdoor temperatures from historic meteorological tables;
- the daily maxima and minima from contemporary measurements (sometimes taken to be the outdoor air temperatures at 6 am and 3 pm) averaged over the period of the survey;
- a weighted moving average (running mean) of the daily mean outdoor temperature up to the day of the survey.

The relative weights of the monthly maximum and minimum monthly mean outdoor temperatures

There has been no research since Humphreys' 1978 study, so far as we know, on the relative importance of the daily maximum and daily minimum temperatures for predicting the temperature for comfort indoors. After exploring the matter, Humphreys concluded that, based on the data then available, a simple average could not be bettered. De Dear and Brager assumed that this was true when calculating the regression line linking the climate to the indoor comfort conditions from the ASHRAE RP-884 database.

The RP-884 database includes the day maximum and minimum temperature in each line of data and so can be used to explore the question. We can use multiple linear regression analysis to predict the comfort temperatures for buildings when operating in the FR mode. For this calculation, we estimated the comfort temperatures exactly as described for the SCATs data, using the individual interview data. We also calculated a univariate regression equation using the simple average of the two temperatures $\{(T_{o\_max} + T_{o\_min})/2\}$ as the predictor variable. The regression equations are:

$$T_{comf} = 14.1 + 0.20T_{o\_max} + 0.35T_{o\_min} \qquad r = 0.806$$

$$T_{comf} = 13.2 + 0.55\{(T_{o\_max} + T_{o\_min})/2\} \qquad r = 0.802$$

Inspection of the regression coefficients suggests that the daily minimum (T_{o_min}) is substantially more influential than the daily maximum (T_{o_max}) in predicting the indoor comfort temperature (T_{comf}). The same is true if we build similar regression equations to predict the indoor operative temperature (T_{op}):

$$T_{op} = 13.6 + 0.21 T_{o\_max} + 0.56T_{o\_min} \qquad r = 0.879$$

$$T_{op} = 8.9 + 0.77\{(T_{o\_max} + T_{o\_min})/2\} \qquad r = 0.870$$

The regression equations attribute much more weight to the minimum temperature than to the maximum, both for the prediction of the indoor mean temperature and for the prediction of the comfort temperature. The result should be treated with caution because the daily maximum and daily minimum temperature in the database are quite highly correlated $(r = 0.82)$ and such co-linearity renders less reliable the estimates of the relative weights (the regression coefficients) attributed to the two variables. For the prediction of the comfort temperature (T_c), the improvement over using the simple average, although statistically significant, is very small, raising the correlation coefficient with the comfort temperature from 0.802 to 0.806, an increase of 0.5 per cent. We made similar calculations using the data from the SCATs database. The correlation between the day maximum temperature and the day minimum temperature for the FR mode was even higher $(r = 0.93)$. This prevented the multiple regression coefficients being reliably estimated.

The relative weight attributed to the maximum and the minimum temperature cannot therefore be reliably established from existing databases, but is not critical because of the high correlation between daily maximum and daily minimum temperature. In this circumstance, it seems sensible to continue to use the simple average of the two, not least for ease of use of the method.

Estimating the value of 'α' for the running mean outdoor temperature series

We mentioned above that an exponentially weighted running mean (T_{rm}) of the outdoor temperature was a useful index of the climate for the purpose of predicting the indoor comfort temperature from the climate. It is most conveniently expressed in the form:

$$T_{rm(tomorrow)} = \alpha * T_{rm(yesterday)} + (1 - \alpha) * T_{m(today)}$$

The day-mean (T_m) of the outdoor temperature is taken to be the average of the maximum and minimum for the day in question. The value of α is ascertained by inspecting the correlation between the comfort temperature and the running mean outdoor temperature, with α chosen to maximise the correlation. The value of α cannot be determined with great accuracy from existing data, nor does great precision seem to be needed. It was first estimated experimentally using survey data from Oxford (southern England) and Aberdeen (northern Scotland).[25] It was re-estimated using the SCATs data, and a similar value (0.8) was found.

It is likely that the value of α represents the thermal inertia of the building together with the delayed behavioural responses of the occupants to temperature changes within the building, so we might expect different values to apply to buildings having different thermal inertias. There was some indication from data collected in Oxford that this was so, but it would require further studies before a quantitative statement could be made.[26]

The half-life (λ) of an exponentially weighted running mean is given by the relation

$$\lambda = 0.69/(1 - \alpha)$$

so a value of α of 0.8 indicates a half-life of some 3.5 days. This indicates that if there were a step change in the outdoor mean temperature, the indoor comfort temperature would take about 3.5 days to move half-way towards its new value, or a week to move three-quarters of the way. After a month, adaptation would be very nearly completed (97 per cent).

Comparison of various metrics of outdoor temperature

The only large body of data currently available for this comparison is the SCATs database. We use all the observations from buildings operating in the FR mode to make the comparison of different values of α and other climate metrics. Table 29.2 gives values of the correlation of the comfort temperature with various metrics of the outdoor temperature, together with the regression coefficient of the comfort temperature on that metric.

The correlation coefficients indicate that the running mean with $\alpha = 0.8$ is (in these data) decisively superior as a predictor of the indoor comfort temperature both to the historic monthly mean from meteorological tables, and to a running mean with $\alpha = 0.96$. It also has the highest regression coefficient, decisively higher than that for the actual hour of the survey interview, and perhaps higher than that of the mean on the day of the survey. The data suggest that this running mean is at present the best available predictor of the indoor comfort temperature. So while no great accuracy should be attributed to the value of 0.8 for α, it should be used until a more

TABLE 29.2 Comparison of different ways of expressing the outdoor temperature.

Metric for the outdoor temperature	Correlation coefficient (r) with neutral temperature (s.e.)	Regression coefficient of comfort temperature on outdoor temperature (s.e.)
Monthly mean from meteorological tables	0.548 (0.019)	0.315 (0.013)
Mean on the day on which the survey took place	0.571 (0.018)	0.304 (0.012)
At the actual hour of the survey interview	0.557 (0.019)	0.259 (0.010)
T_{rm} $\alpha = 0.96$ $2\lambda \approx$ month	0.550 (0.019)	0.317 (0.013)
T_{rm} $\alpha = 0.80$ $2\lambda \approx$ week	0.598 (0.017)	0.331 (0.012)
T_{rm} $\alpha = 0.33$ $2\lambda \approx$ day	0.593 (0.017)	0.318 (0.012)

Notes: N = 1378. The neutral temperature is estimated for each individual interview, using a regression gradient of 0.5 scale units per degree. The standard error (s.e.) of the correlation coefficient is taken as $(1 - r^2)/\sqrt{(N-1)}$.

precise estimate becomes available. It represents chiefly changes that have taken place over the previous week or two. A month is too long a period while a day is too short a period.[27]

Recalling Auliciems's 'thermobile' controls, if the aim were to control the indoor temperature according to the variations of outdoor temperature, then a running mean with $\alpha = 0.8$ would probably be a satisfactory outdoor temperature metric for use within a control algorithm.

The usefulness of the historic monthly data should not be overlooked. For a quick indication of whether a building is likely to operate satisfactorily in the FR mode in a certain month of the year at a certain location, the historic mean is useful.[28] But for a more detailed design assessment, running mean temperatures derived from selected design years should be used. The running means are also convenient to use to estimate comfort temperatures in the thermal simulations of buildings.

Revising the relation between indoor comfort and the climate

In Chapter 28, we introduced and described the database of summary statistics (DBSS) and derived values of the neutral temperature from it. We now use these values to revise the relation between the climate and the temperatures indoors that are thermally neutral. In the DBSS, the neutral temperatures draw no distinction between 'comfortable', if the Bedford scale was used, and 'neutral', if the ASHRAE scale was used. The difference is very small. Also, the mean temperatures from which the neutral temperatures were derived might be either operative temperatures or globe temperatures. They are practically identical. If the survey gave neither mean operative temperature nor mean globe temperature, the mean air temperature was used instead. The error introduced by this substitution is very small. No adjustments have been made for humidity or for air movement. Humidity makes only a very small difference to the neutral temperature. The occupants adjusted their air movement to suit themselves, as they found appropriate for the indoor temperature. It is assumed that fans can be used when needed.

The climate metric in the database differs among the surveys included in the DBSS. The SCATs data use the exponentially weighted running mean with a value for α of 0.8. The surveys in the ASHRAE RP-884 database use the mean of the outdoor temperature over the survey period if available. If this was not available, historic values were obtained from meteorological tables. Other surveys in the DBSS used either contemporary weather data or historic values from tables. We have used the best available metric for each survey. This composite metric we have called the 'prevailing mean outdoor temperature'. It is not ideal, and when it is used to predict the neutral temperature in a regression analysis, it will introduce some error into the predictor variable. However, the range of the outdoor temperatures is very large compared with the likely extent of the error, so the regression line is very little affected by its presence.

Having available such a large database and a more rigorous method of deriving the neutral temperatures considerably increases the robustness and accuracy of the relation between comfort indoors and the climate. We consider first the buildings that are operating in the free-running mode at the time of the survey.

The free-running mode

We selected those surveys in the DBSS that were conducted in buildings in the FR mode of operation (no heating or cooling appliances in use during the survey period), and excluded those with fewer than 20 observations. There are 209 surveys in the database that have 20 or more observations, together representing some 58,000 observations. The scatter plot of the neutral temperatures against the prevailing mean outdoor temperature is shown in Figure 29.4. The mean neutral temperature is 24.9°C (s.d. 3.9 K) at the mean prevailing outdoor temperature of 21.3°C (s.d. 8.5 K). The correlation coefficient between the neutral temperatures and the outdoor prevailing mean is 0.89, and the equation of the regression line is:

$$T_n = 13.8 + 0.53(\pm 0.02)T_o$$

T_n is the neutral indoor operative temperature and T_o the prevailing mean outdoor temperature. It will be noticed that the gradient of the line (0.53) is considerably steeper that those in ASHRAE standard 55 and in EN 15251. It is indistinguishable from that obtained by Humphreys in 1978 (0.534).

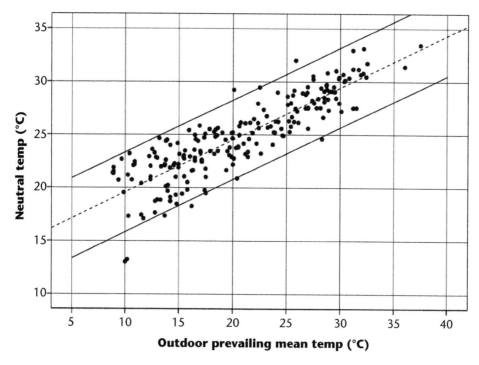

FIGURE 29.4 The neutral temperatures (°C) for buildings in the FR mode against the prevailing mean outdoor temperature (°C).

The scatter of the points about the line (the residual s.d.) is 1.8 K. It was shown in Chapter 28 that the standard error associated with the estimate of a neutral temperature based on 20 observations was some 0.4 K. This is very much smaller than the observed scatter. It follows that the scatter of the points about the line is not attributed chiefly to random error but represents real differences among the neutral temperatures of the various groups of people at any prevailing mean outdoor temperature. It is therefore better to represent the data not by a line, but by a band. The band drawn on the figure includes 95 per cent of the observed neutral temperatures. The comfort temperatures in all the surveys selected (N ≥ 20) from the SCATs database, and from the ASHRAE RP-884 database, fall within the band (Figure 29.5). The regression line should be interpreted as giving the most probable neutral temperature rather than the only possible neutral temperature.

It seems strange at first sight that so rough and simple a measure as the 'prevailing outdoor temperature' is capable of bringing together into a simple correlation such varied and diverse climates, irrespective of, for example, the prevailing outdoor humidity. How can it be that equatorial climates and hot–dry climates are contained within the same single simple relation? The answer is that at the neutral temperatures around 27°C typical of those found in tropical climates, where people are lightly clad

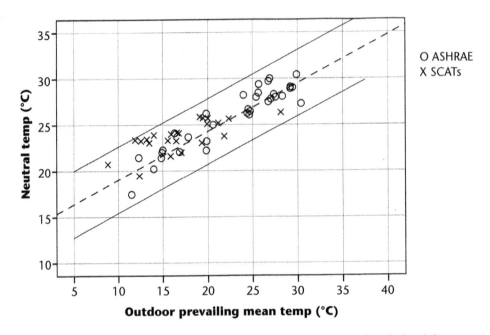

FIGURE 29.5 The location of the SCATs data and the ASHRAE data within the band shown in Figure 29.4.

and have copious air movement, the humidity has but slight influence on the neutral temperatures, so it matters little how humid it is outdoors. A climate of lower humidity but the same prevailing temperature would therefore have much the same indoor neutral temperature. At the very hot prevailing mean outdoor temperatures over 30°C, where the neutral temperature is high enough for us to expect an effect of humidity, there is an absence of inhabited climates. Human populations do not inhabit regions where excessive temperatures coincide with very high humidities, except in air-conditioned buildings or vehicles.

We have not adjusted the neutral temperatures for air speed. It is simpler to note that it is assumed that suitable and effective fans should be available for use when required. By the adaptive principle, we may rely on the people to use them to secure thermal comfort. So at the higher temperatures air movement is also higher. There is, of course, no doubt that the air speed affects the neutral temperature, and in Chapter 20 we considered the nature of its effect, but from an adaptive perspective there is no need for us to enquire into the relation between them. It takes care of itself.

Statistical note: the regression attributed equal weight to each point, irrespective of its number of observations. In the circumstance of systematic differences among the residuals it is inappropriate to weight the regression by the numbers of observations represented by each point. It is better to set a minimum accuracy for a point to be included in the dataset, and above this threshold to give equal weight to each point.

The practical importance of these systematic differences should not be overlooked. They mean that we can no longer attribute a definite 'percentage comfortable' to lines parallel to the regression line, as is currently done in ASHRAE Standard 55 (see Figure 29.2), and as implied in EN 15251 (Figure 29.3). The discomfort depends on the distance from the *actual* neutral temperature, not from the regression estimate of that neutral temperature. This fact limits the 'swing' of temperature that will be acceptable during a working day, as indicated in Chapter 26, while reducing the strictness with which the mean indoor temperature need be close to the regression line. It is for this reason that the adaptive section in the CIBSE Guide suggests placing limits on the day-on-day change in the mean indoor temperature, to ease the occupants' adaptation to the temperature.

Why the relation exists

Prof. Ole Fanger at a Windsor Conference once said of an adaptive chart relating the outdoor temperature to the temperature for comfort indoors: 'I have two problems with this graph . . . the vertical axis and the horizontal axis.' This amusing polemic hides a serious question: why is there any relation at all between the outdoor temperature and the temperature for comfort indoors?

Humphreys had originally explained it in terms of minimising discomfort when moving between indoors and outdoors. There is probably some truth in this explanation, but a better one is available, as we now explain.

According to the adaptive principle, people tend to adjust their clothing, posture, etc. to suit their room temperature and so become comfortable at the temperatures they typically encounter, while at the same time adjusting the room temperature for comfort. As we saw in the previous chapter, this process is very strong, and tends to make their neutral temperature close to the mean temperature they experience. The result of this strong convergence was seen in Figure 28.6 and it applies in both the FR and the HC modes. It is this adaptive process that is fundamental, and the relation between the comfort temperature indoors and the outdoor temperature is a consequence of it. The relation exists because, particularly when a building is in the FR mode, the mean indoor temperature is to a considerable extent determined by the outdoor air temperature, as the following consideration will show.

The air temperature recorded at a meteorological station is the temperature within a Stevenson screen. This is a white-painted box with louvred sides that permit copious through ventilation while excluding direct sunshine from the interior. It may be likened to a white building whose facades exclude all direct sunshine, has copious through ventilation, and which is neither heated nor cooled. The mean temperature within such a building operating in the FR mode, in the absence of internal heat loads, and with external shading that excludes the direct sun, would therefore equal the average outdoor temperature as measured in a Stevenson screen.

The temperature within such a building can be controlled to stay above the outdoor temperature by the admission of solar gain and by limiting the ventilation. It can be stabilised by the provision of adequate 'thermal mass'. It can be kept below the outdoor mean temperature by night-time ventilation – for example by keeping the windows open when it is cooler outdoors than in, and closing them when it is hotter outdoors than indoors.

Figure 29.6 is a scatter-plot of the mean indoor operative temperatures against the prevailing mean outdoor temperature, showing the results of these processes for the buildings in the DBSS operating in the FR mode. The coefficient of determination, R-squared, is 0.76, indicating that 76 per cent of the variation in the indoor mean temperature could be attributed to the variation in the prevailing mean outdoor temperature.

When the outdoor prevailing temperature is low, the use of the building's controls and solar heat gain have combined to raise the indoor temperature above the prevailing outdoor mean. When the outdoor prevailing mean temperature is high, the controls are used to bring the indoor mean temperature closer to the outdoor mean, and in some instances below it.

The real disparity among the comfort temperatures at any particular prevailing outdoor temperature may be explained as follows. If, because of features of its design

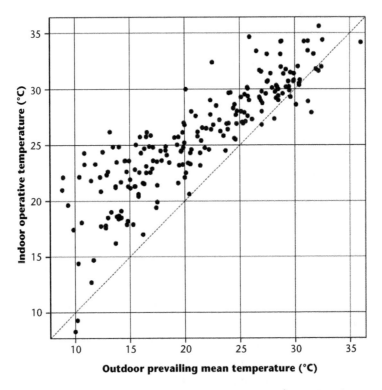

FIGURE 29.6 Scatter plot of the mean indoor operative temperature (°C) against the prevailing outdoor temperature (°C) for buildings in the FR mode.

or operation, a particular building is warmer than the average building for the prevailing outdoor temperature, the occupants will tend over time to adapt to that higher indoor temperature, so raising their neutral temperature above the average for that prevailing outdoor temperature. Over 70 per cent of the residual variation among the neutral temperatures may be explained in this way, as is shown by Figure 29.7.

Figure 29.8 superimposes the neutral temperatures on the mean operative temperatures over the range of prevailing mean outdoor temperature. The large overlap of the zones of the operative temperatures and the neutral temperatures indicates that it is entirely possible to design and operate buildings to run successfully in the FR mode over an extensive range of outdoor prevailing mean temperatures.

The heated-or-cooled mode of operation

From the database of summary statistics, we selected those surveys which took place when the building was being heated or cooled (the HC mode). Again we discarded batches of data having fewer than 20 observations. The HC mode applies to a building while using heating in cold weather, to a building while using cooling in hot weather, and to buildings that are air-conditioned all year round. There are 253 blocks of survey data that fit this HC category, together representing 71,150 observations. Figure 29.9 is a scatter plot of their neutral temperatures against the prevailing outdoor temperature. Again it should be noted that the scatter, for the larger part, represents real differences among the various neutral temperatures, so that we are dealing with a band within which the data may fall rather than a line on which they should lie.

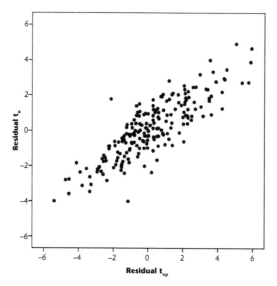

FIGURE 29.7 Relation between elevation of mean operative temperature (K) (horizontal axis, residual t_{op}) and the elevation of the neutral temperature (K) (vertical axis, residual t_n). ($r = 0.85$; horizontal axis restricted to ±6 to exclude outliers) (FR mode).

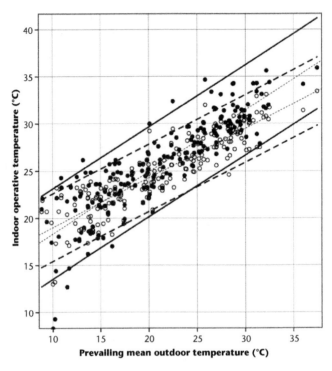

FIGURE 29.8 Scatters of neutral temperatures and mean operative temperatures superimposed. The open points are the neutral temperatures and the filled points the mean operative temperatures. The dashed lines delineate the zone in which lie 95 per cent of the neutral temperatures, and the solid lines the equivalent zone for the mean operative temperatures (FR mode).

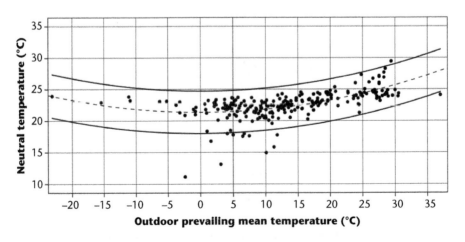

FIGURE 29.9 Scatter of neutral temperatures (°C) and the prevailing mean outdoor temperatures (°C).

A quadratic fit is shown (r = 0.61) and the lines have been set to include 95 per cent of the observations. The residual standard deviation is 1.7 K. The regression equation is:

$$T_n = 21.4 + 0.000T_o + 0.0048(\pm 0.0008)T_o^2$$

The coefficient of T_o is close to zero because the minimum of the curve occurs at about 0°C. We have left the term in T_o in the equation for conceptual complete-ness. The mean of T_n is 22.6°C (s.d. 2.1 K) and the mean of T_o is 13.2°C (s.d. 9.6 K)

The quadratic curve does not represent the data particularly well, but it does show that the neutral temperatures are climate-dependent, have a clear tendency to rise with rising temperature in the zone between 10 and 30°C, and that the relation is non-linear. Data from very cold weather are sparse, but probably indicate a prefer-ence for warmer interiors when it is very cold outside.

The outlying points must be taken seriously, for they are not errors. The lowest neutral temperature (11°C) comes from Rijal's survey of people at home in rural Nepal in the Himalayan winter, where fuel was scarce and people adapted by using heat sparingly and wearing highly insulating clothing.[29] The highest neutral tempera-ture (30°C) is from Williamson's survey of people at home in Northern Australia. They chose to wear little clothing and to use their air conditioning sparingly, lest they lose acclimatisation to their tropical environment.[30] The neutral temperatures for the two modes are superimposed on Figure 29.10.

Again, a relation exists because, even in buildings which are heated or cooled, there is a relation – if only a weak one – between the prevailing outdoor and the indoor temperature. The occupants tend to adapt to the conditions they encounter outdoors as well as indoors and so may wear lighter clothing in hotter weather. So, even in air-conditioned accommodation, people tend to have higher neutral temperatures when the outdoor temperature is high.

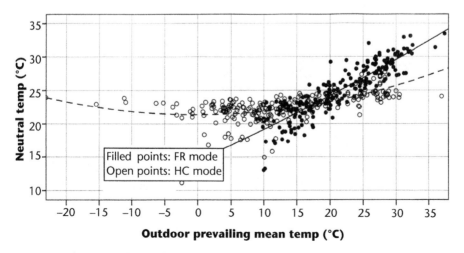

FIGURE 29.10 The FR mode and the HC mode of operation superimposed.

Probable behaviour of mixed-mode buildings

By 'mixed-mode', we mean a building that operates in the free–running mode whenever practical, but has cooling available for use in hot weather. These data suggest that there need be no discomfort if the temperature in a mixed-mode building were allowed to drift seasonally in the free-running mode according to the prevailing outdoor temperature in accordance with Figure 29.4, and that cooling were supplied if the temperature exceeded say 28°C, and heating provided if the temperature fell below say 18°C, in accordance with Figure 29.9. It is likely that these limits would vary from culture to culture, and be affected by the cost and availability of clothing and fuel. It would of course be necessary within this seasonal drift to limit the swings of temperature within a day and ensure that the drift were sufficiently gradual from day to day – a requirement that can be met by thoughtful building design and operation. The seasonal range assumes customs of indoor clothing that allow sufficient seasonal variation of clothing insulation – a custom that has become attenuated in some cultures in recent decades, but could presumably be reinstated were there sufficient reason to do so.

Local models

In the 1990s, Nicol and his co-workers developed a graph specific to Pakistan relating indoor comfort and the climate. The SCATs project produced a graph applicable to Europe. More recently Chinese researchers have produced graphs specific to the various regions of China, and brought together in the Chinese thermal comfort standard for public buildings.[31] Work has proceeded, led by Indraganti, on producing a relation applicable to much of the climate of India.[32] Rijal is currently working on a project to produce a relation applicable to people in dwellings in Japan.[33]

These local guidelines and standards are to be welcomed, provided they are based on sound field research and robust statistical analysis. Local data is to be preferred to worldwide data because of its immediate relevance. There is also a need for continuing work on the worldwide data, for this has the potential, when better understood, of bringing together all the local models into a single relation. The aim of the research is to discover how best to express the world climates in such a way as to bring together, so far as may be possible, the data from different countries, cultures and climates into a single relation.

Concluding comments

The following main points emerge from the analyses presented in this chapter:

- The general pattern of the relation first established in 1978 is confirmed by the ASHRAE database, the SCATs database and again by the analysis we have presented of our extensive worldwide database of summary statistics.

- The preferred metric for the prevailing outdoor temperature is the exponentially weighted running mean of the daily mean outdoor temperature, with a value of α in the region of 0.8. Further work may improve the value of α, and may find that the daily minimum temperature should be given more weight than the daily maximum when calculating the daily mean. The running mean is needed for dynamic thermal simulation of comfort conditions in buildings, while the historic monthly means are valuable for initial design assessments.
- The analysis of our extensive world database of summary statistics finds a greater sensitivity of the indoor comfort temperature to the prevailing outdoor temperature than envisaged in either ASHRAE Standard 44-2004 or EN Standard 15251.
- It is in principle possible to design and operate buildings that provide comfort in the free-running mode, at least within a range of prevailing mean outdoor temperature from 10–30°C.

We finally draw attention to some areas not covered in our analyses. We have not separated different types of occupancy. Most data are from offices, but the DBSS also includes some data from schools and homes, and a few from light industry. Nor have we considered the effect of culture, except to the extent that it is implicit in the prevailing mean outdoor temperature.

Notes

1 This chapter originated as part of a keynote lecture given to the NCEUB Windsor Conference in 2010 (*Adapting to change: new thinking on comfort*, Cumberland Lodge, Windsor, April. London: Network for Comfort and Energy Use in Buildings, http://nceub.org.uk).

2 CIBSE (2006) *The Guide, Section A1, Comfort, Environmental Design*, Chartered Inst. of Bldg. Serv. Engrs, London.

3 EN 15251 (2007) Comité Européen de Normalisation: *Indoor environmental input parameters for design and assessment of energy performance of buildings addressing Indoor Air Quality, Thermal Environment, Lighting and Acoustics*. CEN, Brussels.

4 ANSI/ASHRAE Standard 55-2013: *Thermal environmental conditions for human occupancy*, ASHRAE, Atlanta, 2013.

5 Humphreys, M. A. (1978) Outdoor temperatures and comfort indoors, *Building Research and Practice (J. CIB)* 6(2), 92–105.

6 Auliciems, A. (1981) Towards a psycho-physiological model of thermal perception, *International Journal of Biometeorology* 25, 109–22.

7 de Dear, R. *et al.* (1997) *Developing an adaptive model of thermal comfort and preference*. Final Report on RP 884. Macquarie University, Sydney.

8 de Dear, R. J. and Brager, G. S. (1998) Developing an adaptive model of thermal comfort and preference, *ASHRAE Technical Data Bulletin* 14(1), 27–49.

9 EN 15251 (2007) Comité Européen de Normalisation, Op. cit.

10 McCartney, K. J. and Nicol, J. F. (2002) Developing an adaptive control algorithm for Europe: results of the SCATs project, *Energy and Buildings* 34(6), 623–35.

11 Humphreys, M. A. (1978), Op. cit.

12 Humphreys, M. A. (1975, 1976) *Field studies of thermal comfort compared and applied*. Department of the Environment: Building Research Establishment, Current Paper CP 76/75 (also in: *Physiological*

Requirements on the Microclimate, Institute of Hygiene and Epidemiology, Symposium, Prague, 1975, pp. 115–81, and in *J. Inst. Heat. & Vent. Eng.* 44, 5–27).

13 Humphreys, M. A. (1981) The dependence of comfortable temperature upon indoor and outdoor climate, in: *Bioengineering, Thermal Physiology and Comfort*, Eds: Cena, K. and Clark, J. A., Elsevier, Amsterdam, pp. 229–50.

14 Humphreys, M. A. (1979) The influence of season and ambient temperature on human clothing behavior, in: *Indoor climate*, Eds: Fanger, P. O. and Valbjorn, O., Danish Building Research, Copenhagen, pp. 699–713.

15 Auliciems, A. (1981), Op. cit.

16 Auliciems, A. (1989) Thermal comfort. in: *Building design and human performance*, Ed.: Ruck, N., Van Nostrand Reinhold, New York, pp. 3–28.

17 Hoyt, T. *et al.* (2009) Energy savings from extended air temperature set points and reductions in room air mixing, *International Conference on Environmental Ergonomics*, Boston, July.

18 de Dear, R. *et al.* (1997), Op. cit.

19 de Dear, R. J. and Brager, G. S. (1998), Op. cit.

20 Fanger, P. O. (1970) *Thermal comfort*. Danish Technical Press, Copenhagen.

21 Gagge, A. P. *et al.* (1986) A standard predictive index of human response to the thermal environment, *ASHRAE Transactions* 92(2b), 709–31.

22 EN 15251 (2007) Comité Européen de Normalisation, Op. cit.

23 Thus, strictly speaking, it is a neutral temperature rather than a comfort temperature, a matter explored in the next chapter.

24 Nicol, F. and Humphreys, M. (2010) Derivation of the adaptive equations for thermal comfort in free-running buildings in European Standard EN 15251, *Building and Environment* 45, 11–17.

25 McCartney, K. J. and Nicol, J. F. (2002) Developing an adaptive control algorithm for Europe: results of the SCATs project, *Energy and Buildings* 34(6), 623–35.

26 Nicol, F. and Raja, I. (1996) *Thermal comfort, time and posture: exploratory studies in the nature of adaptive thermal comfort*. School of Architecture, Oxford Brookes University, Oxford.

27 A value of 0.84 for α has recently been found by Liu and colleagues: Liu, W. *et al.* (2014) Occupant time period of thermal adaption to change of outdoor air temperature in naturally ventilated buildings, *Proceedings of 8th Windsor Conference: Counting the Cost of Comfort in a Changing World*. Cumberland Lodge, Windsor, April. Network for Comfort and Energy Use in Buildings, London, http://nceub.org.uk.

28 See Volume 1, Section 6.3.

29 Rijal, H. B. and Yoshida, H. (2006) Winter thermal comfort of residents in the Himalaya region of Nepal, *Proceedings of International Conference on Comfort and Energy Use in Buildings – Getting Them Right*, Windsor. Organised by the Network for Comfort and Energy Use in Buildings (available at nceub.org.uk).

30 Williamson, T. J. *et al.* (1989) Thermal preferences in housing in the humid tropics, *ERDC No 1*, Energy Research and Development Corporation, Canberra.

31 Evaluation Standard for indoor thermal environment in civil buildings (2012). Ministry of Housing and Urban–Rural Development and the General Administration of Quality Supervision, Inspection and Quarantine of the People's Republic of China.

32 Indraganti, M. *et al.* (2014) Field investigation of comfort temperature in Indian office buildings: a case of Chennai and Hyderabad, *Building and Environment* 65, 195–214.

33 Rijal, H. B. (2014) Investigation of comfort temperature and occupant behavior in Japanese houses during the hot and humid season, *Buildings* 4, 437–52, doi:10.3390/buildings4030437. ISSN 2075-5309. www.mdpi.com/journal/buildings/; Rijal, H. B. *et al.* (2013) Investigation of comfort temperature, adaptive model and the window opening behaviour in Japanese houses, *Architectural Science Review* 56(1), 54–69; Rijal, H. B. *et al.* (2014) Development of the adaptive model for thermal comfort in Japanese houses, *Proceedings of 8th Windsor Conference: Counting the Cost of Comfort in a Changing World*. Cumberland Lodge, Windsor, April. Network for Comfort and Energy Use in Buildings, London, http://nceub.org.uk.

30

DO PEOPLE LIKE TO FEEL NEUTRAL? SEMANTIC OFFSETS AND ZERO-ERRORS

The simplicity of the ASHRAE scale of subjective warmth should not lead us to suppose that its interpretation is self-evident or its behaviour invariant. Over the many years of its use in field studies, quite complex behaviour has been noticed. This should be no surprise, for the construction of good semantic scales is no easy matter, as we saw in Chapter 19.

Back in the 1970s, Humphreys suggested, when comparing field studies from many different countries, that people in hot climates might prefer a sensation slightly cooler than neutral, while people in cold climates might prefer a sensation slightly warmer than neutral.[1] The possibility of a discrepancy between thermal neutrality and optimum comfort has implications for the estimation of energy use in buildings. If the assumption that optimal comfort can be equated with thermal neutrality is incorrect, then temperature standards based on the ASHRAE scale will be to some extent faulty. Faulty assumptions about the required temperatures have practical implications both for the interpretation of the result of a thermal simulation of a building and for the estimates of energy requirement obtained from such a simulation.

The need to ascertain more precisely the desired thermal sensation led researchers to supplement the ASHRAE scale with a scale of thermal preference, a scale that asked people whether they would prefer to feel warmer or cooler, or whether they desired no change.[2,3] It is then possible to derive from a survey both the temperature corresponding to thermal neutrality on the ASHRAE scale (the *neutral* temperature) and the temperature corresponding to the 'prefer no change' category on the preference scale (the *preferred* temperature). The difference between the two may be called the 'semantic offset'.

It should be noticed that the discussion assumes that the preference scale obtains the optimum temperature for comfort without bias. This is not wholly true. We pointed out in Chapters 18 and 19 that practical problems can arise when using and analysing scales of thermal preference, and that these can sometimes make the estimate of the preferred temperature unreliable. We will comment on the effect of these problems as they arise during this chapter.

De Dear and his colleagues in their analysis of the ASHRAE RP-884 database of field studies were the first to make a systematic evaluation of the semantic offset. Their method of quantifying the offset was to obtain the neutral temperature by solving the regression equation of the ASHRAE scale value on the operative temperature and then to obtain the preferred temperature from the McIntyre scale by using Ballantyne's method (see Chapter 24). The semantic offset was defined as the neutral temperature minus the preferred temperature. They looked at how the offset might depend on the mean indoor temperature during the survey, and on how it might depend on the outdoor temperature prevailing at the place and time of the survey. Their analysis showed a small offset in the expected direction for people in centrally air-conditioned buildings, both in relation to the indoor mean temperature and the prevailing outdoor temperature. For people in the naturally ventilated buildings, they found no statistically significant relation between the offset and either the outdoor or the indoor temperature.[4,5] However, although not statistically significant, the relation with the outdoor temperature was in the expected direction.

The offset is difficult to quantify. Obtaining a neutral temperature from a regression equation can lead to large errors because of the uncertainties in the regression coefficient, as we saw in Chapter 28. There is also a weakness inherent in Ballantyne's method, apart from random uncertainty. His procedure casts the 'prefer no change' responses at random into the 'prefer warmer' and the 'prefer cooler' categories on the scale. This produces a biased estimate of the preferred temperature if the mean temperature during the survey differs from the preferred temperature, as we noted in Chapter 24. So the estimates of the neutral temperature and of the preferred temperature are each prone to error, and the semantic offset is the difference between them. Now the error variance of the difference between two quantities is the sum of the error variance of each of those quantities, and so the values obtained for the semantic offset are prone to large errors. This is perhaps why researchers may find it difficult to detect a statistically significant relation between the climate and the offset for the naturally ventilated buildings in the database.

Modelling the discrepancy between comfort and neutrality

The discrepancy between thermal neutrality and the preferred condition may be thought of as a moveable 'zero-error' on the ASHRAE scale, whereby the optimum for comfort is not fixed at 'neutral' (zero on the scale), but can sometimes take other values, perhaps 'slightly warm' or 'slightly cool', or even 'warm' or 'cool'. This different approach enables us to build a picture of the behaviour of this moveable zero-error in the ASHRAE scale.

Figure 30.1 is a schematic representation of the behaviour of the zero-error. The grey curves are the hypothesised relation between the actual warmth sensation on the scale of subjective warmth and the preferred warmth sensation on the same scale. Its shape is perhaps more easily explained in terms of the categories of the Bedford scale: much too cool, too cool, comfortably cool, comfortable (neither warm nor cool), comfortably warm, too warm, much too warm. For convenience here we number the scale from −3 to +3 with the central comfort sensation equal to zero. If there were no zero-error on the scale, the grey curves would be entirely flat, and all at the value of zero on the scale of preferred thermal sensation. That is to say, everyone would always desire to be 'neither warm nor cool' (thermally neutral).

We suppose that people's preferred sensations vary from time to time, such that sometimes they like to feel 'comfortable, neither warm nor cool', while at other times they like to feel 'comfortably warm' and at other times 'comfortably cool'. If this is so, we would expect, in accord with the adaptive principle, that they would adjust their clothing and their room temperature so that they actually did feel how they would like to feel. We would therefore expect a positive correlation between the actual warmth sensation and the desired sensation. This adaptive relation is illustrated by the centre region of the figure (warmth sensation −1 to +1). Complete adaptation would give a gradient of unity at the centre, while the gradient would be less if opportunities to adapt were restricted, falling to zero were no adaptation possible.

Outside this region, on the warm side (warmth sensation between +1 and +2 on the figure), a person would probably like to feel a little cooler than they actually felt. The zero-error would reach a maximum in this region. On the cooler side (warmth

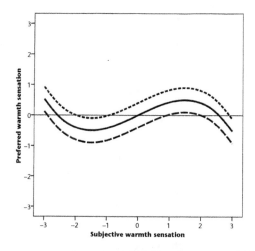

FIGURE 30.1 Schematic diagram of the relation between the warmth sensation and the preferred sensation. The zero-error is the distance between the curve and the horizontal axis (preferred sensation = zero).

sensation between −1 and −2), a person would probably like to feel a little *warmer* than they actually did, giving the trough in the curve.

If the thermal sensation rose to be 'much too warm' (3 on the horizontal axis), it is likely that a person would like to feel 'comfortably cool' (−1 on the preferred sensation scale), while if the thermal sensation fell to be 'much too cool' (−3 on the horizontal axis), it is likely that a person would like to feel comfortable but warm (+1 on the preferred sensation scale). These are the regions of negative gradient towards the margins of the figure.

We might further expect that the solid curve, in its entirety, would move up or down on the graph in relation to the prevailing outdoor temperature. So, in a hot climate, people might like to feel on the whole slightly cooler than neutral (the lower dashed curve), while in a cold climate they might prefer to feel slightly warmer than neutral (the upper dotted curve). In summary, the size of the zero-error depends on just two factors: the current thermal sensation and the outdoor temperature. We now consider the experimental evidence in support of this model of the behaviour of the zero-error.

Obtaining the zero-error by direct enquiry

Humphreys and Hancock used a direct method of obtaining the zero-error.[6] First, they asked their respondents to indicate on the ASHRAE scale how they were feeling. Next they asked them to indicate on the scale their *desired* thermal sensation – how they would have liked to feel. The desired sensation is the zero-error on the ASHRAE scale, measured in the units of the scale. The method requires the use of the scale in its continuous form. Griffiths had used a similar method a number of years previously to check the zero on the ASHRAE scale.[7] The method has the merit of providing a value for the zero-error each time a respondent indicates his subjective warmth sensation.

Humphreys and Hancock obtained their data from studies they made of the thermal comfort of students during university lectures, and also from a separate project that was investigating the thermal comfort of residents in selected dwellings. The data yielded some 900 comparisons of the actual and the desired thermal sensation.

They found that the preferred sensation depended on the current actual sensation, there being a positive correlation between the preferred sensation and the actual sensation in the region between 'neutral' and 'warm', and a negative correlation when the actual thermal sensations were higher. The data therefore confirmed the existence both of the region of positive correlation in the centre region of Figure 30.1, and of the negative correlation when people were uncomfortably warm. However, their data for the cool to cold thermal sensations were too few to confirm the existence of a zone of negative correlation in that region of the figure.

The data also showed something not immediately relevant to ascertaining the behaviour of the zero-error. They found that people differed systematically from one another in the sensation they preferred. While most associated 'neutral' on the ASHRAE scale with optimum comfort, others described their preferred sensation as systematically either warmer or cooler than 'neutral'. So the zero-error on the scale differed from person to person in addition to its dependence on how warm or cool the person may have felt.

This exploratory study shows that there is merit in using the method of direct enquiry. That is to say, whenever we ask a respondent for their sensation on the ASHRAE scale, or some other scale of subjective warmth, we ask them also to indicate on the same scale what sensation they would like best. The procedure dispenses with the need to use a scale of thermal preference, such as the McIntyre scale.

We next consider the evidence on the behaviour of the zero-error that can be obtained from the SCATs database and the ASHRAE RP-884 database. From the field studies conducted in European offices (the SCATs project), as well as from the ASHRAE RP-884 database, Humphreys and Nicol were able to show that the preferred sensation on the ASHRAE scale depended both on the prevailing outdoor temperature and upon the current indoor temperature.[8] They selected those occasions where the respondent had indicated on the McIntyre scale or the Nicol scale that they desired no change in their thermal condition. They then ranked the indoor temperature and the outdoor temperature into quintile bins, and calculated the mean ASHRAE scale value for each. This value was the mean zero-error for the bin, expressed in the units of the ASHRAE scale. They found these mean zero-errors always to be small, lying within the range ±0.3 scale units. The use of the quintile bins avoided the need to assume any particular mathematical form for the relation between the zero-error and the temperatures.

They found that, in both databases, people preferred sensations on the warm side of neutral if it was warm indoors, and on the cool side of neutral if it was cool indoors. The effect was stronger in the ASHRAE data, perhaps because of the higher indoor temperature of its top quintile bin (Figure 30.2). The result does not relate directly to the relation hypothesised in Figure 30.1, because the horizontal axis is the room temperature rather than the warmth sensation on the ASHRAE scale. However, room temperature and the warmth sensation are correlated, so we may use the room temperature as a surrogate for the warmth sensation. The increase in the zero-error with an increase in indoor temperature therefore confirms the shape of the grey line in the centre region of Figure 30.1.

They also found that the zero-error in general decreased with rising outdoor temperatures (Figure 30.3). This corresponded to the raising and lowering of the curves in Figure 30.1 according to the prevailing outdoor temperature. The higher outdoor temperatures were associated with a negative zero-error.

The independent confirmation from the two databases encourages us to believe that the zero-error indeed behaves as hypothesised in Figure 30.1. However, it should be noticed that, in the databases, the indoor temperature is correlated with the outdoor temperature, so trends seen in Figures 30.2 and 30.3 are not completely independent.

Humphreys and Nicol's procedure is open to criticism of two kinds. First, the considerable width of the 'no change' category of the preference scales meant that

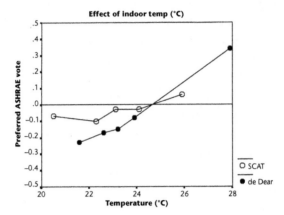

FIGURE 30.2 The preferred point on the ASHRAE scale related to the mean indoor temperature (ASHRAE database and SCATs database).

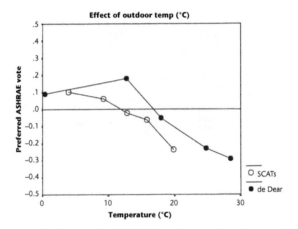

FIGURE 30.3 The preferred point on the ASHRAE scale related to the prevailing mean outdoor temperature (ASHRAE database and SCATs database).

the corresponding value on the ASHRAE scale is prone to bias if the true preference does not coincide with the centre of the 'no change' category. Second, the use of just one category of the preference scale results in a large loss of information. The trends in the data are probably secure, but it would be unwise to derive quantitative estimates from them.

In summary, we have independent data on the behaviour of the zero-error from the two principal databases of thermal comfort surveys, and from two surveys that used the method of direct enquiry. Together they confirm much of the pattern pictured in Figure 30.1. Around thermal neutrality, when people report feeling warmer than neutral it is partly because they *choose* to feel warmer than neutral, while if they feel cooler than neutral it is partly because they *choose* to feel cooler than neutral. Beyond the zone of reasonable comfort, this ceases to apply, and people who are warm desire to feel cooler than neutral. These two zones are supported by the evidence of experiment. There is less evidence from conditions where people feel cool or cold. It is probable that they would like to feel warmer than neutral, but the experimental data are sparse. The systematic departures of the desired sensation from neutrality are not large, amounting to no more than about 0.3 scale units. So at a first approximation, the grey curves on Figure 30.1 can be taken to be flat. We are dealing with just a small correction to the ASHRAE scale.

Semantic offsets in the SCATs database

The exploration above of the behaviour of the zero-error has provided a good understanding of the complex behaviour of the ASHRAE scale, but does not give reliable quantitative adjustments to be applied to the neutral temperature. Recently developed methods of analysis overcome the uncertainties in estimating semantic offsets that de Dear and his colleagues had encountered when using the ASHRAE database, and the problems inherent in the method used by Humphreys and Nicol when exploring the behaviour of the zero-error using the ASHRAE database and the SCATs database. There is therefore the potential to improve the accuracy with which we can estimate the semantic offset. We explain the method and give some preliminary results. The reader should be aware that this is a topic that requires further research, and we do not regard our results as definitive.

In Chapters 25 and 28, we explained how to derive reliable estimates of the neutral operative temperature. First, the regression gradient was established from a pooled within-day analysis of warmth sensation on the operative temperature. Then the neutral temperature was calculated from the joint means of temperature and warmth sensation, using the regression gradient derived from the day surveys. We pointed out that the method could be applied to a single value of the warmth sensation. This gives a low-precision but unbiased value of the neutral temperature. When applied to the data arising from a survey, the method gives a low precision but unbiased value for the neutral temperature that corresponds to each of the individual ASHRAE votes in the survey. The statistics drawn from a large number of unbiased but low-precision observations can be usefully precise.

The surveys in the SCATs database used not the McIntyre three-category scale of thermal preference but the Nicol scale, which has five points. It is therefore possible to obtain unbiased values for the *preferred* temperature from the Nicol scale in the SCATs database using the methods of Chapters 25 and 28. First we obtain the regression gradient from the pooled within-day thermal preferences and the corresponding operative temperatures, as had been done for the ASHRAE scale.

The pooled within-day scatter plot is shown as Figure 30.4. This procedure gave a regression coefficient of 0.190/K. The coefficient was next corrected for the presence of measurement error in the operative temperature, exactly as had been done for the regression of the ASHRAE vote on the operative temperature. This gave a corrected coefficient of 0.238/K. Having done this, we calculated a preferred operative temperature corresponding to each 'preference vote'. The semantic offset is the neutral temperature for each vote minus the corresponding preferred temperature. Each value of the offset is without bias but of low precision.

From the discussion of the behaviour of the zero-error of the ASHRAE scale at the beginning of this chapter, we would expect that the semantic offset would depend both on the thermal sensation and on the prevailing outdoor temperature. We now explore these dependencies. Figure 30.5 is a scatter plot of the semantic offsets obtained from each interview contained in the SCATs database and the mean thermal sensation of the respondents on the interview day. The ASHRAE scale in the SCATs database is numbered from one (cold) to seven (hot) and this numbering is retained in Figure 30.5. The striped appearance of the data arises from the categorisation of the ASHRAE scale and the Nicol scale.

A number of wild values of the semantic offset were found. These extreme values arise from the occasional mistakes respondents make when using a preference scale, whereby its polarity becomes reversed, as we noted in Chapter 19. Offsets greater than ±6 K were therefore discarded as outliers. There were 59 such observations out of a total of 4654, just 1.3 per cent of the data.

The solid line on Figure 30.5 is a locally weighted regression line that follows any local trends there may be in the data (a Loess smoothed regression line). It shows

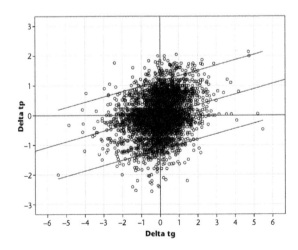

FIGURE 30.4 The pooled within-day regression of preference (delta tp) against the operative temperature (delta tg). PP mode (part air conditioned and part not) omitted. (Points outside ±2 can arise because the within-day mean in general differs from zero.)

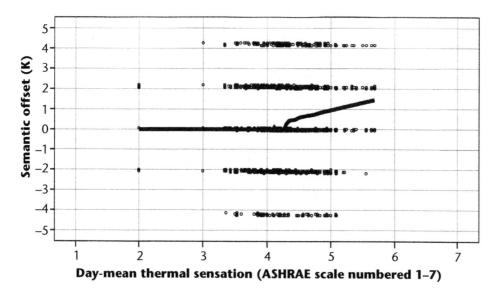

FIGURE 30.5 Scatter plot of the semantic offset (neutral temperature minus preferred temperature) against the day-mean thermal sensation.

that the semantic offset was zero when the day-mean thermal sensation was below neutral (4 on the scale). So in this region the preferred temperature and the neutral temperature did not differ systematically from each other. Above thermal neutrality, and disregarding the small kink in the trend-line, the offset rose in a near-linear manner, reaching some 1.5 K at the highest day-mean thermal sensations. At a day-mean thermal sensation of 5 on the ASHRAE scale (slightly warm), the neutral temperature was about a degree above the preferred temperature.

The semantic offset in warm outdoor temperatures: Figure 30.6 is a similar scatter plot, but this time the offset is plotted against the prevailing mean outdoor temperature, expressed as an exponentially weighted mean of the day-mean outdoor air temperature. Between −3°C and +13°C, the semantic offset is zero. As the outdoor prevailing mean rises above 13°C the offset increases in a near-linear fashion until it reaches some 1.8 K at 29°C. In this region the semantic offset is positive, showing that the neutral temperature is a little above the preferred temperature.

   The region of interest is therefore when the prevailing mean outdoor temperature is above 13°C and the day-mean thermal sensation on the ASHRAE scale is warmer than neutral. These data were selected for a further analysis using multiple regression.
   In the region of interest the trends are near linear, so a linear regression of the offset against the two variables should be an adequate statistical model. We further note that the outdoor temperature is correlated with the day-mean thermal sensation,

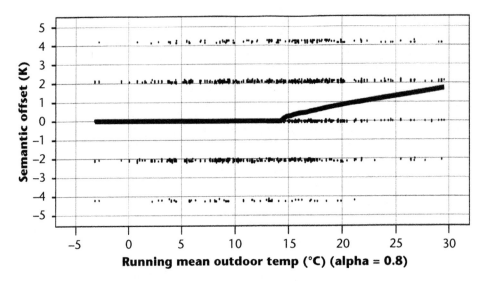

FIGURE 30.6 Scatter plot of the semantic offset (neutral temperature minus preferred temperature) against the prevailing mean outdoor temperature.

because the temperature outdoors influences the indoor temperature, particularly when a building is operating in the free-running mode. This correlation leaves open the question of whether the offset is attributable to the day-mean thermal sensation, to the outdoor temperature, or to both of them. We use multiple regression analysis to help us to answer this question.

The statistics are as follows. There are 1491 occasions when the outdoor temperature is above 13°C and the day-mean thermal sensation above 4 scale units. In these selected data the day-mean thermal sensations have a grand mean of 4.43 scale units and a standard deviation of 0.36. The mean outdoor temperature is 17.1°C, standard deviation 3.2 K. The correlation between the day-mean thermal sensations and the outdoor temperatures is 0.24, high enough to necessitate the use of multiple regression but not so high that co-linearity becomes a problem. The regression equation is:

$$T_{offset} = 0.49(0.12)TS_{day\text{-}mean} + 0.061(0.014)T_o - 2.5$$

T_{offset} is the semantic offset (K), $TS_{day\text{-}mean}$ the day-mean thermal sensation in ASHRAE scale units, and T_o the outdoor prevailing mean air temperature (°C). The numbers in parentheses are the standard deviations of the estimates of the regression coefficients.

The equation is informative. The two predictor variables contribute about equally to the variation of the semantic offset when their respective ranges are taken into account (beta values of 0.104 and 0.116 respectively). Both regression coefficients are highly significant statistically (p < 0.001) and usefully precise.

It is the effect of the outdoor temperature that we desire to quantify for the thermal design of buildings. The range of outdoor temperature covered by the equation is from about 13°C to about 29°C, some 16 K. According to the equation, traversing this range while holding constant the day-mean thermal sensation will increase the semantic offset by just 1.0 K. It is evident that the variation of the prevailing mean outdoor temperature encountered in the SCATs project has only a small effect on the semantic offset, after we have allowed for the contribution of the day-mean thermal sensation.

Concluding note

The question of whether people desire to feel neutral has long been a concern in the field evaluation of thermal comfort, the concern first being raised, we believe, in 1972 during the discussion of Webb's data from Roorkee and Baghdad, data showing people comfortable in room temperatures over 30°C (Chapter 5). In this chapter, we have tried to understand the meaning of the neutral point on the ASHRAE scale, and have explored how it changes meaning in relation to a changing thermal sensation and a changing outdoor temperature. We have shown that there is good evidence that the semantic offset depends on both these variables.

A new method of estimating the semantic offset has been applied to the SCATs data. This unbiased method of analysis strongly suggests that the effect of the prevailing mean outdoor air temperature on the systematic difference between the neutral temperature and the preferred temperature is no more than 1 K over the range of prevailing mean outdoor temperature covered by the data (−3°C to + 29°C). Evaluation of its effect outside this range must await further research and analysis, but it is clear that the semantic offset is only a small correction to the neutral temperature. The large values of the semantic offset occasionally reported from a single survey may perhaps be attributed to idiosyncrasies of the scales that were used in that survey (see the discussion of scaling irregularity in Chapters 18 and 19).

Notes

1 Humphreys, M. A. (1975) *Field studies of thermal comfort compared and applied.* Department of the Environment, Building Research Establishment, Watford. Current Paper CP 76/75, 1975 (also in *J. Inst. Heat. Vent. Eng.* 44, 5–7, and *Physiological Requirements on the Microclimate*, Symposium, Prague, 1975).
2 Fox, R. H. *et al.* (1973) Body temperatures in the elderly: a national study of physiological, social and environmental conditions, *British Medical Journal* 1, 200–06.
3 McIntyre, D. A. (1980) *Indoor climate.* Applied Science Publishers, London.
4 de Dear, R. *et al.* (1997) *Developing an adaptive model of thermal comfort and preference.* Final report on Project RP-884, Maquarie University, Sydney.
5 de Dear, R. J. and Brager, G. S. (1998) Developing an adaptive model of thermal comfort and preference, *ASHRAE Transactions* 104(1), 27–49.

6 Humphreys, M. A. and Hancock, M. (2007) Do people like to feel 'neutral'? Exploring the variation of the desired sensation on the ASHRAE scale, *Energy and Buildings* 39(7), 867–74.

7 Griffiths, I. D. (undated, circa 1990) *Thermal comfort in buildings with passive solar features: field studies*, Report to the Commission of the European Communities ENS3S-090-UK, Department of Psychology, University of Surrey, Guildford.

8 Humphreys, M. A. and Nicol, J. F. (2004) Do people like to feel 'neutral'? Response to the ASHRAE scale of subjective warmth in relation to thermal preference, indoor and outdoor temperature, *ASHRAE Transactions* 110(2), 569–77.

31

CLOTHING AND THERMAL BEHAVIOUR

From the viewpoint of thermal comfort research, clothing is seen to be a means of adding extra thermal insulation to the human body so that the person wearing it can be thermally comfortable in a cooler thermal environment. We begin this chapter by considering the several other purposes that clothing serves, as these can influence people's thermal behaviour.

The diverse functions of clothing

Clothing is used to display wealth and social status. Portraits of kings and queens and the nobility of centuries past show the wealthy splendidly dressed. Their fine and costly clothing was often adorned with furs and jewels, as can be seen in even the portraits of young children of noble families. There were 'sumptuary laws' in Europe that prohibited, among other things, the lower social classes from wearing clothes of certain colours and of fine materials. Fine clothing was a mark of social distinction. Even a wealthy person of a lower social class was prohibited from wearing the type of clothing associated with the nobility. To wear the finest clothing, one had not only to be able to afford it, but also to be considered somehow worthy of wearing it. A family had to have been wealthy for a generation or two before being accepted into a higher social class. These laws proved difficult to enforce, and eventually fell into disuse.

Clothing is still used to display wealth and to mark social distinctions. A wealthy businessman might have many suits, each tailored specially to fit him and made of the finest cloth. He will not wear the same suit two days running, lest he should appear even slightly dishevelled, and lest it be thought that he could not afford to wear a suit freshly cleaned and pressed every day. Fine shoes too are used to display the status of the wearer, rather than to improve the comfort of the feet.

It is not only the wealthy who care about how they dress. Almost everyone does. It is from the clothes that a first impression of character and status is obtained. Young people choose to dress differently from their parents, and clothing must be appropriate to the age of the person, lest we seem to dress 'too young' or 'too old'. People have conflicting desires: they wish to conform to the social norm for the place in society that they occupy, while also wishing somehow to be distinctive. So people choose with care what they will wear, having regard for what is suitable for the occasion and suitable to their personality. Much of this choice happens at a level a little below conscious thought. We somehow 'know' in our home culture what clothing is suitable, and feel embarrassed when we get it wrong.

In most cultures, the clothing of men and women differs, and it would not be fitting for a man to dress as a woman would, nor a woman to dress as a man would. In Europe the boundaries are not as rigid as once they were, and many items of clothing are the same for women and men, but it is still possible with reasonable certainty to identify most articles of clothing as intended for women or for men.

Clothing is also used for modesty, hiding certain parts of the body while allowing other parts to be seen. What is considered modest or decent varies from culture to culture, and changes over the generations. Even in the relaxed attitudes that prevail in many contemporary cultures, there are limits. Few cultures accept total nudity in public places, even when it would be sensible from a thermal viewpoint. The origin of ideas of modesty and decency are ancient and obscure, but even in 'liberated' societies, they remain socially powerful, whether or not they have the force of law.

The increased frequency and ease of travel in recent decades, together with our ability to see on television and the internet how other nations dress, has been a force towards international uniformity. In Europe, it is no longer possible at a glance to identify a person's nationality by their everyday dress, as it was until as recently as the middle of the twentieth century. Traditional national costumes are now worn only on special occasions and for special events.

Clothing is used to indicate a person's function in society. A soldier is identified by the uniform, while distinctions of rank are indicated by differences in the uniform. The same applies to the police, and in many workplaces dress indicates function, even when there is no obvious dress-code or uniform.

Clothing is often used as a statement of religious belief and of the devoutness of the wearer. Examples are the wearing of distinctive dress by ministers of religion and those who have devoted themselves to a religious life, such as monks. Well-known examples can be found in most of the religions of the world.

Clothing also protects the body. This is obviously true of the protective clothing worn for various tasks and sports and by various occupations, but it is also true of everyday clothing. It serves as a barrier against abrasion and against painful or unpleasant contact with hot and cold surfaces.

It is important to remember this wide range of functions noted above when we are treating clothing as thermal insulation. The thermal insulation function may not be the one that matters most to a particular person at a particular time. One of its other functions may take precedence over comfort. A jacket may be discarded to give improved freedom of arm movement, or kept on to avoid giving offence on a formal occasion, disregarding any resulting thermal discomfort.

The complexity of clothing

Even when considered from the viewpoint of its thermal function, clothing behaviour is complex. The thermal insulation of the fabric is only one of the factors that contribute to the insulation of a complete ensemble. Much of the insulation is provided by air trapped between and within the layers of fabric, and so the insulation of the ensemble changes if this air is disturbed when people move, or when the air is moving. Loose clothing may provide high insulation when a person keeps still, but a much-reduced insulation when the person moves, or when a ceiling fan is being used. Thus the air permeability of each layer also affects the dynamic insulation of the whole ensemble.

In hot arid climates, the clothing serves to protect against the heat of the sun, and its colour is then important too. Black clothing is a more effective barrier to the direct sun, being opaque to the sunshine, but it becomes hot because it absorbs almost all the incident radiation. So the choice of colour depends on the circumstances – white is not always preferable, and can allow the solar radiation to penetrate the layers of clothing. The thermal insulation depends on the radiant, conductive and convective heat exchanges within the layers of the ensemble, between it and the body, and between the ensemble and the thermal environment. Quite distinct from preventing the loss of heat to a cold environment, in a hot, and particularly in a dry environment, the clothing may be creating an acceptable microclimate around the body by reducing heat gain while the loose garments encourage air movement and allow the evaporation of sweat.

Nor is insulation the only characteristic of importance to the thermal properties of clothing. In hot conditions and when the metabolic rate is high, it is essential that sweat produced can evaporate to cool the body. Some modern clothing items limit the passage of moisture (sometimes purposely so), and a lack of water-vapour permeability of the clothing can cause discomfort and thermal stress.

A particular item of clothing can also be used in a variety of ways. Indraganti and her co-workers[1] demonstrated at the 2014 Windsor Conference the different ways in which the Indian sari can be used according to climate and context – yet the basic insulation of each alternative would be hard to determine, and the change from one to another hard to record in a field survey context.

Given this complexity it is not surprising then that the measurement of the thermal insulation of a clothing ensemble obtained using a thermal manikin can sometimes be a less than accurate indicator of the practical thermal benefit of the ensemble. Other ways of assessing the thermal insulation of a clothing ensemble may also fail to capture its practical insulation in daily life. For example, the summing of the insulation of each layer to obtain the total provides only a rough value. Rather less reliable is the relation between the weight of the ensemble and its overall thermal insulation – light down-filled garments can be highly insulating.

Nor is it surprising that people are quite often dressed in ways that do not seem appropriate to the thermal comfort researcher. Despite the adaptive principle, people do not always have thermal comfort as their priority.

Clothing and room temperature

The large databases of thermal comfort provide a general picture of the way clothing insulation varies with the room temperature. Figure 31.1 is the distribution of the clothing insulation found in the surveys included in the ASHRAE RP-884. In this database, each line of data gives the clothing insulation of a single respondent on the occasion of the interview, so the distribution is of individual values rather than group-mean values.

The range of thermal insulation is wide. The distribution departs from Normal because there is a theoretical lower limit of zero for the clothing insulation, but no fixed upper limit. The database contains surveys from some cold winter interiors in northern Pakistan, where several layers of clothing were needed to reduce cold discomfort.

People tend to choose their clothing to be comfortable at the temperature of the room, so we would expect some of the variation of the clothing insulation to be attributable to the different room temperatures. This dependence is evident in

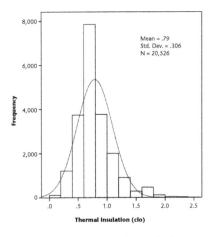

FIGURE 31.1 Distribution of the thermal insulation of the clothing (ASHRAE RP-884 database). The values include the insulation of the chair where appropriate.

Figure 31.2, a scatter plot of the thermal insulation of the clothing against the room operative temperature. The points on the scatter plot are spot-values rather than group-means. The grey trend line is Loess-smoothed.

The clothing insulation decreases as the indoor temperature rises, as expected. The trend line is almost straight until about 22°C, the clothing insulation reducing from 2.0 clo at 5°C to 1.0 clo at 19°C. The reduction is insufficient to compensate for the temperature change (see Chapter 20 on the simple heat exchange model), and some of the respondents in the winter Pakistan surveys felt cold despite their clothing, as we noted in the discussion of adaptation to seasonal variations of temperature in Chapter 27. The rate of decrease becomes smaller as the temperatures rise above about 25°C. People in the warmer rooms did not all compensate fully for the raised temperatures, presumably because their social circumstances did not permit them to remove as many layers of clothing as would have been required for complete comfort.

We also see in Figure 31.2 that, at any particular temperature, there is a very wide range of clothing insulation. For example, at 20°C the range is from 0.4 clo to 2.0 clo. This diversity is partly because people use their clothing as a thermal comfort control. So people will wear more clothes when they are too cold, and shed them when they later become too warm. This control action can even reverse the expected relation between clothing and the sensation of warmth in the short term. Heavy clothing may

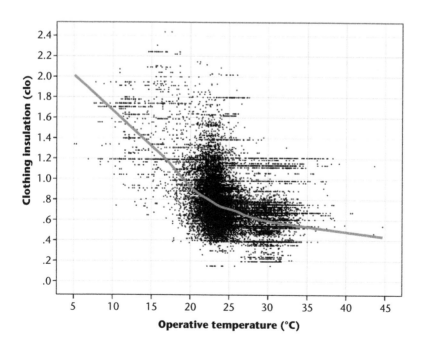

FIGURE 31.2 Scatter plot of the thermal insulation of the clothing and the room operative temperature. Each point is a single observation (ASHRAE RP-884 data).

indicate that the person is feeling cold, while light clothing may indicate that she is feeling hot. A further reason for the wide spread is the uncertainty in the estimation of the clothing insulation in field studies, where it is almost always obtained from a checklist of the garments. These factors make it impossible to predict with accuracy a spot-value of the clothing insulation from the room temperature alone.

Next we consider the data from the SCATs project, the year-round investigation of thermal comfort of office workers in five European countries. Figure 31.3 shows the distribution of the clothing insulation. We notice the absence of a 'tail' in the distribution. Offices in Europe are heated in winter, so there was not the same need for highly insulating clothing.

Figure 31.4 is the scatter plot of the clothing insulation and the room temperature, again showing the Loess-smoothed trend line. We have kept the axes the same as in Figure 31.2 so that the scatter plots may be compared at a glance, and the degree of smoothing is unchanged so that their trend lines may also be directly compared. The range of room temperatures in the SCATs data is much smaller than in the ASHRAE data.

Further information on clothing insulation in relation to the room temperature is available from the database of summary statistics (DBSS) we used in Chapters 28 and 29. It will be recalled that the DBSS includes the surveys of the ASHRAE database and the SCATs database in summary form. We have excluded them from Figure 31.5 to make the three scatter plots entirely independent. We have also excluded a single outlying value where the clothing insulation was estimated by weight as 5 clo when the mean temperature was 8°C. Unlike either the ASHRAE database or the SCATs database, the DBSS consists of group-mean values (rather than single discrete

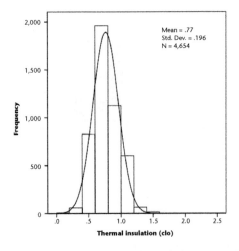

FIGURE 31.3 Distribution of the thermal insulation of the clothing (SCATs database). The values include the insulation of the chair where appropriate.

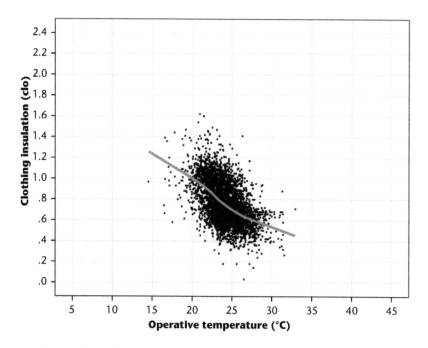

FIGURE 31.4 Scatter plot of the thermal insulation of the clothing and the room operative temperature. Each point is a single observation (SCATs data).

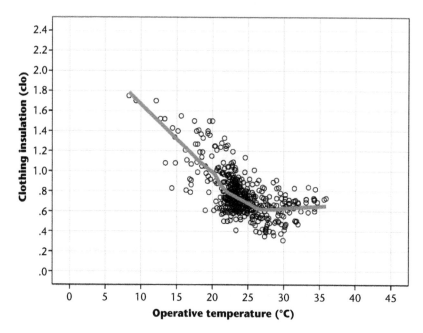

FIGURE 31.5 Scatter plot of the thermal insulation of the clothing and the room operative temperature. Each point is a mean value from a survey of thermal comfort. Database of summary statistics (DBSS) excluding ASHRAE and SCATs data, and also a single outlier (5 clo at 8°C).

TABLE 31.1 Comparison of the mean clothing insulation in the three databases.

Room temp (°C):	5	10	15	20	25	30	35	40
Clo (ASHRAE):	2.0	1.7	1.3	0.9	0.7	0.6	0.5	0.5
Clo (SCATs):			1.2	1.0	0.7	0.5		
Clo (DBSS)*:		1.7	1.3	1.0	0.7	0.6	0.7	

* Excluding the ASHRAE and the SCATs data.

observations) and this reduces the scatter of the values of the clothing insulation at any particular room temperature, and also of the room temperature for any particular value of the clothing insulation.

The trend line again has the same degree of Loess smoothing. A slight upturn in the trend line at the highest temperatures is probably attributable to the presence of surveys where reduced clothing was not socially acceptable.

Table 31.1 compares the mean clothing insulation found in the three databases across the range of room temperature. The values are all within 0.1 clo, except for a discrepancy of 0.2 clo at 35°C, the place of the slight upturn in the DBSS data. The agreement between the databases is surprising, and suggests we can predict the likely overall-mean clothing insulation (though not the individual values) for a particular room temperature.

This is not as useful as might first be thought. The trend line does not correspond to the behaviour of any particular group of people, nor does it enable us to predict with useful accuracy the clothing an individual will wear, or even the mean clothing insulation for any particular survey included in a database. We will show later in the chapter, when looking at clothing adjustment as a thermal comfort control mechanism, why it is to be expected that at any particular room temperature there is likely to be a very wide range of values for the clothing insulation.

The influence of outdoor temperature

We have focussed on the relation with the indoor temperature rather than with the outdoor temperature. The indoor temperature is the relevant variable because people spend almost all their time indoors in most societies today. It is the indoor temperature that governs human heat exchange of people indoors, and hence also the indoor clothing insulation that is needed for comfort. The effect of outdoor temperature on the indoor clothing is therefore indirect. (The effect of outdoor temperature on *outdoor* clothing is a separate field of research.)

The expected indirect influence of the outdoor temperature on the clothing indoors is for people to wear slightly less clothing indoors in warm weather than in cold, for the same room temperature. Estimating the size of this effect is difficult because the indoor temperature and the outdoor temperature are correlated in the databases, and because the effect of the outdoor temperature on the clothing insulation is likely to be non-linear.

To estimate the size of the effect, we ranked the outdoor temperatures into quintile bins (20 per cent of the data in each bin), separately for each database. We did the same for the room temperature. Then in each temperature bin we compared the clothing insulation for the upper and lower quintiles of outdoor temperature. (Statistical note: we used a univariate general linear model with clothing insulation as the dependent variable and the indoor and outdoor temperature quintiles as the predictors.)

Of the three databases, the most uniform is the SCATs data, because, except for the language spoken, the same research protocol was used in all five countries all year round. So the effect of the outdoor temperature should appear most consistently from this database. Figure 31.6 shows the result of the analysis. The horizontal axis shows the quintile groups of the indoor globe temperature (tg in the figure), and the lines show the estimated clothing insulation for the quintiles of the running-mean outdoor temperature (tr80 in the key of the figure).

The clothing insulation indoors was 0.16 clo lower when the outdoor temperature was in the top quintile of the outdoor temperature. For the DBSS data, the corresponding value was 0.15 clo, so the two are in close agreement. However, the corresponding value for the ASHRAE database was higher at 0.23 clo, and the next-highest quintile had a still larger effect (0.34 clo). The inversion reduces confidence in quantifying the effect. Perhaps we should say that the range of outdoor temperature found in the databases may be expected to cause a shift of about 0.2 clo in the indoor clothing insulation, in addition to the effect of any change in the indoor temperature. It is small compared with the overall range of the clothing insulation in the databases, and equivalent to a shift of about 1 K in the temperature for thermal neutrality for sedentary activities. A more precise statement is not possible.

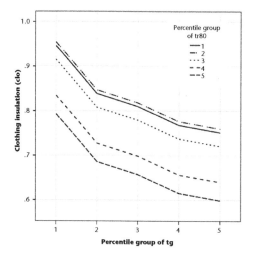

FIGURE 31.6 Estimating the influence of the outdoor running-mean temperature on the thermal insulation of the clothing indoors (univariate general linear model).

Some of the effect of the outdoor temperature is transitory. In summertime when someone lightly clad and feeling hot enters a cool building, the first impression is very pleasant. It may take an hour or two before the person begins to feel cold in the light clothing. The reverse effect has been noticed in wintertime on the London underground railway system. People on the streets wear heavy clothing but often still feel the cold. When they enter the underground system, the walkways, platforms and carriages are warm. The warmth is welcome at first while people are warming up. Only after some time on the crowded trains do people become too hot in their winter clothing. This transitory effect means that people feel hotter in summer on the underground than they do in winter, although they wear less clothing and the temperature deep underground does not change with the season.

Modelling the dynamics of clothing behaviour

We turn next to consider the dynamics of the clothing behaviour. From research over many years into the use of clothing to achieve adaptation to the thermal environment, two 'rules of thumb' can be stated.

- Within a day at the workplace or at school there is almost no change of clothing in response to room temperature change. (This may not apply to people at home, where there are fewer constraints on the clothing.)
- From month to month the change in clothing is sufficient for complete adaptation, unless the month-mean indoor temperature requires wearing clothing ensembles that are socially or culturally unsuitable.

Can we be more precise than this in tracing the rate at which the clothing changes in response to a changing room temperature?

Because there is little change within the day at work, it seems sensible to express the rate of change of clothing in terms of the change from one day to the next in response to day-to-day changes in room temperature. If a mismatch between clothing and room temperature made us uncomfortable at work yesterday, we try to correct it today by wearing different clothing. For example, if we were a little too warm yesterday, we would choose to wear slightly lighter clothing today. If seriously too warm yesterday we would wear even less clothing today. So the greater the mismatch, the greater would be the change in the clothing.

This behaviour can be modelled by saying that the clothing insulation we choose today depends on the clothing insulation we wore yesterday, and also on yesterday's room temperature. A linear model will suffice.

Consider the relation:

$$C_d = a + bC_{(d-1)} + cT_{(d-1)} + \varepsilon \tag{31.1}$$

where C_d is today's clothing, $C_{(d-1)}$ is yesterday's clothing, and $T_{(d-1)}$ yesterday's mean room temperature; a, b and c are constants (parameters to be estimated from the data), and ε is a random error term. The equation says that what is worn today depends on what was worn yesterday and on yesterday's temperature. So if we have data from many pairs of consecutive days, it is possible to evaluate the parameters by fitting a multiple regression equation to the data. No long runs of consecutive days are needed. This is useful, because field thermal comfort data do not often include long runs of consecutive days. A survey of thermal comfort that includes estimates of clothing may be very short – just a few days. For longer surveys, the sequence of the data is usually broken at the weekend, though Monday's clothing might be influenced to some extent by the memory of Friday. So the possibility of obtaining estimates from many pairs of consecutive days rather than from a very long unbroken sequence is of great practical value.

The particular index of clothing does not matter. It could be the thermal insulation estimated as a clo-value, or a simple count of the number of layers worn, or some other variable likely to be linearly related to the clothing insulation. This again is useful, because it is not always practicable to estimate the clo-value accurately in field data. However, within any block of data from which an estimate of the parameters is to be obtained, the same index of clothing must be used. It will be shown that equation 31.1 is equivalent to saying that the clothing insulation today depends on an exponentially weighted series of the room temperature over a long period of previous consecutive days. We show the working in full for the benefit of readers who are not familiar with the mathematics of geometric progressions and exponential series.

Consider the clothing yesterday, day $(n-1)$. It would depend on the clothing and temperature the day before yesterday, day $(d-2)$. We would have:

$$C_{(d-1)} = a + bC_{(d-2)} + cT_{(d-2)} + \varepsilon$$

We can enter this value of $C_{(d-1)}$ into the first equation, obtaining

$$C_d = a + b(a + bC_{(d-2)} + cT_{(d-2)} + \varepsilon) + cT_{(d-1)} + \varepsilon$$
$$= a + ab + b^2C_{(d-2)} + cT_{(d-1)} + bcT_{(d-2)} + \varepsilon + b\varepsilon$$

But the clothing on day $(d-2)$ would depend on its value on the day before that, day $(d-3)$, so we would have:

$$C_{(d-2)} = a + bC_{(d-3)} + cT_{(d-3)} + \varepsilon$$

If we enter this value in the new equation, we obtain:

$$C_d = a + ab + b^2(a + bC_{(d-3)} + cT_{(d-3)} + \varepsilon) + cT_{(d-1)} + bcT_{(d-2)} + \varepsilon + b\varepsilon$$

$$= \{a + ab + ab^2\} + \{cT_{(d-1)} + bcT_{(d-2)} + b^2cT_{(d-3)}\} + \{\varepsilon + \varepsilon b + \varepsilon b^2\} + b^3C_{(d-3)}$$

A pattern is now emerging, and we can continue to repeat the procedure for each successive previous day indefinitely. We then obtain (if N is an indefinitely large integer):

$$C_d = a\{1 + b + b^2 + b^3 + b^4 + b^5 + \ldots b^N\}$$

$$+ c\{T_{(d-1)} + bT_{(d-2)} + b^2T_{(d-3)} + b^3T_{(d-3)} + b^4T_{(d-4)} + b^5T_{(d-5)} + \ldots b^N T_{(d-N)}\}$$

$$+ \varepsilon\{1 + b + b^2 + b^3 + b^4 + b^5 + \ldots b^N\}$$

$$+ b^N C_{(d-N)} \tag{31.2}$$

We can deduce from the second line of the equation that the parameter b must be less than unity. The room temperature yesterday surely has more influence on today's clothing than does the room temperature on a day in the remote past. Therefore b^N must be smaller than b (true only if b is less than unity). The higher powers of b are then close to zero, and we may say for practical purposes that $b^N = 0$. The last line of the equation is therefore zero and can be omitted.

We now consider the first line of the equation. The portion in parentheses is the sum (S) of the terms of a geometric progression. Since b is less than unity it has a finite sum equal to $1/(1 - b)$.

Mathematical note: the proof of this is surprisingly neat and simple:

Let $\quad S \quad = 1 \; + \; b \; + \; b^2 \; + \; b^3 \; + \; b^4 \; + \; b^5 \; + \quad \ldots \quad b^N$

then $\quad Sb = \quad\quad\quad b \; + \; b^2 \; + \; b^3 \; + \; b^4 \; + \; b^5 \; + \quad \ldots \quad b^N + b^{N+1}$

Subtracting Sb from S we obtain:

$S - Sb = 1 - b^{N+1}$

but $b^{N+1} \to 0$, *so* $S - Sb = 1$

hence $S = 1/(1 - b)$

We therefore have for today's clothing:

$$C_d = a/(1 - b)$$

$$+ c\{T_{(d-1)} + bT_{(d-2)} + b^2T_{(d-3)} + b^3T_{(d-3)} + b^4T_{(d-4)} + b^5T_{(d-5)} + \ldots b^N T_{(d-N)}\}$$

$$+ \varepsilon/(1 - b)$$

The second line of the equation is of great interest. It is the series of the temperatures of past days, weighted according to a geometrical progression with multiplying constant b. If we divide it by the sum of the weights (which is equal to $1/(1-b)$) it becomes what is known as the exponentially weighted moving average (or exponential running mean) of the temperatures of the previous succession of days. So we may write:

$$C_d = a/(1-b)$$
$$+ c/(1-b)\left[(1-b)\{T_{(d-1)} + bT_{(d-2)} + b^2T_{(d-3)} + b^3T_{(d-3)} + \ldots b^NT_{(d-N)}\}\right]$$
$$+ \varepsilon/(1-b)$$

The expression enclosed between square brackets [. . .] in the second line is now the exponentially weighted running mean of the temperatures of the succession of previous days. The clothing today (C_d) can therefore be described by the running mean of the daily sequence of room temperature.

The third line of this equation is a random-error term. We have carried it through the calculations in order to show that the presence of random error in the data does not invalidate the procedure. It does not appear in the expression for the running mean temperature, and remains finite despite the successive inclusion of innumerable past days.

It is worth pausing at this point to reflect on the result. From the simple statement that people tend to correct any discomfort that arises from a mismatch between their clothing and the room temperature, we have deduced that the clothing depends on an exponentially weighted running mean of the room temperatures over the series of previous days. Thus the use of exponential running means in adaptive comfort theory is a logical rather than a merely empirical choice.

It is easy to update the running mean temperature from one day to the next. Today's running mean is:

$$T_{rm(d)} = (1-b)\{T_{(d-1)} + bT_{(d-2)} + b^2T_{(d-3)} + b^3T_{(d-3)} + \ldots b^NT_{(d-N)}\}$$

So tomorrow's running mean will be (marking 'tomorrow' as day $(d+1)$):

$$T_{rm(d+1)} = (1-b)\{T_d + bT_{(d-1)} + b^2T_{(d-2)} + b^3T_{(d-3)} + b^4T_{(d-3)} + \ldots b^{N+1}T_{(d-N)}\}$$
$$= (1-b)\{T_d\} + b\left[(1-b)\{T_{(d-1)} + bT_{(d-2)} + b^2T_{(d-3)} + b^3T_{(d-3)} + \ldots b^NT_{(d-N)}\}\right]$$

The term in square brackets [. . .] is seen to be *today's* running mean, $T_{rm(d)}$. So we have:

$$T_{rm(d+1)} = (1-b)\{T_d\} + b\{T_{rm(d)}\}$$

This property of the running mean makes it very simple to use, because it can be updated every day in response to the new temperature.

Evaluating *b* from survey data

The value of b indicates how rapidly the clothing changes in response to the day-to-day change in room temperature. If, for example, b were zero, only yesterday's temperature would affect today's clothing, while if b were unity all previous days would be equally influential no matter how far back in time they were. The response to a temperature change would be indefinitely slow. The exponential time constant is $(1 - b)$. The half-life (the time taken for the clothing to go half-way to its final value after a step-change in the room temperature) is $0.69/(1 - b)$.

A value of b can be obtained from Equation 31.1 by applying linear regression analysis to survey data from pairs of successive days using the clothing and the mean room temperature. Early evaluations of b for secondary school children were obtained from observations of children's clothing and the mean room temperature. There were 120 pairs of days (rows of data) from which to estimate the parameters of the regression equation.[2] The value of b came to 0.15 ± 0.014. The estimate of the exponential time constant $(1 - b)$ is therefore 0.85, giving a half-life $(0.69/(1 - b))$ of 0.81 days. It may be said that equilibrium is reached after five half-lives have elapsed, so if there were a step change in the mean room temperature it would take about four days for the clothing to reach its new equilibrium value. This would be the time the children would take to adapt fully to the new room temperature. Data were also available for younger children, but the day-on-day changes in room temperature were so small that the value of b, although similar (0.11 ± 0.07), was not statistically significant.

There were also data for people outdoors, shopping or visiting the zoo, giving 56 pairs of days for the regression analysis, and the value of b was 0.14 ± 0.05. This again indicated that, were there to be a step change in the outdoor daily mean temperature, people would take about four days fully to adapt to it.

These time constants apply to conditions where the required clothing is readily available. If the adaptation required a more radical change in thermal behaviour, then much longer time constants would apply. For example, if the temperature were so cold that normal indoor clothing would not suffice, a cultural change would be required and people would take longer to adapt to the new circumstance. This matter was discussed in Chapter 6 when considering the response to the lowered office temperatures at the UK Building Research Station (BRS) during the winter of 1972–73.

Figure 31.7 illustrates the response of the clothing of a class of school children were there to be a step change in the classroom temperature. Suppose that the classroom temperature had been steady for several days and that the children had adapted their clothing to suit that temperature. Between days 2 and 3 on the figure,

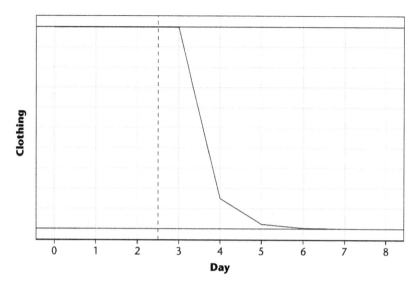

FIGURE 31.7 The response of the clothing to a step change in room temperature.

the temperature changes overnight to a higher level, and stays at the new level for several days. On day 3, the children would find they were hot. On the next they would respond by wearing less clothing, and so on day by day until they had adapted to their new classroom temperature. The figure is for a value of b equal to 0.15. It is seen that the children would be well adapted by the third day after the change (day 5 on the figure), and completely adapted by day 7.

This simple illustration is, of course, far from reality. A classroom operating in the free-running mode (neither heated nor cooled) will have a temperature that changes from day to day in a way that depends on the vagaries of the weather. The children could adapt to gradual seasonal trends in classroom temperature, but not to quasi-random differences from day to day.

Figure 31.8 illustrates the difficulty. We suppose that, at the start of the series of days, the children have clothing suitable to 20°C, this being the room temperature to which they are currently adapted. Then over the next school days the classroom temperature changes at random. (The temperatures are a random selection from a Normal distribution with a mean of 20 and a standard deviation of 3. A rather large standard deviation has been chosen to make the effect clear.) On day 0, they encounter a temperature of 17.3°C. They come next day clothed for 17.7°C, only to find the temperature has changed to 19.9°C. Next day they manage better. They have come clothed for 19.6°C and encounter 19.2°C.

The open circles in Figure 31.8 show the randomly changing room temperatures, while the crosses show the temperature that the children would be adapted to that day, according to the equation for the running mean. There is no convergence. On days 4, 6, 7, 8 and 9 the children would experience quite serious thermal

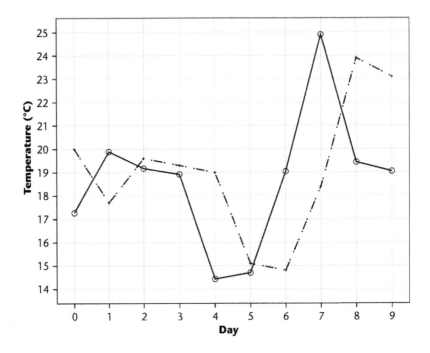

FIGURE 31.8 Response to a random series of daily mean temperatures.

discomfort, as the classroom temperature would be four or more degrees different from the temperature for which they were clothed. Their attempt to adapt is ineffective because the temperature is unpredictable. The children might eventually learn the need to come to school with clothing that was easy to alter during the school day. But it is inconvenient carrying extra garments in case the room happens to be cold. Adaptation can be seen as a learning process, and it is not possible to learn from a series of random events, as it contains no information. It follows that the building should be designed so that the indoor temperature changes little from day to day. This can be achieved by, for example, providing sufficient thermal inertia in the structure.

An experiment on bedclothing and comfort (see Chapter 8) also provided data from which to estimate b. The clothing index in this field experiment was the number of layers of bedclothes, and the bedroom temperatures were recorded during the night. There were data from 26 people from summer through to winter, giving in all some 1800 pairs of nights upon which to base the estimate. The median value of b was 0.53, giving a value of 1.47 days for the half-life of the exponential series, and five half-lives as 7.4 nights, but the speed of response differed among the respondents. The time taken for the median person to adapt fully to a step change in the bedroom temperature would be about a week.[3] It seems then that night-to-night variations in bedroom temperature were not as quickly adapted to as were day-to-day changes in school classroom temperature. Since those days, the blanket-and-eiderdown system of bedclothing has been largely superseded in the UK by the duvet (continental quilt). As far as we know, there have been no systematic experiments on rapidity of response when changing from a winter-weight duvet to a summer one, or vice-versa.

More sources of data for estimating the clothing response

The ASHRAE RP-884 database contains dated information on the clothing and temperature of numerous individuals. Most of the surveys in the database are of cross-sectional design, where each person gives just one set of responses. These cannot tell us anything about the speed of adjustment of clothing. However, a few of its surveys are of longitudinal design, where the same person gives responses on many consecutive days. It would be possible to derive time constants for clothing adaptation from these data, but as far as we know this has not been attempted. The SCATs project also collected longitudinal data on numerous people over periods from weeks to months. They are held in a separate SCATs database, and these data too could be used. Most useful would be data from spring and autumn from lightweight buildings operating in the free-running mode, where room temperatures might be expected to show considerable day-to-day variation.

Obtaining thermal comfort information from the observation of clothing

If we assume that people choose their clothing to optimise their thermal comfort, it is possible to deduce room temperatures for comfort from observations of the clothing alone. This can be a useful method in circumstances where the administration of questionnaires would be intrusive or inconvenient for other reasons. The next sections explain the theoretical basis of the method.[4]

It is easiest to explain the method by considering a group of people sharing the same thermal environment, as for example a class of school children in their classroom (see Chapter 4). In the UK, many but not all schools have a uniform that the children are expected to wear when at school. This usually includes a sweatshirt that the children may wear over a shirt or dress. Because of this it is simple to identify visually those children who have removed their sweatshirt. We imagine a class of children in a cold classroom. All the children will be wearing their sweatshirts. Now let the room warm up gradually, and assume that the children take off their sweatshirts if they are too warm. As the room continues to warm up, more and more of the children will be without sweatshirts, until the room is so warm that no children will be wearing them. The proportion without sweatshirts, if plotted against the room temperature, would form a sigmoid curve. Curve B in Figure 31.9 represents this condition. It may be defined more strictly as the percentage of the population (in this case a class of children) who, if wearing

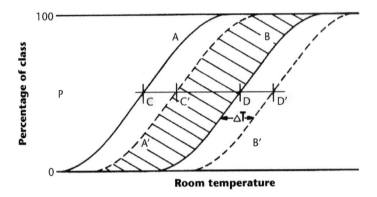

FIGURE 31.9 Idealised clothing behaviour.

Source: M. A. Humphreys 1973.

clothing of thermal resistance R_c (in this example a sweatshirt over their dress or shirt) would be suffering warmth discomfort.

For the same class, still with their sweatshirts on, consider now the percentage who are 'not too cold' (i.e. they are comfortable or too warm). At any temperature (except at the extremes) this percentage will be greater than that shown by curve B, because it includes both the children who are too warm and those who are comfortable. The percentage 'not too cold' thus lies to the left of curve B on a curve parallel to it. This is curve A in Figure 31.9. It may be defined as the percentage of the population who, if wearing clothing of thermal resistance R_c, would be 'not too cold', that is, who were comfortable or too warm.

The curves A' and B' are the same as curves A and B except that they apply to a class of children *without* sweatshirts. The temperature displacement between curve A and A' is equal to that between B and B', and is the warming effect of the sweatshirt. (This temperature difference can be estimated from the clo-value of the sweatshirt and the metabolic rate of the children, as explained in Chapter 20.)

We now move from our thought experiment into the real world. In a real classroom in a building that is operating in the free-running mode (no heating or cooling), the classroom temperature will fluctuate from day to day and vary within the day. We now show that the experimental observations of the percentages without sweatshirts should lie within the shaded zone of the figure.

Consider a point (t', p') anywhere on curve A'. The population is without sweatshirts. By definition, a proportion p' are not suffering cold discomfort. Therefore (1 − p') must be uncomfortably cold. Now let the population be free to put on extra clothing. At least (1 − p') would do so. Hence only p' or less would remain without sweatshirts. Therefore if one were observing a free population, the observed proportions without sweatshirts would all lie on or below the curve A'. (A 'free population' may be defined as one whose members choose between two or more weights of clothing to avoid discomfort.)

Now consider a point (t, p) anywhere on the curve B. The population is wearing sweatshirts. By definition a proportion p suffers warmth discomfort. Now let the population be free to take off their sweatshirt. At least p would do so. Hence p or more would be without sweatshirts. So if one were observing a free population, the observed proportions without sweatshirts would all lie on or above the curve B. Therefore all observations of the proportions without sweatshirts would lie between curves A' and B. This is the shaded area in Figure 31.9. Curves A' and B are thus the envelope of the observed proportions without sweatshirts. The positions of A' and B can therefore be estimated from observations of the proportion of the population without sweatshirts. Further, if we now subtract curve B from curve A we obtain a bell curve for the proportion comfortable when wearing sweatshirts. Subtracting curve B' from A' gives the proportion comfortable without sweatshirts. From the bell curves we can read off the optimum temperatures for comfort for children with and without sweatshirts.

We can therefore deduce quantitative thermal comfort information simply by repeatedly observing the clothing of a group of people at a variety of room temperatures, if we may assume that the people select their clothing to avoid discomfort.

Fitting the data

Deriving the sigmoid curves from the data is not straightforward. The curves A' and B are the envelope of a complete set of observations, observations that should all fall within the shaded zone on the figure. But because of statistical variation the zone has 'fuzzy edges', and so we do not know precisely where to place the curves. The analysis aims to find a pair of parallel lines which, for a given separation, enclose the greatest number of experimental observations, due weight being given to each observation. The separation of the lines must be such that any observations that fall outside the lines might reasonably be expected to do so.

The appropriate statistical procedure is to apply a Probit transformation to the data to render straight the sigmoid curves, and then to regress the room temperature upon the clothing Probit. The edges of the zone are then placed at ±1.5 times the residual standard deviation of the room temperature. This in effect assumes that some 85 per cent of the observations should lie within the zone. This way of obtaining the boundaries of the zone works quite well in practice, but lacks a theoretical foundation.

The proportion of observations that we would expect by chance to fall outside the zone will depend on the number of observations in each batch. The usual class size was about 30 children, giving a reasonable degree of accuracy to each observation. Another source of error lies in the inconsistency of the children's responses. Not only is there variation among the children, which is accounted for in the model, but also each child is unlikely to respond consistently to the room temperature – he or she will not always remove the sweatshirt at the same room temperature.

There is good reason to regress the room temperature on the clothing Probit rather than applying the iterative Probit technique, which at first glance would be the obvious method to adopt. The clothing tends to be invariant to within-day temperature variation, as we have noted previously. A rise or fall of room temperature during the school day does not therefore result in a corresponding change of clothing. The changes take place chiefly from day to day. This gives a strong 'horizontal grain' to the observations within the zone, and the presence of this structure must be represented in the way the data are analysed.[5]

It would be incorrect to apply the usual iterative Probit analysis procedure to the data arising from clothing observations, because its assumptions about the error structure do not apply to our data. Its application leads to inconsistent results among incomplete sets of data – data where the range of temperature is insufficient to give a full range of clothing response, from cool conditions where everyone wears a sweatshirt through to warm conditions where nobody does.

Constraints on the clothing behaviour

The derivation of thermal comfort conditions from the observation of the clothing depends on the assumption that people choose their clothing to avoid thermal discomfort. We noted at the beginning of this chapter that clothing serves a variety of purposes, and that people may tolerate some degree of discomfort rather than wear thermally suitable clothing that is for some reason socially inappropriate. For this reason it is wise, when using clothing observations to estimate comfort temperatures in conditions where such constraints might apply, to supplement the observations of clothing with brief structured interviews. These can be with a small random subsample of the experimental population, the interview being designed to discover what factors had led to the choice of clothing. The presence and severity of the constraints may thus be evaluated.

The behavioural model we have just described can be re-cast to apply to such behaviour as the use of fans and the opening and closing of windows. We discussed these applications in Volume 1. When a person or group has more than one way by which to restore thermal comfort (or achieve some other goal such as improving the air quality) they will tend to choose first the most convenient of the ways, and only if that is unsuccessful or too slow will they try another way. There will therefore be no fixed hierarchy of adaptive actions.

Constraints on actions can be expressed in terms of temperature differences. For example, a group may tolerate a temperature rise of perhaps three degrees before opening a window on a noisy façade. We can speak of this as a three-degree constraint.

Concluding comments

From the point of view of heat-exchange models of thermal comfort it is the physical properties of clothing that dominate. Laboratory research has been devoted to measuring the thermal insulation and the vapour permeability of more and more garments and clothing ensembles. Thermal manikins have become more sophisticated, better representing the human body, and some are able to move and 'sweat'. This research has improved our estimates of the clothing insulation in daily life. Particularly welcome is the wider range of ensembles that have been measured, which now include the everyday wear of a number of non-European cultures.

The adaptive approach to thermal comfort makes use of these results, but behavioural questions are its primary concern. How do people use their clothing to achieve thermal comfort in everyday life? How quickly do they respond to a changing temperature by changing their clothing? How much and in what circumstances do the non-thermal purposes of clothing constrain people's thermal behaviour?

In this chapter we have considered how such matters can be investigated, and have described the theoretical models that we have developed. Our hope is that this chapter will stimulate interest in understanding and quantifying people's thermal clothing behaviour. Such understanding will be valuable for the design and operation of buildings. Buildings must enable and encourage types of thermal behaviour that lead to thermal comfort. This must be done in a way that is consistent with minimising the use of energy. Further consideration of these matters must await Volume 3 in our series.

Notes

1 Indraganti, M. *et al.* (2015) Thermal adaptation and insulation opportunities provided by draping Indian saris on the body, *Architectural Science Review* 58(1), 87–92.
2 These examples are taken from: Humphreys, M.A. (1979) The influence of season and ambient temperature on human clothing behaviour, in: *Indoor climate*, Eds: Fanger, P. O. and Valbjorn, O., Danish Building Research, Copenhagen, pp. 699–713. The analysis presented in the 1979 paper rested on a slightly different mathematical model, but the difference does not systematically affect the values extracted for the exponential time constants.
3 The analysis considered only the speed of night-to-night changes, and does not tell us about adjustments during a single night, such as turning back the covers to get cooler or pulling them up again to get warmer.
4 The method was first given in: Humphreys, M.A. (1973) Clothing and thermal comfort of secondary school children in summertime, in: *Thermal comfort and moderate heat stress*, Eds: Langdon, F. J. *et al.*, HMSO, London, and further developed in: Humphreys, M.A. (1973) Classroom temperature, clothing and thermal comfort – a study of secondary school children in summertime, *J. Inst. Heat. & Vent. Eng.* 41, 191–202.
5 This behaviour helps explain why, at any particular room temperature, there is characteristically a wide range of values for the clothing insulation.

32

INTERACTIONS AMONG ENVIRONMENTAL VARIABLES, ADAPTATION AND OVERALL COMFORT

In these volumes, we focus on the thermal dimension of comfort. However, the thermal landscapes we occupy are just one aspect of our local environment that can contribute to overall comfort. If the indoor environment is to satisfy its occupants, not only must the thermal conditions within it be good, but so too must be the visual and acoustic aspects of the rooms and spaces. The quality of the air also needs to be sufficiently good, both with regard to its chemical composition and to the particulate matter suspended in it. In this chapter, we consider the ways in which these different aspects of the indoor environment might be combined, how they can interact with one another, and what role adaptation plays in the overall perception, management and resulting quality of the indoor environment.

Interactions

By an interaction, we mean that the response of a person to the environment may be affected not only by the individual components, such as temperature and sound level, but also by their combination. Thus the response to one aspect of the environment (perhaps a noise) might depend on the level of another (such as room temperature).

When this is so, it affects the statistical models used to predict the responses of the occupants of the environment. We can no longer use a simple regression model for prediction, such as the response, Y, to the environmental variables X_1 and X_2 as in the equation:

$$Y = aX_1 + bX_2 + \varepsilon$$

If an interaction is present we must include a further term, the product of X_1 and X_2, so the predictive equation would become:

$$Y = aX_1 + bX_2 + cX_1X_2 + \varepsilon$$

Whether there is an interaction between X_1 and X_2 can be discovered by checking whether the coefficient c of the term X_1X_2 differs significantly from zero.

Note: The number of possible interactions is very large. If we suppose that we have n environmental variables to consider, the number of first-order interactions is given by

$$(n^2 - n)/2$$

so if there were 20 aspects of the environment to be considered, the number of potential interactions would be 190. Work on uncovering and classifying interactions continues. It would be out of place in a volume on adaptive thermal comfort to enumerate these interactions and explain their potentially serious effects on comfort and wellbeing. ASHRAE Guideline 10 classifies the interactions that have so far been indentified, and the information is continually updated.[1]

Overlapping concepts

The concepts available in a particular language may not map perfectly onto the concepts used in the sciences of the indoor environment. For example in English, as spoken in the UK, the concept of the warmth of the environment is not independent of the concept of the 'freshness' of the air. A cold day will be described as 'a bit fresh'. The words commonly used to describe the quality of the air overlap in meaning with those used to describe its temperature. Thus an apparent interaction between two aspects of the environment may have a linguistic rather than a physical explanation. So researching interactions requires precise wording of the semantic scales we use to quantify the subjective responses of the population (see Chapters 18 and 19).

The impossibility of an overall comfort index

Within the thermal environment, indices have been developed that combine the several aspects of the thermal environment into a single number. A simple and useful example is the globe temperature, which combines the effect of thermal radiation and of the air temperature. More complete are the indices that combine the effects of the air temperature, the mean radiant temperature, the air movement, the humidity, the clothing insulation and the metabolic rate. The best known of these are the PMV (predicted mean vote) and the SET (standard effective temperature) indices we have already described.

These indices do not comprehensively describe the effect of the thermal environment on a person. Two environments can have the same PMV but feel quite different. Wearing light clothing in a warm environment feels quite different from wearing heavy clothing in a cool environment, even if the two conditions are equal on the scale of PMV or SET. A room heated by warm air feels different from a room heated by a radiant source, although they might be equal on the scale of globe temperature, PMV or SET. Further, the air movement is a vector quantity, having both magnitude and direction, and the direction in which the air is moving relative to the body affects how we feel. The same applies to thermal radiation. Even the complex indices are therefore incomplete, since they consider only the overall heat exchange between the person and the thermal environment.

What has been said of the thermal environment applies also to the other aspects of the environment, each of which is complex. There exists no truly comprehensive index of the visual environment, of the acoustic environment, or of the quality of the air. There can therefore exist no comprehensive index of the overall quality of the environment.

The relative importance of the different aspects for comfort also depends on the task being performed. For example, a task that entails sustained close visual work may demand a high level of illumination on the task. The visual environment is then very important, while the acoustic and thermal environments might be less so. A restaurant may require a very different visual environment from that normal for close visual work. Generally lower levels of illumination may be used to create a relaxed mood for an evening meal. Examples could be multiplied.[2] Such complexities mean that it is impossible to combine the different aspects of an environment into a single number that describes the overall environment in terms of its comfort or quality.

The role of adaptation in multifaceted environments

How then can we ensure that people are satisfied with or comfortable in their overall environment? Adaptation can come to our rescue. People adapt not only to optimise their thermal environment, but other aspects of their environment as well. Furthermore, they will, if given the opportunity, adjust the levels of these various features of the environment to suit themselves. Thus, in addition to optimising their thermal environment, people will, given the opportunity, adjust the lighting to their current preference, using the lighting controls, blinds and shades.[3] They will adjust the acoustic environment by supplying background sound of the chosen type and volume, opening or closing windows to admit or exclude outdoor sounds, and perhaps reduce the reverberation time of the room with carpeting and other soft furnishings. People will also adapt behaviourally to any effects of interactions among the various facets of the environment, adjusting themselves to the environment and the environment to themselves to optimise their experience so far as may be practicable.

The concept of *adaptive opportunity* is therefore valid not only for the thermal environment, but also for the other aspects of the environment, and for balancing the aspects to provide an overall optimum that fits the task or activity. For example, when writing this chapter at home in his study, Humphreys chose not to heat the room despite it being 18°C, and but rather to wear warm clothing. This was one of a series of interconnected choices. The casement windows were opened a crack for ventilation to maintain good air quality, the curtain was part-drawn to prevent glare on the computer screen, and the brightness of the screen was set for visual comfort. Fixed levels for each of those aspects of the indoor climate would have precluded adjusting them to suit the particular task he was performing. Rather these adjustments should be easy to make, and be of sufficient effect to enable optimisation for comfort and task performance.

Other aspects might have entered into awareness and required action. Discomfort from sitting too long in one position causes adaptation by changing the posture, while sound from the loud speakers at the nearby cattle-auction sometimes requires the window to be closed. The ever-changing nature of personal requirements and an ever-changing outdoor environment lead to a succession of adaptations to optimise the indoor environment.

Limitation of optimising the environment by adaptive actions

People can adapt only to that which their bodies perceive. For example, carbon monoxide is toxic yet odourless. People cannot detect it in the atmosphere, and so they do not take the appropriate action to avoid it. It is therefore necessary to prevent harmful levels of carbon monoxide in the room air by the appropriate design, operation and maintenance of any equipment that produces it. The same applies to any harmful aspect of the environment that cannot be perceived by the body.

Compromises and conflicts

Sometimes it is not possible to satisfy all the requirements at the same time. For example there can be a conflict between having windows open to maintain good air quality and needing to close them to exclude noise from outside. A compromise is needed – perhaps opening the windows for a few minutes to air the room and then keeping them closed for a longer period. The compromise adopted will depend on the relative importance of quietness and of sufficient ventilation, and that in turn depends on the task and on personal preference.

Sometimes the design of the building and its services causes unnecessary conflicts. Automated window controls that aim to keep the indoor temperature constant in buildings operating in the free-running mode can be noisy in operation and open and close unnecessarily often. The noise annoyance may outweigh the benefit of the improved temperature control. Window design may be such that daylight is optimised but at the same time the room is caused to overheat, or ventilation cannot be obtained without excessive air speeds in the occupied zone, or the action of opening windows can prevent the effective use of blinds.

Problems of this kind have been identified by Post–Occupancy Evaluation (POE) and Building Performance Evaluation (BPE),[4,5] which enable building managers and occupants to fine tune, over time, the interaction between the requirements of the buildings, their services and the people in them. All three requirements evolve together in the context of the local climate and its micro-climates in a relationship that is almost impossible to capture with building simulation models. Achieving low-cost and low-energy comfort relies on the adaptation of the building, its systems and its occupants to optimise the comfort conditions within it. Key to achieving this is to deal with the various clashes between the systems and the environment that have been documented by Bordass and Leaman of the Usable Buildings Trust.[6]

Such conflicts proliferate in buildings with complex control systems, but they can also result in serious discomfort even in simple buildings. The problem can be avoided only by understanding the fundamentals of designing for comfort and performance, and by applying them with careful attention to the details of design. How this might be achieved and the comfort challenge faced by designers is the subject of Volume 3 in our trilogy on adaptive thermal comfort.

Notes

1 Interactions affecting the achievement of acceptable indoor environments. *ASHRAE Guideline 10-2011,* American Society of Heating, Refrigerating and Air-Conditioning Engineers, Atlanta.
2 The SCATs project, besides measuring the thermal environment, measured the visual and acoustic environment and the quality of the air. That these aspects cannot be combined as a single number is shown in: Humphreys, M. A. (2005) Quantifying occupant comfort: are combined indices of the indoor environment practicable? *Building Research & Information* 33(4), 317–25.
3 Nicol, F. *et al.* (2006) Using field measurements of desktop illuminance in European offices to investigate its dependence on outdoor conditions and its effect on occupant satisfaction, productivity and the use of lights and blinds, *Energy and Buildings* 38(7), 802–13.
4 Preiser, W. and Vischer, J. (2005) *Post occupancy evaluation.* Harper Collins, New York; and Baird, G. (2010) *Sustainable buildings in practice – what the users think.* Routledge, Abingdon.
5 Baird, G. *et al.* (1996) *Building evaluation techniques.* McGraw Hill, New York.
6 http://www.usablebuildings.co.uk/.

33

DRAWING THE THREADS TOGETHER

At the close of this second volume on adaptive thermal comfort, it is time to draw the various threads together and consider what has been achieved. The foundations of the adaptive approach have been securely laid, and the approach has taken its place as a recognised environmental discipline, contributing to national and international standards. It has become a worldwide venture with research in progress in all continents, a growing international fellowship of researchers working together on the science, the systems and the application of the approach. The data from the numerous surveys of recent years, bringing together the work of numerous research teams and backed by funding from various bodies including ASHRAE, are currently being collated into a comprehensive new database that will be truly worldwide in its scope.

All this has come from very small beginnings. Charles Webb noticed, when he was conducting field studies of thermal comfort in the hot summers of North India and Iraq, that his respondents felt little warmer than did people in the much cooler interiors of the UK. Back in the UK, he began a year-round data-logging project at the Building Research Station (BRS). When analysing its results, Humphreys and Nicol found that there was barely any difference in how warm people felt in winter and summer, despite the seasonal variation in the room temperatures. People did, however, respond to fluctuations of the indoor temperatures that occurred in their room during the course of a month. It seemed then that people somehow *adapted* to the seasonal drifts in the indoor temperature, but did not adapt so much to short-term fluctuations of the room temperature.

This observation laid the foundation for the adaptive approach and deflected interest from improving the formulation of indices of the combined effects of the thermal variables (air temperature, radiant temperature, air movement, humidity) that had been the prime focus of thermal comfort research in previous decades.

Thermal comfort, then, could be viewed as a self-regulating system. People, given the opportunity, would achieve thermal comfort by regulating their thermal environment to suit themselves, and by choosing clothes suitable for that thermal environment. The focus of interest therefore shifted away from the processes of heat exchange between a person and the environment, and towards a person's thermal behaviour. To understand thermal comfort, it was necessary to understand human thermal behaviour.

Comparison of field results from around the world led to two conclusions. First, that across the world people could achieve comfort over a vast range of thermal environments. Second, that in the hottest of these environments there was a clear discrepancy between the predictions of the latest and best thermal indices and how warm people actually felt. People did not feel as hot as the indices predicted, as was finally demonstrated beyond reasonable doubt by de Dear and colleagues. This led to a lively debate between proponents of the adaptive approach and proponents of the thermal-index and heat-exchange approach to comfort.

From the adaptive viewpoint, changing the clothing insulation was seen to be the principal but not the sole means by which people adapted to their thermal environment. An understanding of thermal clothing behaviour would lead to an understanding of the most basic adaptive behavioural mechanism. Studies of the clothing behaviour of children at school and of people outdoors at leisure or shopping found that people changed their clothing little during a day, changes being made chiefly from day to day. This finding has since been widely confirmed and shows that, if people are to adapt successfully, changes of room temperature should be modest within a day and gradual from day to day. The model of clothing behaviour was built on the assumption that people chose their clothing to avoid discomfort, and the principles of the model have also been applied to the control of the thermal environment by opening and closing windows and by switching fans on or off.

At the heart of the adaptive approach is the adaptive principle: if a change occurs such as to produce discomfort, people react in ways that tend to restore their comfort. Put more positively we might say that people are comfort seekers, and that anything that hinders the adaptive process is likely to cause discomfort. Modelling this comfort-seeking behaviour led to the conclusion that the clothing would follow an exponentially weighted running mean of the daily-mean indoor temperature. These running mean temperatures are also useful in describing the rate at which the comfort temperature within a building changes in response to a sequence of day-mean outdoor temperatures. These ideas have led to the inclusion of logical models of human thermal behaviour in the computer thermal simulation of buildings.

Our early understanding of the adaptive approach was too individualistic. There came a growing realisation that the social and cultural context influences the adaptive actions people take, and in some circumstances can prevent them. The culture controls what clothing is socially acceptable, and within the culture what is appropriate for a particular social setting. It is this cultural effect that limits the seasonal variation of indoor temperature that is acceptable in any particular society. If the indoor temperature becomes so extreme that social norms would have to be breached or new habits acquired, then adaptation either does not occur, or takes much longer to occur. To be thermally comfortable in a foreign society and in a different climate is therefore likely to entail adopting the culture and customs that prevail in that society and climate – that is to say, to adopt a new pattern of thermal behaviour.

So far in this summary we have not mentioned what is usually taken to be the adaptive model. The current ASHRAE standard defines an adaptive model as a chart relating the prevailing outdoor temperature to the room temperature likely to be most comfortable indoors. Such a relationship was first established in the 1970s and has gone through several revisions with the acquisition of ever-increasing quantities of data.

We do not view such charts fundamental to the adaptive approach. They work because people tend to adapt to the indoor temperatures they have, and because these temperatures, particularly in the free-running mode, are related to the outdoor temperature through the physics of the building and the thermal behaviour of the occupants. The relation depends on the building design, materials and layout and it seems that on average buildings today are warmer at a given prevailing outdoor temperature than they were forty years ago. This is unfortunate, particularly in view of the predictions of global warming in the coming century. Several factors contribute to these higher indoor temperatures. In many countries building codes require a highly insulating façade, more electrical equipment is being used indoors, and glass has become a favoured building material.

More fundamental than the charts relating the prevailing outdoor temperature to the indoor comfort temperature is the correlation between the mean indoor temperature in a building and the temperature that people find to be comfortable. They are highly correlated, and this strong correlation arises directly from the adaptive principle. If the indoor temperature is well within the social norms for the climate and season, then by the adaptive principle the occupants will adjust themselves to it. If the adaptation is complete, then the comfort temperature must equal the mean temperature. This way of thinking moves thermal comfort into the sphere of the social sciences, without losing its ultimate dependence on the physics of heat exchange and the thermal physiology of the human body.

The adaptive approach does not lend itself readily to expression in the form of a standard for indoor temperatures. It does not predict what temperatures are comfortable, since it regards those temperatures as contingent upon what temperatures are normally provided in that society. Its ultimate expression in standards is likely to be in terms of the provision a building makes to facilitate people's adaptation – the extent of adaptive opportunity it provides. One of the future aims of adaptive research will surely be to quantify such opportunities in a way that can be expressed in standards, and in such a way that the lessons of adaptive comfort can be expressed in building regulations and embodied in the design of buildings.

The adaptive approach seeks to understand human adaptive behaviour and where practicable to quantify it. Such a behavioural understanding enables us to suggest principles for the design, construction, and operation of buildings that enable the occupants to achieve comfort, while at the same time make few demands on the world's resources. That will be the topic of the third volume in this series.

REFERENCES

Abdelrahman, M. A. (1990) Field evaluation of a comfort meter, *ASHRAE Transactions* 96(2), 212–15.

Adebamowo, M. A. (2006) Thermal comfort for naturally ventilated houses in Lagos metropolis, *Proceedings of the 4th Windsor Conference,* NCEUB, pp. 71–84.

Akande, O. K. and Adebamowo, M. A. (2010) Indoor thermal comfort for residential buildings in hot-dry climate of Nigeria, *Proceedings of the 6th Windsor Conference,* NCEUB, pp. 86–96.

Alexandre, J., Friere, A., Teixeira, A., Silva, M. and Rouboa, A. (2011) Impact of European Standard EN15251 in the certification of services buildings – a Portuguese case study, *Energy Policy* 39(10), 6390–9.

Ambler, H. R. (1955) Notes on the climate of Nigeria with reference to personnel, *Journal of Tropical Medicine and Hygiene* 58, 99–112.

ANSI/ASHRAE Standard 55-2013: *Thermal environmental conditions for human occupancy,* American Society of Heating, Refrigerating and Air–Conditioning Engineers (ASHRAE), Atlanta, Georgia, USA, 2013.

Arens, E., Humphreys, M. A., de Dear, R. and Zhang, H. (2010) Are 'class A' temperature requirements realistic or desirable? *Building and Environment* 45(1), 4–10.

ASHRAE (1998) Field studies of thermal comfort and adaptation, *Technical Bulletin* 14(1), 27–49.

ASHRAE (2011) Interactions affecting the achievement of acceptable indoor environments. *ASHRAE Guideline* 10-2011, American Society of Heating, Refrigerating and Air-Conditioning Engineers, Atlanta, Georgia, USA.

Auliciems, A. (1969) Some group differences in thermal comfort, *Heating and Ventilating Engineer and Journal of Air-Conditioning* 71, 562–564.

Auliciems, A. (1969) Thermal requirements of secondary school children in winter, *Journal of Hygiene* 67(1), 59–65.

Auliciems, A. (1973) Thermal sensations of secondary school children in summertime, *Journal of Hygiene,* 71, 453–8.

Auliciems, A. (1975) Warmth and comfort in the subtropical winter: a study in Brisbane schools, *Journal of Hygiene* 74, 339–43.

Auliciems, A. (1977) Thermal comfort criteria for indoor design in the Australian winter, *Architectural Science Review* 20, 86–90.

Auliciems, A. (1981) Towards a psycho-physiological model of thermal perception, *International Journal of Biometeorology* 25, 109–22.

Auliciems, A. (1983) Psycho-physiological criteria for global thermal zones of building design, *International Journal of Biometeorology* 26(Supplement), 69–86.

Auliciems, A. (1986) Air conditioning in Australia III: thermobile controls, *Architectural Science Review* 33, 43–48.

Auliciems, A. (1989) Thermal comfort, in: *Building design and human performance*, Ed.: Ruck, N., Van Nostrand Reinhold, New York, pp. 3–28.

Bailie, A. P., Griffiths, I. D. and Huber, J. W. (1987) *Thermal comfort assessment: a new approach to comfort criteria in buildings.* Dept of Psychology, ETSU S-1177, University of Surrey, Guildford.

Baird, G. (2010). *Sustainable buildings in practice – what the users think.* Routledge, Abingdon.

Baird, G., Gray, J., Isaacs, N., Kernohan, D. and McIndoe, G. (1996) *Building evaluation techniques.* McGraw Hill, New York.

Ballantyne, E. R., Hill, R. K. and Spencer, J. W. (1977) Probit analysis of thermal sensation assessments, *International Journal of Biometeorology* 21(1), 29–43.

Bedford, T. (1936) *The warmth factor in comfort at work*, Medical Research Council Report no. 76, HMSO, London.

Bellos, D. (2011) *Is that a fish in your ear? Translation and the meaning of everything.* Particular Books (Penguin), London.

Black, F. and Milroy, E. (1966) Experience of air conditioning in offices, *J. Inst. Heat. & Vent. Eng.*, September, 188–96.

Bouden, C. and Ghrab, N. (1999) *Thermal comfort in Tunisian buildings: results of an enquiry.* Final Report, Solar Energy Laboratory, ENIT (in French).

Bouden, C. and Ghrab, N. (2001) Thermal comfort in Tunisia: results of a one year survey. in: *Moving thermal comfort standards into the 21st century.* Windsor Conference Proceedings compiled by Kate McCartney, OCSD Architecture Group, Oxford Brookes University, Oxford, pp. 197–207.

Bouden, C., Ghrab-Morcos, N., Nicol F. and Humphreys M. (1998) *A thermal comfort survey in Tunisia.* EPIC 98, Lyon, France. Proceedings ACTES, pp. 491–6.

Brissman, J., Kranz, M., Persson, P-G., Nicol, F., McCartney, K., Stoops, J., Pavlou, C., Tsangrassoulis, A. and Santamouris, M. (2001) *Conducting monitoring and comfort surveys to verify the algorithm in practice: air conditioned buildings.* SCATs final report of task 7.

BSI (2007) BS EN 15251: 2007 *Indoor environmental input parameters for design and assessment of energy performance of buildings addressing indoor air quality, thermal environment, lighting and acoustics.* Comité Européen de Normalisation, Brussels.

Burton, A. C. and Edholm, O. G. (1955) *Man in a cold environment.* Arnold, London.

Busch, J. F. (1990) Thermal responses to the Thai office environment, *ASHRAE Transactions* 96(1), 853–58.

Caird, G. B. (1963) *The Gospel of St. Luke.* The Pelican Gospel Commentaries, A490, Penguin, London.

Cao, B., Zhu, Y., Ouyang, Q., Zhou, X. and Li, H. (2009) Field study of human thermal comfort and thermal adaptability during summer and winter in Beijing, in: *The 6th International symposium on heating, ventilating and air conditioning*, Nov. 6–9, 2009, Nanjing, China, pp. 1207–16.

Cena, K., Spotila, J. R. and Avery, H. W. (1986) Thermal comfort of the elderly is affected by clothing, activity and psychological adjustment, *ASHRAE Transactions* 92(2), 329–42.

Cena, K. and de Dear, R. (1998) *Field study of occupant comfort and office thermal environments in a hot-arid climate.* Final Report on ASHRAE RP-921. ASHRAE, Atlanta, Georgia, USA.

Cena, K. M., Ladd, P. G. and Spotila, J. R. (1990) A practical approach to thermal comfort surveys in homes and offices: discussion of methods and concerns, *ASHRAE Transactions* 96(1), 853–8.

Cena, K. M., Spotila, J. R. and Ryan, E. B. (1988) Effect of behavioral strategies and activity on thermal comfort of the elderly, *ASHRAE Transactions* 94(1), 83–103.

Chan, D. W. T., Burnett, J., de Dear, R. J. and Ng, S. C. H. (1998) A large scale survey of thermal comfort in office premises in Hong Kong, *ASHRAE Technical Data Bulletin* 14(1), 76–84 (see also *ASHRAE Transactions* 104(1)).

Cheng, C. L. and Van Ness, J. W. (1999) *Statistical regression with measurement error.* Kendall's Library of Statistics, Volume 6. Arnold, London.

Chrenko, F. A. (1953) Probit analysis of subjective reactions to thermal stimuli – a study of radiant panel heating in buildings, *British Journal of Psychology*, General Section 44(3), 248–56.

CIBSE (2006) *Guide A Environmental design*. Chartered Institution of Building Services Engineers, London.

CIBSE (2013) *CIBSE technical memorandum TM52: the limits of thermal comfort: predicting overheating in European buildings*. Chartered Institution of Building Services Engineers, London.

Coldicutt, S., Williamson, T. J. and Penny, R. E. C. (1991) Attitudes and compromises affecting design for thermal performance of housing in Australia, *Environment International* 17, 251–61.

Cooper, I. (1982) Comfort and energy conservation: a need for reconciliation? *Energy and Buildings* 5(2), 83–7.

Davies, A. D. M. (1972) *Subjective ratings of the classroom environment: a sixty–two week study of St. George's School Wallasey*. University of Liverpool, Liverpool.

Davies, M. G. and Davies, A. D. M. (1973) Warmth ratings and temperature ratings in a classroom, in: *Thermal comfort and moderate heat stress*, Eds: Langdon, F. J., Humphreys, M. A. and Nicol, J. F., *Proceedings of the CIB Commission W45 (Human Requirements) Symposium*, Building Research Station, 13–15 September 1972. HMSO, London, pp. 79–85.

de Dear, R. J. (1994) Outdoor climatic influences on indoor thermal comfort, in: *Thermal comfort: past, present and future*, Eds: Oseland, N. A. and Humphreys, M. A., Building Research Establishment Report, Watford, pp. 106–32.

de Dear, R. J. and Brager, G. S. (1998) Developing an adaptive model of thermal comfort and preference, *ASHRAE Technical Data Bulletin* 14(1), 27–49.

de Dear, R. J. and Brager, G. S. (2002) Thermal comfort in naturally ventilated buildings: revisions to ASHRAE Standard 55, *Energy and Buildings* 34(6), 549–61.

de Dear, R., Brager, G. and Cooper, D. (1997) *Developing an adaptive model of thermal comfort and preference*. Final Report on RP-884. Macquarie University, Sydney.

Deutscher, G. (2010) *Through the language glass: how words colour your world*. Heinemann, London.

Doherty, T. J. and Arens, E. (1988) Evaluation of the physiological bases of thermal comfort models, *ASHRAE Transactions* 94(1), 1371–85.

Edwards, A. L. (1957) *Techniques of attitude scale construction*. Appleton Century Crofts Inc., New York. (A standard textbook reprinted many times; most recently: 1983, Irvington, New York.)

EN 15251 (2007) *Indoor environmental input parameters for design and assessment of energy performance of buildings addressing indoor air quality, thermal environment, lighting and acoustics*. Comité Européen de Normalisation, Brussels.

Errors-in-variables (2013) en.wikipedia.org/wiki/Errors-in-variables_models (accessed on 12 March 2013).

Evaluation standard for indoor thermal environment in civil buildings (2012). Ministry of Housing and Urban–Rural Development and the General Administration of Quality Supervision, Inspection and Quarantine of the People's Republic of China.

Fanger, P. O. (1970) *Thermal comfort*. Danish Technical Press, Copenhagen.

Fanger, P. O. (1994) How to apply models predicting thermal sensation and discomfort in practice, in: *Thermal comfort: past, present and future*, Eds: Oseland, N. A. and Humphreys, M. A., Building Research Establishment Report, Watford, pp. 11–4.

Fanger, P. O. and Toftum, J. (2001) Thermal comfort in the future – excellence and expectation, *Proceedings: Moving Thermal Comfort into the 21st Century*, Cumberland Lodge, Windsor.

Fanger, P. O. and Toftum, J. (2002) Prediction of thermal sensation in non-air-conditioned buildings in warm climates, *Proceedings of the 9th International Conference on Indoor Air Quality and Climate*, Monterey, California.

Feriadi, H. and Wong, N. H. (2004) Thermal comfort for naturally ventilated houses in Indonesia, *Energy and Buildings* 36(7), 614–26.

Finney, D. J. (1947, 1952, 1971) *Probit analysis*. Cambridge University Press, Cambridge.

Fisekis, K., Davies, M., Kolokotroni, M. and Langford, P. (2002) Prediction of discomfort glare from windows, *Lighting Research and Technology* 35(4), 360–71.

Fisher, R. A. (1925) *Statistical methods for research workers*. Oliver and Boyd, Edinburgh and London. (The book went through 14 editions, the last appearing in 1970.)

Fishman, D. S. and Pimbert, S. L. (1982) The thermal environment in offices, *Energy in Buildings* 5(2), 109–16.

Folk, G. E. (1969) *Introduction to environmental physiology*. Lee and Febiger, Philadephia.

Fox, R. H., Woodward, P. M., Exton-Smith, A. N., Green, M. F., Donnison, D. V. and Wicks, M. H. (1973) Body temperatures in the elderly: a national study of physiological, social and environmental conditions, *British Medical Journal* 1, 200–06.

Gagge, A. P., Fobolets, A. P. R. and Berglund, L. G. (1986) A standard predictive index of human response to the thermal environment, *ASHRAE Transactions* 92(2b), 709–31.

Gillard, J. W. (2006) *An historical overview of linear regression with errors in both variables*. Technical Report, School of Mathematics, Cardiff University.

Goldsmith, R. (1960) Use of clothing records to demonstrate acclimatisation to cold in man, *Applied Physiology* 15, 776–80.

Goromosov, M. S. (1963/1965) *The microclimate in dwellings*. State Publishing House for Medical Research, Moscow. (English translation: BRS Library Communication No 1325, 1965.)

Gray, P. G. and Corlett, T. (1952) A survey of lighting in offices. Appendix 1 of: *Postwar Building Research* No. 30. HMSO, London.

Gribbin, J. (1990) *Hothouse earth*. Bantam Press, London.

Griffiths, I. D. (undated, circa 1990) *Thermal comfort in buildings with passive solar features: field studies*. Report to the Commission of the European Communities ENS3S-090-UK, Department of Psychology, University of Surrey, Guildford.

Grivel, F. and Barth, M. (1981) Thermal comfort in office spaces: predictions and observations, in: *Building energy management*, Eds: de Fernandes, O., Woods, J. E. and Faist, A. P., Pergamon Press, Oxford, pp. 681–93.

Gross, D. M. (2006) *The secret history of emotion. From Aristotle's rhetoric to modern brain science*. University of Chicago Press, Chicago.

Guilford, J. P. (1954) *Psychometric methods*. Second Edition. McGraw-Hill, New York.

Haldi, F. and Robinson, D. (2010) On the unification of thermal perception and adaptive actions, *Building and Environment* 45(11), 2440–57.

Harrison, K. D. (2007) *When languages die. The extinction of the world's languages and the erosion of human knowledge*. Oxford University Press, Oxford.

Heidari, S. (2000) *Thermal comfort in Iranian courtyard housing*. PhD Thesis, University of Sheffield, UK.

Hickish, D. E. (1955) Thermal sensations of workers in light industry in summer. A field study in southern England, *Journal of Hygiene* 53(1), 112–23.

Hindmarsh, M. E. and MacPherson, R. K. (1962) Thermal comfort in Australia, *Australian Journal of Science* 24, 335–9.

Howell, W. C., Kennedy, P. A. (1979) Field validation of the Fanger thermal comfort model, *Human Factors* 21(2), 229–39.

Hoyt, T., Kwang, H. L., Zhang, H., Arens, E. and Webster, T. (2009) Energy savings from extended air temperature setpoints and reductions in room air mixing, *International Conference on Environmental Ergonomics*, Boston, July.

Humphreys, M. A. (1970) A simple theoretical derivation of thermal comfort conditions, *J. Inst. Heat. & Vent. Eng.* 38, 95–8.

Humphreys, M.A. (1973) Classroom temperature, clothing and thermal comfort – a study of secondary school children in summertime, *J. Inst. Heat. & Vent. Eng.* 41, 191–202.

Humphreys, M. A. (1973) Clothing and thermal comfort of secondary school children in summertime, in: *Thermal comfort and moderate heat stress*, Eds: Langdon, F. J., Humphreys, M. A. and Nicol, J. F., HMSO, London.

Humphreys, M. A. (1975) *Field studies of thermal comfort compared and applied*. Department of the Environment: Building Research Establishment, CP 76/75. (Reissued in: *J. Inst. Heat. & Vent. Eng.* 44, 5–27, 1976, and in *Physiological Requirements on the Microclimate*, Symposium, Prague, 1975.)

Humphreys, M. A. (1977) A study of the thermal comfort of primary school children in summer, *Building and Environment* 12, 231–40.

Humphreys, M. A. (1977) Clothing and the outdoor microclimate in summer, *Building and Environment* 12, 137–42.

Humphreys, M. A. (1977) The optimum diameter for a globe thermometer for indoor use, *Annals of Occupational Hygiene* 20(2), 135–40.

Humphreys, M. A. (1978) Outdoor temperatures and comfort indoors, *Building Research and Practice (J. CIB)* 6(2), 92–105.

Humphreys, M. A. (1979) The influence of season and ambient temperature on human clothing behaviour, in: *Indoor climate*, Eds: Fanger, P. O. and Valbjørn, O., Danish Building Research, Copenhagen, pp. 699–713.

Humphreys, M. A. (1981) The dependence of comfortable temperature upon indoor and outdoor climate, in: *Bioengineering, thermal physiology and comfort*, Eds: Cena, K. and Clark, J. A., Elsevier, Amsterdam, pp. 229–50.

Humphreys, M. A. (1992) Thermal comfort in the context of energy conservation, in: *Energy efficient building*, Eds: Roaf, S. and Hancock, M., Blackwell, Oxford, pp. 3–13.

Humphreys, M. A. (1992) Thermal comfort requirements, climate and energy, in: *Renewable energy, technology and the environment*, Ed.: Sayigh, A. A. M., Pergamon Press, Oxford, pp. 1725–34.

Humphreys, M. A. (1994) An adaptive approach to the thermal comfort of office workers in North West Pakistan, *Renewable Energy* 5(5–8), 985–92.

Humphreys, M. A. (1994) Field studies and climate chamber experiments in thermal comfort research, in: *Thermal comfort: past, present and future*, Eds: Oseland, N. A. and Humphreys, M. A., Building Research Establishment Report, Watford, pp. 52–72.

Humphreys, M. A. (1995) Thermal comfort temperatures and the habits of hobbits, in: *Standards for thermal comfort*, Eds: Nicol, F., Humphreys, M., Sykes, O. and Roaf, S., E. & F. N. Spon (Chapman & Hall), London, pp. 3–13.

Humphreys, M. A. (2005) Quantifying occupant comfort: are combined indices of the indoor environment practicable? *Building Research and Information* 33(4), 317–25.

Humphreys, M. A. (2008) 'Why did the piggy bark?' Some effects of language and context on the interpretation of words used in scales of warmth and thermal preference, in: *Proceedings of International Conference on Air-Conditioning and the low Carbon Cooling Challenge*, Windsor, July. Organised by the Network for Comfort and Energy Use in Buildings (NCEUB).

Humphreys, M. A. and Hancock, M. (2007) Do people like to feel 'neutral'? Exploring the variation of the desired sensation on the ASHRAE scale, *Energy and Buildings* 39(7), 867–74.

Humphreys, M. A. and Nicol, J. F. (1970) An investigation into thermal comfort of office workers, *J. Inst. Heat. & Vent. Eng.* 38, 181–9.

Humphreys, M. A. and Nicol, J. F. (1995) An adaptive guideline for UK office temperatures, in: *Standards for thermal comfort*, Eds: Nicol, F., Humphreys, M., Sykes, O. and Roaf, S., E. & F. N. Spon (Chapman & Hall), London, pp. 190–5.

Humphreys, M. A. and Nicol, J. F. (1996) *Conflicting criteria for thermal sensation within the Fanger predicted mean vote equation*, CIBSE/ASHRAE Joint National Conference Papers, Harrogate.

Humphreys, M. A. and Nicol, J. F. (1998) Understanding the adaptive approach to thermal comfort, *ASHRAE Transactions* 104(1), 991–1004.

Humphreys, M. A. and Nicol, J. F. (2000) Outdoor temperature and indoor thermal comfort – raising the precision of the relationship for the 1998 ASHRAE database of field studies, *ASHRAE Transactions* 106(2), 485–92.

Humphreys, M. A. and Nicol, J. F. (2002) The validity of ISO-PMV for predicting comfort votes in everyday life, *Energy and Buildings* 34, 667–84.

Humphreys, M. A. and Nicol, J. F. (2004) Do people like to feel 'neutral'? Response to the ASHRAE scale of subjective warmth in relation to thermal preference, indoor and outdoor temperature, *ASHRAE Transactions* 110(2), 569–77.

Humphreys, M. A. and Nicol, J. F. (2007) Self-assessed productivity and the office environment: monthly surveys in five European countries, *ASHRAE Transactions* 113(1), 606–16.

Humphreys, M. A., Nicol, J. F. and McCartney, K. J. (2002) *An analysis of some subjective assessments of indoor air-quality in five European countries*, Indoor Air 2002, Monterey.

Humphreys, M. A., Nicol, J. F., and Raja, I. A. (2007) Field studies of thermal comfort and the progress of the adaptive model, *Advances in Building Energy Research* 1, 55–88.

Humphreys, M. A., Rijal, H. B. and Nicol, J. F. (2013) Updating the adaptive relation between climate and comfort indoors; new insights and an extended database, *Building and Environment* 63, 40–55. (http://dx.doi.org/10.1016/j.buildenv.2013.01.024).

Humphreys, M. A., McCartney, K. J., Nicol, J. F. and Raja, I. A. (1999) *An analysis of some observations of finger–temperature and thermal comfort of office workers*, Indoor Air 1999, Edinburgh.

Hunt, D. R. G. and Gidman, M. I. (1982) A national survey of house temperatures, *Building and Environment* 17(2), 107–24.

Indraganti, M., Lee, J., Zhang, H. and Arens, E. (2015) Thermal adaptation and insulation opportunities provided by draping Indian saris on the body, *Architectural Science Review* 58(1), 87–92.

Indraganti, M., Ooka, R. and Rijal, H. B. (2013) Field investigation of comfort temperature in Indian office buildings: a case of Chennai and Hyderabad, *Building and Environment* 65, 195–214.

ISO (2005) *ISO 7730: Ergonomics of the thermal environment – analytical determination and interpretation of thermal comfort using calculation of the PMV and PPD indices and local thermal comfort criteria*. ISO, Geneva.

Jamy, G. N. (1995) Towards new indoor comfort temperature standards for Pakistani buildings, in: *Standards for thermal comfort*, Eds: Nicol, F., Humphreys, M., Sykes, O. and Roaf, S., E. & F. N. Spon (Chapman & Hall), London, pp. 14–21.

Kato, T., Yamagishi, A. and Yamashita, Y. (1996) Difference between winter and summer of the indoor thermal environment and residents' thinking of detached houses in Nagano city, *J. Archit. Plann. Environ. Eng.* 481, 23–31 (in Japanese with English abstract).

Kwok, A. G., (1998) Thermal comfort in tropical classrooms, *ASHRAE Technical Data Bulletin* 14(1), 85–101.

Lane, W. R. (1965) *Education, children and thermal comfort*. University of Iowa, Iowa.

Langdon, F. J. and Loudon, A. G. (1970) Discomfort in schools from overheating in summer, *J. Inst. Heat. & Vent. Eng.* 37, 265–79.

Langdon, F. J., Humphreys, M. A. and Nicol, J. F. (1973) *Thermal comfort and moderate heat stress*. HMSO, London.

Li, H., Zhou, Y., Ouyang, Q. and Cao, B. (2009) Measurement and field survey of indoor thermal comfort in rural housing of northern china in winter, in: *The 6th International Symposium on Heating, Ventilating and Air-Conditioning*, Nov. 6–9, Nanjing, China, pp. 245–51.

Likert, R. (1932) A technique for the measurement of attitudes, *Archives of Psychology* 140, 1–55.

Liu, W., Wargocki, P., and Xiong, J. (2014) Occupant time period of thermal adaption to change of outdoor air temperature in naturally ventilated buildings, in: *Proceedings of 8th Windsor Conference: Counting the cost of comfort in a Changing World*, Cumberland Lodge, Windsor, pp. 10–13. Network for Comfort and Energy Use in Buildings, London, (http://nceub.org.uk).

Loudon, A. G. (1968) Window design criteria to avoid overheating by excessive solar gains, *Building Research Station Current Paper CP 4/68*.

Lundqvist, G. R. (1973) Thermal field measurements in schools, in: *Thermal comfort and moderate heat stress. Proceedings of the CIB Commission W45 (Human Requirements) Symposium*, Eds: Langdon, F. J., Humphreys, M. A. and Nicol, J. F., HMSO, London.

Malama, A., Sharples, S., Pitts, A. C. and Jitkhajornwanich, K. (1998) An investigation of the thermal comfort adaptive model in a tropical upland climate, *ASHRAE Technical Data Bulletin* 14(1), 102–11. (see also: *ASHRAE Transactions* 104(1)).

Mallick, F. H. (1992) Thermal comfort in tropical climates: an investigation of comfort criteria for Bangladeshi subjects, *PLEA Conference Proceedings*, pp. 47–52.

Matthews, J. (1993) *An investigation into the thermal comfort of factory workers in North India using the Probit method of analysis*. MSc Dissertation, University of East London, London.

Matthews, J. and Nicol, J. F. (1994) Thermal comfort of factory workers in Northern India. in: *Standards for thermal comfort*, Eds: Nicol, F., Humphreys, M., Sykes, O. and Roaf, S., E. & F. N. Spon. (Chapman & Hall), London, pp. 227–33.

McCartney, K. J. and Nicol, J. F. (2001) Developing an adaptive control algorithm for Europe: results of the SCATs project, in: *Moving thermal comfort standards into the 21st century*, Ed.: McCartney, K., Oxford Brookes University, Oxford, pp. 176–97.

McCartney, K. J. and Nicol, J. F. (2002) Developing an adaptive control algorithm for Europe: results of the SCATs project, *Energy and Buildings* 34(6), 623–35.

McIntyre, D. A. (1980) *Indoor climate*. Applied Science Publishers, London.

Nakamura, Y., Yokoyama, S., Tsuzuki, K., Miyamoto, S., Ishii, A., Tsutsumi, J. and Okamoto, T. (2008) Method for simultaneous measurement of the occupied environment temperature in various areas for grasp of adaptation to climate in daily life, *Journal of Human and Living Environment* 15(1), 5–14 (in Japanese with English abstract).

Nakaya, T., Matsubara, N. and Kurazumi, Y. (2005) A field study of thermal environment and thermal comfort in Kansai region, Japan: Neutral temperature and acceptable range in summer. *J. Environ. Eng.* 597, 51–6 (in Japanese with English abstract).

NCEUB 2014 Windsor Conference: http://nceub.org.uk//W2014/webpage/W2014_index.html.

Nevins, R. G., Rohles, F. H., Springer, W. and Fayerherm, A. M. (1966) A temperature–humidity chart for thermal comfort of seated persons, *ASHRAE Transactions* 72(2), 283–91.

Nicol, J. F. (1974) An analysis of some observations of thermal comfort in Roorkee, India and Baghdad, Iraq, *Annals of Human Biology* 1(4), 411–26.

Nicol, F. (1993) *Thermal comfort: a handbook for field studies toward an adaptive model*. University of East London, London.

Nicol, J. F. (1995) Thermal comfort and temperature standards in Pakistan, in: *Standards for thermal comfort*, Eds: Nicol, F., Humphreys, M., Sykes, O. and Roaf, S., E. & F. N. Spon (Chapman & Hall), London, pp. 149–56.

Nicol, J. F. (2001) Characterising occupant behaviour in buildings: towards a stochastic model of occupant use of windows, lights, blinds heaters and fans, *Proceedings of the Seventh International IBPSA Conference*, Rio, Vol. 2. International Building Performance Simulation Association, pp. 1073–1078.

Nicol, J. F. and Humphreys, M. A. (1972) Thermal comfort as part of a self–regulating system, in: *Proc. Symposium: Thermal comfort and moderate heat stress*, Eds: Langdon, F. J., Humphreys, M. A. and Nicol, J. F., CIB Commission W45, HMSO, London.

Nicol, J. F. and Humphreys, M. A. (2002) Adaptive thermal comfort and sustainable thermal standards for buildings, *Energy and Buildings* 34(6), 563–72.

Nicol, J. F. and Humphreys, M. A. (2004) A stochastic approach to thermal comfort, occupant behaviour and energy use in buildings, *ASHRAE Transactions* 110(2), 554–6.

Nicol, J. F. and Humphreys, M. A. (2007) Maximum temperatures in European office buildings to avoid heat discomfort, *Solar Energy* 81(3), 295–304.

Nicol, J. F. and Humphreys, M. A. (2009) New standards for comfort and energy use in buildings, *Building Research and Information* 37(1), 68–73.

Nicol, J. F. and Humphreys, M. A. (2010) Derivation of the equations for comfort in free-running buildings in CEN Standard EN15251, *Buildings and Environment* 45(1), 11–7.

Nicol, F. and McCartney, K. (1997) Modelling temperature and human behaviour in buildings – field studies 1996–97, *Proceedings of BEPAC/EPSRC Miniconference: Sustainable Building*, Abingdon.

Nicol, J. F. and McCartney, K. J. (1999) Assessing adaptive opportunities in buildings, in: *Engineering in the 21st century – the changing world*, Chartered Institution of Building Services Engineers, London, pp. 219–29.

Nicol, F. and McCartney, K. (2001) Final report (public) smart controls and thermal comfort (SCATs) (also subsidiary reports to Tasks 2 and 3 and contributions to reports to Tasks 1, 6 and 7), *Report to the European Commission of the smart controls and thermal comfort project* (Contract JOE3-CT97-0066), Oxford Brookes University, Oxford (available at www.nceub.org.uk).

Nicol, J. F. and Parsons, K. (guest editors) (2002) *Energy and Buildings* 34(6).

Nicol, F. and Raja, I. (1996) *Thermal comfort, time and posture: exploratory studies in the nature of adaptive thermal comfort*. School of Architecture, Oxford Brookes University, Oxford.

Nicol, F. and Roaf, S. (2005) Post occupancy evaluation and field studies of thermal comfort, *Building Research and Information* 33(4), 338–46.

Nicol, F. and Roaf, S. (guest editors) (2005) *Building Research and Information* (BRI), 33(4).

Nicol, F., Raja, I., and Alaudin, A. (1996) *Thermal comfort in Pakistan II.* Report to the Overseas Development Administration. School of Architecture, Oxford Brookes University, Oxford.

Nicol, F., Wilson, M. and Chiancarella, C. (2006) Using field measurements of desktop illuminance in European offices to investigate its dependence on outdoor conditions and its effect on occupant satisfaction, productivity and the use of lights and blinds, *Energy and Buildings* 38(7), 802–13.

Nicol, F., Humphreys, M., Sykes, O. and Roaf, S. (Eds) (1995) *Standards for thermal comfort*, E. & F. N. Spon (Chapman & Hall), London.

Nicol, J. F., Raja, I. A., Allaudin, A. and Jamy, G. N. (1999) Climatic variations in comfort temperatures: the Pakistan projects, *Energy and Buildings* 30(3), 261–79.

Nicol, F., Roaf, S., Brotas, L. and Humphreys, M. (Eds) (2014) *Proceedings of the 2014 Windsor Conference*, NCEUB, London (nceub.org.uk). See: www.windsorconference.com.

Nicol, F., Jamy, G. N., Sykes, O., Humphreys, M., Roaf, S. and Hancock, M. (1994) *A survey of thermal comfort in Pakistan: towards new indoor temperature standards.* Final Report, School of Architecture, Oxford Brookes University, Oxford.

Olesen, B. W. and Parsons, K. C. (2002) Introduction to thermal comfort standards and to the proposed new version of EN ISO 7730, *Energy and Buildings* 34(6), 537–48.

Oppenheim, A. N. (1992) *Questionnaire design, interviewing and attitude measurement.* Second edition, Continuum, London.

Oseland, N. A. (1994) A comparison of the predicted and reported thermal sensation vote in homes during winter and summer, *Energy and Buildings* 21(1), 45–54.

Oseland, N. A. (1994) Predicted and reported thermal sensation votes in UK homes, in: *Banking on design?* Ed.: Seidal, A., *Proceedings 25th Annual Conference of the Environmental Design Association*, St Antonio. EDRA, pp. 175–9.

Oseland, N. A. (1995) Predicted and reported thermal sensation in climate chambers, offices and homes, *Energy and Buildings* 23(2), 105–15.

Oseland, N. A. (1997) *Thermal comfort: a comparison of observed occupant requirements with those predicted and specified in standards.* PhD thesis, Cranfield University, UK.

Oseland, N. A. (1998) Acceptable temperature ranges in naturally ventilated and air-conditioned offices, *ASHRAE Technical Data Bulletin* 14(1), 50–62 (also: *ASHRAE Transactions* 104(1)).

Oseland, N. A. and Humphreys, M. A. (1994) Trends in thermal comfort research, *Building Research Establishment Report* No. 266, BRE Watford, Herts.

Oseland, N. A. and Humphreys, M. A. (Eds) (1994) *Thermal comfort: past, present and future.* Building Research Establishment Report, Watford.

Oseland, N. A. and Raw, G. (1990) Thermal comfort in starter homes in UK, *BRE note PD 186/90, Proceedings of Environmental Design Research Association 22nd Annual Conference on Healthy Environments.*

Palmai, G. (1962) Thermal comfort and acclimatisation to cold in a subantarctic environment, *The Medical Journal of Australia*, January, 9–12.

Parsons, K. (2003) *Human thermal environments.* CRC Press (Taylor and Francis), London.

Parsons, K. C., Webb, L. H., McCartney, K. J., Humphreys, M. A. and Nicol, J. F. (1997) A climatic chamber study into the validity of Fanger's PMV/PPD thermal comfort index for subjects wearing different levels of clothing insulation, *CIBSE National Conference Proceedings*, pp. 193–205.

Peccolo, C. (1962) *The effect of thermal comfort on learning.* Digest of PhD Thesis, University of Iowa, Iowa.

Penwarden, A. D. and Wise, A. F. E. (1975) Wind environment around buildings, *Building Research Establishment Report*, Department of the Environment. HMSO, London.

Pirsel, L. (1989) Which temperature is being felt comfortable in dwellings? *Proc. CIB 1989* 2, pp. 167–72.

Pitts, A. (2006) The languages and semantics of thermal comfort, *Proceedings of the NCEUB Conference: Comfort and Energy Use in Buildings – Getting Them Right*, Cumberland Lodge, Windsor.

Preiser, W. and Vischer, J. (2005) *Post occupancy evaluation*. Harper Collins, New York.

Raja, I. A. (1996) *Solar energy resources of Pakistan*. Oxford Brookes University, Oxford.

Raja, I. A. and Nicol, J. F. (1997) A technique for postural recording and analysis for thermal comfort research, *Applied Ergonomics* 28(3), 221–5.

Raja, I. A., Nicol, J. F. and McCartney, K. J. (1998) Natural ventilated buildings: use of controls for changing indoor climate, *Renewable Energy* 15, 391–4.

Ramasodi, L. (1993) *Thermal comfort and the tropical summer index*. MSc Dissertation, University of East London, London.

Ramirez-Esparza, N., Gosling, S. D., Benet-Martinez, V., Potter, J. P. and Pennebaker, J. W. (2006) Do bilinguals have two personalities? A special case of cultural frame switching, *Journal of Research in Personality* 40, 99–120.

Rijal, H. B. (2014) Investigation of comfort temperature and occupant behavior in Japanese houses during the hot and humid season, *Buildings* 4, 437–52. doi:10.3390/buildings4030437. www.mdpi.com/journal/buildings/.

Rijal, H. B. and Yoshida, H. (2006) Winter thermal comfort of residents in the Himalaya region of Nepal, *Proceedings of International Conference on Comfort and Energy Use in Buildings – Getting Them Right*, Windsor, Network for Comfort and Energy Use in Buildings (NCEUB).

Rijal, H. B., Humphreys, M. A. and Nicol J. F. (2009) How do the occupants control the temperature in mixed–mode buildings? *Building Research and Information* 37(4), 381–96.

Rijal, H. B., Humphreys, M. A. and Nicol, J. F. (2014) Development of the adaptive model for thermal comfort in Japanese houses, *Proceedings of 8th Windsor Conference: Counting the Cost of Comfort in a Changing World*, Cumberland Lodge, Windsor, April. Network for Comfort and Energy Use in Buildings, http://nceub.org.uk, London.

Rijal, H. B., Yoshida, H. and Umemiya, N. (2002) Investigation of the thermal comfort in Nepal, *Proceedings of International Symposium on Building Research and the Sustainability of the Built Environment in the Tropics*, Indonesia, pp. 243–62.

Rijal, H. B., Honjo, M., Kobayashi, R. and Nakaya, T. (2013) Investigation of comfort temperature, adaptive model and the window opening behaviour in Japanese houses, *Architectural Science Review* 56(1), 54–69.

Rijal, H. B., Toshiaki, O., Humphreys, M. A., Nicol, J. F. (2012) A comparison of the winter thermal comfort of floor heating systems and air conditioning systems in Japanese homes, *5th International Building Physics Conference (IBPC)*, Kyoto, Japan.

Rijal, H. B., Tuohy, P., Humphreys, M. A., Nicol, J. F., Samuel, A. (2012) Considering the impact of situation-specific motivations and constraints in the design of naturally ventilated and hybrid buildings, *Architectural Science Review* 55(1), 35–48.

Rijal, H., Tuohy, P., Humphreys, M. A. Nicol, F., Samuel, A. and Clarke, J. (2007) Using results from field surveys to predict the effect of open windows on thermal comfort and energy use in buildings, *Energy and Buildings* 39(7), 823–36.

Rijal, H., Tuohy, P., Humphreys, M. A., Nicol, F., Samuell, A., Raja, I. A. and Clarke, J. (2008) Development of adaptive algorithms for the operation of windows, fans and doors to predict thermal comfort and energy use in Pakistani buildings, *ASHRAE Transactions* 114(2), 555–73.

Roaf, S. C. (1988) *The windcatchers of Yazd*. PhD thesis, Oxford Polytechnic, Oxford.

Sà, P. (1938) *Conforto Termico*, Departmento de Estatistica e Publicidade, Rio de Janeiro.

Saito, M. (2009) Study on occupants' cognitive temperature scale for their environmental controls behaviors: In the case of University laboratories in summer in Sapporo, *J. Environ. Eng.* 74(646), 1291–7 (in Japanese with English abstract).

Sangowawa, T., Adebamowo, M. A. and Godwin, J. (2008) Cooling, comfort and low-energy in a warm humid climate: the experience of Lagos, Nigeria, *Proceedings of the 5th Windsor Conference*, NCEUB, pp. 35–49.

Sawachi, T., Matsuo, Y. (1989) Daily cycles of activities in dwellings in the case of housewives: Study on residents' behavior contributing to formation of indoor climate (Part 2), *J. Archit. Plann. Environ. Eng.* 398, 35–46 (in Japanese with English abstract).

Schiller, G. E. (1990) A comparison of measured and predicted comfort in office buildings, *ASHRAE Transactions* 96(1), 609–22.

Sharma, M. R. and Ali, S. (1986). Tropical summer index – a study of thermal comfort in Indian subjects, *Building and Environment* 21(1), 11–24.

SIB (Anon) (1967) *Teachers' opinions of classroom climate – a questionnaire survey.* Statens Institut for Byggnadsforskning, Report No. 31, Stockholm.

Siegal, M. (2008) *Marvellous minds: the discovery of what children know.* Oxford University Press, Oxford.

Sliwowski, L. Z., Cena, K. and Sliwinska, E. (1983) Indoor climate problems in Polish apartment blocks, in: *PLEA Conference*, Ed.: Yanas, S., Pergamon, Oxford.

Spindler, K. (1995) *The man in the ice.* Trans: Osers E. Phoenix, London.

Standeven, M. A. and Baker, N. V. (1995) Comfort conditions in PASCOOL surveys, in: *Standards for thermal comfort*, Eds: Nicol F., Humphreys M., Sykes O. and Roaf, S., E. & F. N Spon (Chapman & Hall), London, pp. 161–8.

Stevenson, F. and Rijal, H. B. (2008) *Post-occupancy evaluation of the Stewart Milne Group's Sigma® House.* Stewart Milne Group (Final report).

Stoops, J. L. (2000) *The thermal environment and occupant perception in European office buildings.* Licentiate thesis, Chalmers University, Sweden.

Suzuki, N. and Shukuya, M. (2009) Field and semi-field surveys on thermal-environment experience and its associated acquired cognition by family members, parents and children, *26th conference on Passive and Low Energy Architecture*, Quebec City, Canada, 22–24 June.

Teli, D., Jentch, M. F., James, P. A. B. and Bahaj, A. S. (2012) Field study on thermal comfort in a UK primary school, *Proceedings of 7th Windsor Conference: The Changing Context of Comfort in an Unpredictable World*, Cumberland Lodge, Windsor, April. Network for Comfort and Energy Use in Buildings, http://nceub.org.uk.

Thurstone, L. L. (1927) A law of comparative judgement, *Psychological Review* 34, 273–86.

Tobita, K., Matsubara, N., Kurazumi, Y., Nakaya, T. and Shimada, R. (2009) Difference of the thermal sensation votes by the scales in the field study of houses during winter, *J. Environ. Eng.*, 74(646), 1291–7 (in Japanese with English abstract).

Tobita, K., Nakaya, T., Matsubara, N., Kurazumi, Y. and Shimada, R. (2007) Calculation of neutral temperature and acceptable range by the field study of houses in Kansai area, Japan, in winter, *J. Environ. Eng.* 614, 71–7 (in Japanese with English abstract).

Tolkein, J. R. R. (1937) *The hobbit.* Allen and Unwin, London.

Tsukida, A., and Gupta, M. R. (2011) *How to analyse paired comparison data.* UWEE Technical Report, URL: http://www.ee.washington.edu.

Tuohy, P., Humphreys, M. A., Nicol, F., Rijal, H. and Clarke, J. (2009) Occupant behaviour in naturally ventilated and hybrid buildings, *ASHRAE Transactions* 115(1), 16–27.

Turnquist, R. O. and Volmer, R. P. (1980) Assessing environmental conditions in apartments of the elderly, *ASHRAE Transactions* 86(1), 536–40.

Umemiya, N., Okura, R. and Tanaka, Y. (2008) Setting temperature and clothing insulation in student rooms in summer, *Proceedings of the 29th Air Infiltration and Ventilation Centre Conference* 3, pp. 79–84.

Usable Buildings Website: http://www.usablebuildings.co.uk.

Webb, C. G. (1959) An analysis of some observations of thermal comfort in an equatorial climate, *British Journal of Industrial Medicine* 16(3), 297–310.

Webb, C. G. (1960) Thermal discomfort in an equatorial climate, *J. Inst. Heat. & Vent. Eng.* 27, 297–303.

Webb, C. G. (1964) Thermal discomfort in a tropical environment, *Nature* 202(4938), 1193–4.

Wierzbicka, A. (2006) *English: meaning and culture.* Oxford University Press, Oxford.

Williamson, T. J., Coldicutt, S. and Penny, R. E. C. (1989) Thermal preferences in housing in the humid tropics, *ERDC No 1*, Energy Research and Development Corporation, Canberra.

Williamson, T. J., Coldicutt, S. and Penny, R. E.C. (1989) *Thermal comfort and preferences in housing: south and central Australia.* University of Adelaide, Adelaide.

Wilson, M. and Nicol, J. F. (2001) Noise in offices and urban canyons, *Proceedings of the Institute of Acoustics* 23(5), 41–4.

Wilson, M. P. and Nicol, J. F. (2003) *Some thoughts on acoustic comfort: a look at adaptive standards for noise.* Paper to Institute of Acoustics Conference 'Soundbite', Oxford 5–6 November.

Woolard, D. S. (1979) *Thermal habitability of shelters in the Solomon Islands.* PhD Thesis, University of Queensland, Australia.

Woolard, D. S. (1981) The graphic scale of thermal sensation, *Architectural Science Review* 24(4), 90–3.

Woolard, D. S. (1981) Thermal sensations of Solomon Islanders at home, *Architectural Science Review* 24(4), 94–7.

Woolgar, C. M. (2006) *The senses in late medieval England.* Yale University Press, London.

Yamashita, H., Umemiya, N. and Okura, R. (2009) Setting temperature, air temperature and thermal comfort in student rooms of university, *Proceedings of the Kinki Chapter of the Society of Heating, Air-Conditioning and Sanitary Engineers of Japan (SHASE)* 38, 25–8 (in Japanese with English abstract).

Yang, L., Yan, H., Xu, Y. and Lam, J. C. (2013) Residential thermal environment in cold climates at high altitudes and building energy use implications, *Energy and Buildings* 62, 139–45.

INDEX

Printed and bound by CPI Group (UK) Ltd, Croydon, CR0 4YY

24/10/2024

01778298-0002